ml

 UNIVER

D1454294

30125

s are to be returned on or before
the last date below.

1 8 APR 2007

JAN 1999

PROPAGATION OF RADIOWAVES

PROPAGATION OF RADIOWAVES

**Edited by
M. P. M. Hall,
L. W. Barclay and
M. T. Hewitt**

The Institution of Electrical Engineers

Published by: The Institution of Electrical Engineers, London,
United Kingdom

© 1996: The Institution of Electrical Engineers

The Institution of Electrical Engineers,
Michael Faraday House,
Six Hills Way, Stevenage,
Herts. SG1 2AY, United Kingdom

British Library Cataloguing in Publication Data

A CIP catalogue record for this book
is available from the British Library

ISBN 0 85296 819 1

Printed in England by Short Run Press Ltd., Exeter

Contents

Preface

This book is based on 19 lectures given at the fifth IEE Vacation School on Radiowave Propagation which was held on 17–22 April 1994. Sixty 'students' attended the course which covered a wide range of subject material on the different aspects of the propagation medium and different service requirements.

The subject of radio communication continues to change rapidly, with a succession of new questions arising from the planning of new services. In particular, the rising requirement for wider bandwidths and the rapid rise in point-to-multipoint and mobile communication (and consequent need for frequency reuse) lead to specific demands for improved prediction methods — all requiring many aspects to be examined.

The book is intended for a wide readership: for those engaged in postgraduate studies or research; for those active in planning radio systems (in operating organisations or industry); and for others also wishing to broaden their familiarity with the subject matter. An understanding of radiowave propagation is fundamental to the design of communication, radar and broadcasting systems.

Considerable emphasis occurs in the book on texts of the Radiocommunication Sector of the International Telecommunication Union (ITU-R) (of which a large part was previously the CCIR). Such texts generally cover material widely accepted in the discipline after extensive study reaching international concurrence. Much of it is updated progressively from new knowledge and research, though some criteria and basic ideas remain on record without change. The texts are mentioned in the book as, for example, *ITU-R Recommendation PN.452-6*. The reference to the Recommendation listed at the end of the book gives its title and the volume where it is to be found; in this case in the PN series. The suffix in the text numbering indicates the number of times that the text has been revised. These texts are published by the ITU, and are given in the *List of Publications* issued by the ITU Sales Service: CH-1211, Geneva 20, Switzerland. Where not otherwise specified, the latest edition of a Recommendation is assumed. In particular, the provisional contents of the 1995 versions of certain Recommendations were known at the time of preparation of this book, and these are expected to be published in 1996.

The editors wish to thank the authors for employing their expertise with enthusiasm, as well as thanking the IEE for its part in producing the book.

Martin P.M. Hall
Leslie W. Barclay
M. Tim Hewitt
July 1995

Contributors

Chapters 1 and 13
Mr M P M Hall
Rutherford Appleton Laboratory
Chilton, Didcot
Oxon OX11 0QX

Chapter 2
Dr M J Mehler
BT Laboratories
Martlesham Heath
Ipswich
Suffolk IP5 7RE

Chapter 3
Prof L W Barclay
12 St Stephens Road
Cold Norton
Chelmsford
Essex CM3 6JE

Chapters 4 and 18
Dr D F Bacon
Radiocommunications Agency
South Quay Three
189 Marsh Wall
London E14 9SX

Chapter 5
Mr P A Miller
Cellnet
260 Bath Road
Slough
Berkshire SL1 4DX

Chapters 6 and 8
Dr K H Craig
Rutherford Appleton
   Laboratory
Chilton, Didcot
Oxon OX11 0QX

Chapters 7 and 9
Mr J W F Goddard
Rutherford Appleton
   Laboratory
Chilton, Didcot
Oxon OX11 0QX

Chapter 10
Prof P A Matthews
University of Leeds
Department of Electronic &
   Electrical Engineering
Leeds LSR 9JT

Chapter 11
Mr G D Richman
BT Laboratories
Martlesham Heath
Ipswich
Suffolk IP5 7RE

Chapter 12
Dr M A Beach
University of Bristol
Department of Electrical &
  Electronic Engineering
Merchant Venturers Building
Woodland Road
Bristol BS8 1UB

Mr S A Allpress
AT&T Bell Laboratories
Whippany Road
New Jersey 07981
USA

Chapter 14
Prof H Rishbeth
University of Southampton
Department of Physics
Highfield, Southampton
Hants SO17 1BJ

Chapter 15
Mr J D Milsom
GEC-Marconi Research
  Centre
West Hanningfield Road
Great Baddow
Chelmsford
Essex CM2 8HN

Chapters 16 and 17
Mr P A Bradley
Rutherford Appleton
  Laboratory
Chilton, Didcot
Oxon OX11 0QX

Chapter 19
Dr C J Gibbins
Rutherford Appleton
  Laboratory
Chilton, Didcot
Oxon OX11 0QX

Chapter 1

Overview of radiowave propagation

M.P.M. Hall

1.1 Introduction

The success of any communication depends on the influence of the propagation medium, be it dependent on sound waves, radio waves or light. Although the Radio Regulations [1] of the International Telecommunication Union (ITU) limit the term 'radio waves' to electromagnetic waves of frequencies 'arbitrarily lower than 3000 GHz' (i.e. including 'submillimetric waves'), some reference may be made here to potential interest in higher frequencies and so encompass the electromagnetic spectrum 'from DC to light', as the phrase is.

We are concerned here with propagation in space rather than guides such as wires, transmission lines, coaxial cables, waveguides or optical fibres. However, the waves may well be guided by the ionosphere, by layers of intense refractive-index change or by the Earth's surface. In addition to these, many other influences of the atmosphere and terrain will be considered. The degree to which these influences affect propagation depends on the frequency of the wave, and in some cases also on the polarisation.

Table 1.1 shows the *wavebands* expressed in decades, with the descriptive term used by ITU-R (the Radiocommunication Sector of the ITU), such as 'myriametric waves' or 'millimetric waves', and also abbreviations for the commonly adopted terms, such as 'very high frequency' or 'extremely low frequency'. No frequency assignments have been made in the UK below 9 kHz and few above 60 GHz.

1.2 Wave propagation in the atmosphere

The frequency range from 1 GHz to about 30 GHz is often referred to as the 'microwave' band (paralleling the notations 'long wave', 'medium wave' and 'short wave' most often used in relation to the broadcasting bands within LF, MF and HF, respectively).

Table 1.1   Part of the electromagnetic spectrum

Frequency range	Wavelength	Descriptive designation	
Below 3 kHz	Above 100 km		ELF
3–30 kHz	10–100 km	Myriametric waves	VLF
30–300 kHz	1–10 km	Kilometric waves	LF
300–3000 kHz	100–1000 m	Hectometric waves	MF
3–30 MHz	10–100 m	Decametric waves	HF
30–300 MHz	1–10 m	Metric waves	VHF
300–3000 MHz	10–100 cm	Decimetric waves	UHF
3–30 GHz	1–10 cm	Centimetric waves	SHF
30–300 GHz	1–10 mm	Millimetric waves	EHF
300–3000 GHz	0.1–1 mm	'Sub-millimetric waves'	
3–30 THz	10–100 μm	'Far-infrared waves'	
30–430 THz	0.7–10 μm	'Near-infrared waves'	
430–860 THz	0.35–0.7 μm	'Optical waves'	

Perhaps the most frequently encountered distinction of atmospheric influences is between those of the troposphere and those of the ionosphere (to be discussed later). The *troposphere* is the lowest region of the atmosphere, in which temperature generally decreases with height. Its upper height limit (the base of the tropopause, in which the temperature ceases to decrease with height, is about 17 km at the equator and 9 km at the poles. The height of the tropopause varies with the atmospheric conditions: for instance, at medium latitudes it is about 13 km in high-pressure zones (anticyclones), declining to less than 7 km in depressions. In clear air, the radio refractive-index of the troposphere is slightly greater than unity (typically about 1.0003). The way in which the refractive-index changes with height has much consequence for radiowave propagation at frequencies greater than about 30 MHz, and, at frequencies greater than about 10 GHz, the absorption and scatter due to clouds, rain, snow etc. which occur in this region also have much effect. Figure 1.1 illustrates some of these effects, separating the clear-air effects from those due to clouds and precipitation.

Above 30 MHz the wavelength is comparable with the distance over which variations of atmospheric refractive-index occur in the troposphere. These variations are due to changes of temperature, pressure and humidity. The refractive-index of the troposphere generally decreases with height. This leads to a slight downwards *refraction* of radio rays, which can be very important for communication at VHF, UHF and SHF. If the rate of refractive-index decrease with height is sufficiently large and extends over a sufficient height interval and horizontal extent, it may give rise to atmospheric *ducts* which guide radio energy

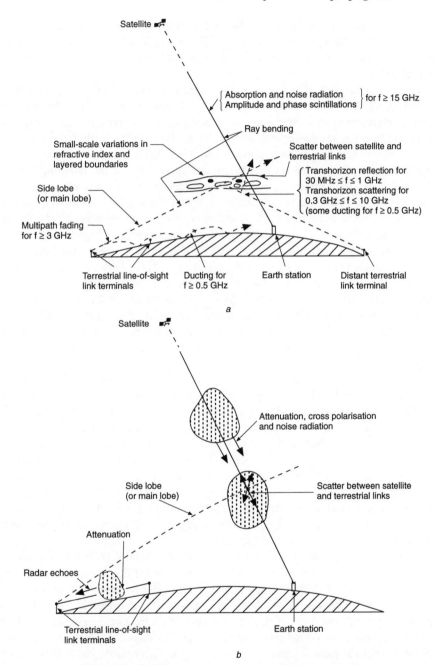

Figure 1.1   *Some effects of the troposphere on radiowave propagation*
      a Effects of atmospheric gases (clear air) and associated refractive-index changes
      b Effects of cloud and precipitation (above 3 GHz)

far beyond the normal horizon.   Normal duct heights are such that complete trapping within them occurs only at centimetric wavelengths, but partial trapping (and very rarely total trapping) may also be found at the shorter metric wavelengths.

If, over a large horizontal area, the refractive-index decreases with height abruptly, this may lead to partial *reflection* of radio energy. Both ducting and partial-reflection mechanisms may cause multipath interference on line-of-sight or interpath interference on beyond-horizon links. Randomly distributed small-scale spatial fluctuations of refractive-index about the local mean value cause weak signal levels always to be present at large distances beyond the horizon. This is due to *tropospheric scattering* from the irregularities, and such scattering may be used to provide radio communication over several hundred kilometres. On line-of-sight paths, these refractive-index fluctuations may cause significant *scintillation* (rapid fading), which is larger in magnitude for longer range or higher frequency.

For two terrestrial radio terminals within line-of-sight, there may often be a ground-reflected ray in addition to the direct ray. There may also be a reflected ray (or more than one) from layers of abrupt change of refractive-index with height. According to their relative phase, the '*multipath*' contributions may give rise to slow signal enhancement or fading. For UHF, the abruptness of change with height must be greater than for VHF, but the reflecting surface area may be smaller.   The latter means that relatively small terrain features may cause significant reflection at SHF.

The effect of obstacles such as hills, buildings or trees depends on the wavelength considered. Such obstacles cause reflection (and multipath), scatter, diffraction and absorption. In a built-up area this results in a wide variation in field strength.   The losses caused by absorption and scattering increase with frequency, until, at frequencies above UHF, walls or masonry more than about 20 cm thick may be regarded as opaque, together with buildings (except those of very light construction) and woods which are visually opaque.

If one (or both) of the terminals is *mobile*, or vehicles are moving nearby, the relative phases (and amplitudes) of multipath contributions cause a rapid fading, and there is also some *Doppler shift* of frequency. Such multipath propagation is more prevalent in city areas than rural areas.

At frequencies greater than 3 GHz, *rain* or *cloud* may give rise to significant *absorption* and scattering of radio energy. Such absorbing media also generate *thermal noise* which may be important in Earth–space systems. Scattering from cloud or precipitation (hydrometeors) may cause considerable interference between systems, but even in clear air the absorption by *atmospheric gases* alone, and the associated noise radiation, are important for frequencies greater than about 20 GHz. (For the radio scientist, 'clear air' does not necessarily imply high optical visibility, as 'hazy' conditions with high water vapour attenuation and scintillation are included whilst fog and other hydrometeor conditions are not.) Figure 1.2 gives an indication of the relative importance of rain, fog and atmospheric gases in the centimetric wavelength to visible regions of the spectrum. The onset of these effects is quite rapid, but now, as a result of the continuing demand for radio frequency bandwidth, much effort is going into exploring the potential use of the 'windows' (and even the absorption peaks) in the millimetric- and submillimetric-wave range. Optical visibility in fog or cloud for 0.1 and 1 gm^{-3} (as shown in Figure 1.2) is about 250 and 50 m, respectively. The three levels of rainfall rate indicated correspond to light drizzle, moderate rain and very intense rain; there is very large variation in the different regions of the world for the time percentages for which such rainfall rates may be expected.

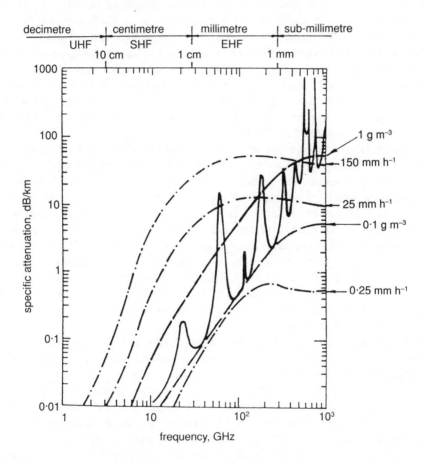

Figure 1.2   *Comparison of specific attenuation due to gaseous constituents,*
fog and precipitation near the Earth's surface
——— gaseous attenuation calculated assuming 1013 mb, 20 °C, 7.5 g/m³
– · – · – rain attenuation for rates shown
– – – fog attenuation for water content shown, at 20 °C

The *ionosphere* is the region of the atmosphere in which ionisation of gases (i.e. the concentration of free electrons) is particularly intense, and it extends from heights of approximately 60 km to 600 km. Figure 1.3 shows regions of the atmosphere in terms of the temperature profile. It also shows the layers of the ionosphere. This region takes a major control on radiowaves at frequencies below about 30 MHz, and radio paths operating at frequencies up to a few hundred megahertz can occasionally encounter severe disruption from ionospheric effects. The ionisation is caused by ultraviolet radiation from the sun acting in a height region where the molecular density is sufficiently low that recombination is not rapid. There are three main ionisation layers, designated by Sir Edward Appleton by the symbols D, E and F, in ascending height order. They vary in height and intensity according to the time of day, the time of year and the latitude. The typical diurnal variations of the E and F layers are indicated in Figure 1.4. The D layer is not directly comparable since it does not display reflection at near-vertical incidence, and so its height (about 70 km by day and 90 km by night) cannot be expressed in terms of a 'virtual height', the height from which rays appear to be reflected, (i.e. if the velocity were the same as in free space). Basic physics of the ionosphere is considered in Chapter 14.

The *effect of the ionosphere* on wave propagation is considered in Chapter 16; it is very dependent on frequency. The ionosphere may be regarded as a low-conductivity dielectric with refractive-index always less than unity, decreasing as electron density increases and/or frequency decreases. At ELF and VLF, the wavelength is very much larger than the separation between the D region of the ionosphere and the Earth's surface, and electromagnetic waves may propagate in this space as in a waveguide. Indeed, for frequencies less than 20 kHz the wave propagates in a single waveguide mode (with horizontal magnetic field and vertical electric field). At MF and HF the propagation may be represented by rays refracted and/or reflected according to the frequency, the electron density and elevation angle, while within the ionosphere, radio waves experience dispersion and a change of polarisation.

In general at frequencies between about 3 kHz and 30 MHz, two types of propagation enter into consideration: *ground wave* propagation along the Earth's surface and indirect *sky wave* (or 'ionospheric wave'). The attenuation with distance of a ground-wave signal depends on the characteristics of the ground and the radio frequency. This attenuation is lower over surfaces with a high conductivity, such as the sea. At VLF, ground waves are propagated to large distances, whereas they can be used over only very short distances at HF. Some of the behaviour of sky waves at HF is indicated in Figure 1.5. Below a certain 'penetration' frequency, and for a given launch angle, rays will be reflected back to Earth. Ionospheric refraction is less for higher frequencies and for higher elevation angles. As the elevation angle is increased, the height of reflection is increased, reaching a maximum for vertical incidence.

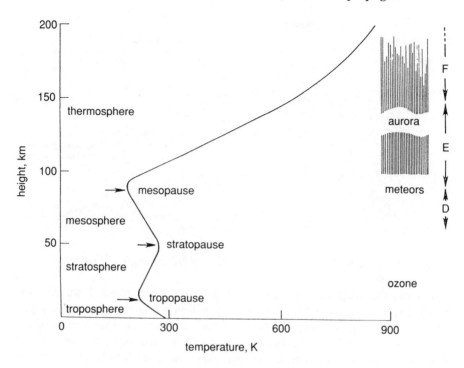

Figure 1.3   Regions of the Earth's atmosphere, showing the mean temperature
             profile, approximate heights of ionospheric regions and other
             features

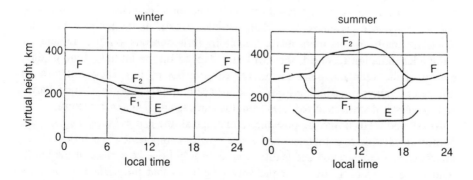

Figure 1.4   Typical diurnal variation of ionospheric layers

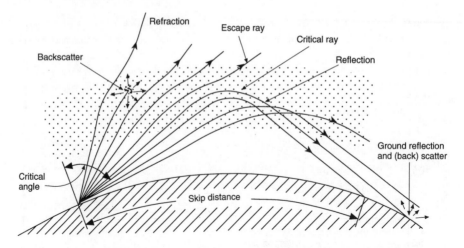

Figure 1.5   Dependence of ionospheric rays at HF on angle of incidence

If, now, the frequency is increased, the height of reflection increases, and eventually, at the *critical frequency*, the ray will penetrate the ionospheric layer, this critical frequency being dependent on the maximum electron density of the layer. For frequencies somewhat below the critical frequency, rays transmitted with a zenith angle below a certain *critical angle* will be reflected back to ground. Some energy is scattered from the ground back along the tracks to the transmitter site, and some energy is scattered forward (or reflected forward). 'High rays' and 'low rays' may arrive together at distances beyond a skip distance within which no signals are received. In addition, some energy is scattered back from ionospheric irregularities, and some is absorbed.

For transmission from one point on the Earth's surface to another, there will be a *maximum usable frequency*, less than the 'penetration frequency' and increasing with path distance. Because these characteristics depend on the ionisation density along the path in the ionosphere, which in turn depends on the amount of ultraviolet radiation received, they will also depend on the latitude, time of day, season of the year and point in the 11-year sunspot cycle. They will also be affected by bursts of radiation from solar surface eruptions, known as solar flares, and by bursts of streams of particles from the sun causing ionospheric storms.

At HF and VHF it may be possible to distinguish the *ground wave* as having a Norton *surface-wave* component and a *space-wave* component. The surface wave travels close to the ground and is diffracted by it to follow the surface of the Earth. 'Close' here means in terms of the wavelength, so that the surface wave may occupy the space below the ionosphere at VLF. By contrast, the surface wave is very near the surface and highly attenuated at UHF (and above), and if the antennas are a wavelength or more above ground, as is normal at these frequencies, the surface wave may be neglected in comparison with the space wave. When considering VHF line-of-sight effects in Chapter 4, it is mentioned that the space wave is made up from a *direct wave* and a *ground-reflected wave*. At MF it is not useful to separate the surface wave, direct wave and ground-reflected wave, but the *ionospheric (sky) wave* is distinct in leaving the transmitter in an upwards direction. A space wave is not distinguishable at low frequencies when the antenna height is low compared with the wavelength. The direct and ground-reflected waves then have equal magnitude and opposite phase and so give an almost zero resultant.

The amount of diffraction of the *surface wave* around the Earth depends on the ratio of the wavelength to the radius of the Earth, decreasing steadily as the wavelength is decreased. This wavebend is strengthened by the wavefront having a slight forward tilt due to a proportion of the radio energy propagating into, and being absorbed, by currents induced in the Earth. The rate of absorption, and so the rate of attenuation of the wave travelling above the Earth's surface, depends on the conductivity and permittivity of the ground. These characteristics differ considerably for different types of ground. At low frequencies the surface wave is mainly dependent on conductivity and is strongest for high conductivity. At higher frequencies a high permittivity is the important factor in giving a strong surface wave. Thus for all frequencies surface-wave propagation is best over sea and worst over dry land.

Although (as mentioned above) the direct wave and ground-reflected wave almost cancel for an antenna low to the ground (compared with the wavelength), there is a pronounced phase delay to the reflected ray (relative to the direct ray) as the antenna is raised above the ground. The space components become increasingly important, and the resultant field must be calculated as the vector sum of the space and surface waves. Although there may be only a slight increase in signal at first, there comes a region at which the rate of signal increase with height is almost constant before it falls off again. The rate of signal increase with height is referred to as the *height-gain factor*. It is a function of the ground conductivity and permittivity, the polarisation and the frequency. It is larger over sea than over land, and larger at VHF than HF. In addition, relative phase changes of the direct and reflected rays produce a series of maxima and minima.

Above 30 MHz the ionosphere becomes increasingly permeable to the sky wave as the frequency rises, and, except for occasional sporadic E layer effects, it may be considered that the ionosphere ceases to act as a reflector. However, space-to-Earth paths do continue to experience some ionospheric-scintillation and Faraday-rotation effects.

1.3   Characteristics of the wavebands

In this Section, consideration is given to each decade waveband, taking into account not only the propagation aspects but also a few of the system considerations that apply, and the resulting radiocommunication usage of the spectrum (current and proposed, primarily in the UK). Here 'radiocommunication' is taken in the broad sense of the ITU Radio Regulations, i.e. 'any transmission, emission or reception of signs, signals, writing, images and source of intelligence of any nature by means of radio waves'. Throughout the wavebands, considerations such as antenna design and bandwidth requirements are as important as those of propagation in determining this usage.

1.3.1 ELF (below 3 kHz) and VLF (myriametric waves, 3–30 kHz)

Atmospheric influences at ELF:	ionosphere forms upper boundary to waveguide propagation
Terrain influences at ELF:	Earth surface forms lower boundary to propagation
System considerations for ELF:	transmission of radio power requires enormous lengths of cable well earthed down at each end; very low information rate
Typical ELF services:	short-range (e.g. in water, between divers) and long-range (to water) submarine communications; ultrastable worldwide communication, mine and subterranean communication; remote sensing under ground.

At extremely low frequency (ELF), taken here to include frequencies up to 3 kHz, waves travel within the concentric-sphere cavity between the Earth and ionosphere, and also propagate deep into the Earth or sea. At 100 Hz, the attenuation in sea water is 0.3 dB/m (a third of that for a wave propagating in the Earth–ionosphere waveguide. This becomes 1 dB/m at 1 kHz, and more than 30 dB/m between 1 MHz and the optical window). Attenuation through average ground is about a tenth of that through sea water, a hundredth for dry ground. However, the transmission rate may be only about 1 bit/s, i.e. one of 1024 coded messages could be identified in 10 s. In addition, antenna noise and atmospheric noise become problems, the latter having a background level due to lightning strokes from major centres around the world which is added to by very large impulses from any nearby lightning. There are no frequency allocations in this frequency range.

Atmospheric influences at VLF:	D region of ionosphere forms upper boundary to propagation

Terrain influences at VLF:	Earth's surface forms lower boundary to propagating wave
System considerations for VLF:	even 100 m height for antennas is only a small fraction of a wavelength; difficult to make transmitter antennas directional; only low data rates
Typical VLF services:	world-wide telegraphy with ships; fixed services over long distances; navigational aids (e.g. Omega); electrical storm warning systems; time standards
Comments:	very few channels are available, and it is difficult to design efficient antennas, so the band is reserved for a few high power transmitters with massive antenna systems; frequencies are not allocated below 9 kHz.

At very low frequency (VLF), a wave may travel over long distances guided between the D layer and the Earth's surface. The wave is attenuated by loss of energy due to imperfect conductivity at the two surfaces, but only to the extent of about 2×10^{-3} dB/km at 10 kHz over sea, about 3×10^{-3} dB/km over land and 5×10^{-2} dB/km over ice. The D layer changes its height fairly regularly between 70 km by day and 90 km by night, and in view of the stability of the propagation conditions (and low attenuation), these frequencies are suitable for long-distance radionavigation systems, such as Omega, and standard-frequency emissions. Owing to their depth of penetration, these waves are used for underground communication and underwater communication to submarines (especially the lower frequencies); also for geological surveying applications. There are no frequency allocations below 9 kHz.

1.3.2 LF (kilometric waves, 30–300 kHz) and MF (hectometric waves, 300–3000 kHz)

Atmospheric influences at LF:	wave still below D region up to about 100 kHz; sky wave becomes distinct from ground wave above about 100 kHz, and gives a second contribution
Terrain influences at LF:	ground wave follows Earth curvature

System considerations for LF: even 100 m height for antennas is only a fraction of a wavelength; difficult to make transmitter antennas directional

Typical LF services: long-distance communication with ships; fixed services over long distances; broadcasting; radionavigational aids

Comments: antennas dictate that only vertical polarisation is realistic.

At frequencies up to about 100 kHz in the low-frequency (LF) band, the change in electron density from zero to maximum occurs within a distance which is small compared with a wavelength, and the layer may be regarded as an abrupt discontinuity acting as an almost perfectly reflecting surface. At frequencies above about 100 kHz, a sky wave may be separated from a ground wave, the ground wave showing more attenuation with distance at the higher frequencies. Fading occurs at shorter distances due to interference between the ground wave and sky wave, and at longer distances due to ionospheric fluctuations. The frequencies in this range are particularly suitable for medium-distance radionavigation systems (Decca and Loran) and for radio beacons.

Atmospheric influences at MF: sky wave separate from ground wave; surface wave for short distances and lower frequencies; ionospheric wave for longer distances, and upper frequencies, but also stronger at night, even at shorter distances

Terrain influences at MF: reflection

System considerations for MF: half-wavelength tower at 1 MHz is 150 m high (quarter-wave folded unipole is 75 m high) antennas can be directional using multi-elements; inverted L or T aerial (height 5 m), or ferrite-loaded coil in receiver.

Typical MF services: broadcasting; radionavigation; some land, maritime and aeronautical mobile; some fixed service

Comments: antennas quite large, but quite efficient.

The ground wave becomes severely attenuated at medium frequency (MF), but can still provide usable signals. By day the sky wave is much absorbed in the lower region of the ionosphere, but after sunset there is less absorption, with the result that the sky wave becomes predominant even at fairly short distances from the transmitter. MF waves are largely employed for national broadcasting, a station of moderate power having a service radius of about 100 km (at 500–1500 kHz) provided by ground wave. Beyond this distance comes a zone in which fading occurs, and these stations are not normally receivable in daylight beyond a distance of perhaps 250 km over land or 1000 km over water. The ionospheric wave is strongly absorbed in the D layer during the daytime. During darkness, however, when ionisation is low, the ionospheric wave is only slightly attenuated, and reception by the E or F layers is possible over longer distances. The frequency band from 1.5 to 3 MHz is unsatisfactory for distant communication because of large absorption in the daytime and too little ionisation at night.

1.3.3 HF (decametric waves, 3–30 MHz)

Atmospheric influences:	ionospheric wave only beyond skip distance, especially 3–6 MHz; surface wave only at short distance (but more over sea), especially 6–30 MHz
Terrain influences:	reflection (and scatter)
System considerations:	log-periodic array antennas (vertical or horizontal), vertical whip antennas, or horizontal dipole arrays
Typical services:	fixed point-to-point; land (within skip distance), maritime and aeronautical mobile; long-distance broadcasting
Comments:	beamed communication services.

Except at very close ranges to the transmitter, sky waves provide the principal propagation mechanism at high frequency (HF). Frequencies are selected for each particular requirement by means of 'maximum usable frequency' predictions. The region between 3 and 6 MHz is used mainly for surface-wave communication within the limits of a continent, while the part between 6 and about 30 MHz (depending on the time in the sunspot cycle) is used for ionospheric-wave long-distance intercontinental services. Within the skip distance, the ground wave is used for mobile land communication. Communication to great distances, including all round the world, is possible by successive ionospheric reflections.

1.3.4 VHF (metric waves, 30–300 MHz) and UHF (decimetric waves, 300–3000 MHz)

Atmospheric influences at VHF:	refraction and reflection by refractive-index irregularities producing transhorizon paths; some sporadic E and ionospheric-scatter transhorizon effects, and Faraday rotation and ionospheric scintillation on Earth–space paths
Terrain influences at VHF:	screening by major hills, but some diffraction into valleys; surface reflections off large areas (sea, lakes, flat ground) causing multipath effects on line-of-sight paths
System considerations for VHF:	multi-element dipole (Yagi) antennas, or slots, helixes etc.; several MHz per radio channel where required
Typical VHF services:	sound and (outside UK) television broadcasting (to about 100 km); land, aeronautical and marine mobile; portable and cordless telephones; aeronautical radionavigation beacons
Comments:	line-of-sight terrestrial transmissions and somewhat beyond.

At very high frequency (VHF), refractive-index effects in the troposphere become important, e.g. reflections from low layers causing multipath effects (often much less than multipath reflection effects from the Earth's surface). Reflections from higher layers may cause transhorizon interference. Ionospheric effects are very limited, but sporadic E-layer reflections in the ionosphere can cause transhorizon interference, at lower frequencies, up to distances of the order of 200 km at 60 MHz.

The surface wave is rapidly attenuated at VHF, and communication is by the space wave within the optical horizon, and slightly beyond. Diffraction allows short-range reception into built-up areas, though mobile systems are subject to screening by hills and to multipath effects caused by scatter or reflections off obstacles. However, the diffraction loss around hills is less than for UHF.

Atmospheric influences at UHF:	refraction effects; reflection from layers at lower frequencies; ducting possible at higher frequencies; refractive-index fluctuations — forward scatter beyond horizon above 500 MHz
Terrain influences at UHF:	screening by hills and collections of buildings
System considerations for UHF:	multi-element dipole (Yagi) antennas; wide bandwidths available; parabolic dishes for higher frequencies
Typical UHF services:	television broadcasting; some aircraft navigation, landing etc.; most surveillance and secondary radars; fixed (terrestrial point-to-point); mobile manpacks and vehicles; satellite mobile; satellite tracking, telemetry and command network; cellular radio; cordless telephones
Comments:	line-of-sight and very slightly beyond; also tropospheric scatter transhorizon for higher frequencies.

There are more severe screening effects from obstacles at ultra high frequency (UHF) than at VHF. At frequencies above about 500 MHz, tropospheric scatter provides a limited degree of reception at ranges up to about 300–600 km. Wider bandwidth per channel and more channels per waveband is attractive for television, as well as higher gain antennas. High gain antennas and waveguides practical for radar at higher frequencies, and still free from rain effects at 3 GHz, though some ducting problems.

1.3.5 SHF (centimetric waves, 3–30 GHz)

Atmospheric influences:	rain, hail, snow etc. cause very variable attenuation with frequency; refraction and ducting; refractive-index fluctuations cause scintillation
Terrain influences:	diffraction around buildings; screening close to buildings; scatter and reflection off elements of buildings, terrain and trees; sea reflection depends on wave height

System considerations: high-gain parabolic dishes and horns; waveguides; large numbers of channels on each carrier

Typical services: fixed (terrestrial point-to-point carrying multiple voice channels and several television channels); fixed satellite; radar; mobile services; future satellite mobile; remote sensing from satellites

Comments: utilisation still being increased, including above about 15 GHz (where atmospheric effects are worst).

SHF offers large numbers of wideband channels on each carrier, with opportunities for versatility in use of channels for multipath voice, TV or high-speed data. Extensive terrestrial line-of-sight networks have developed as well as Earth–space routes, usually with frequency sharing between services. Ducting on transhorizon paths may be a severe cause of interference, and multipath effects may cause severe fading on near-horizontal paths. Site-shielding from interference signals may employ hills or even groups of buildings (according to frequency). Absorption by rain, fog and cloud, as well as atmospheric gases, rapidly becomes a very severe constraint for system reliability at higher frequencies (see Figure 1.2), on both terrestrial and Earth–space paths.

1.3.6 EHF (millimetric waves, 30–300 GHz) and submillimetric waves (300–3000 GHz)

Atmospheric influences at EHF: rain, hail, snow etc., cause very severe attenuation and scatter; cloud, mist cause very variable attenuation with frequency; dust, smoke have some effect; refractive-index gradient; refractive-index fluctuations cause scintillation; absorption by atmospheric oxygen and water vapour

Terrain influences at EHF: screening by buildings or dense nearby trees

System considerations for EHF: paraboloid dish antennas become small

Typical EHF services: short line-of-sight communications, both fixed and mobile; some satellite applications; remote sensing from satellites

Comments:	frequency band developing as equipment elements become available, planning around atmospheric effects; allocations for terrestrial and satellite services up to 275 GHz.

The extremely high-frequency (EHF) region of the spectrum is now being developed as new technology becomes available. Precipitation, clouds and fog and atmospheric gases become a severe problem, though some 'windows' remain (see Figure 1.2) In metropolitan areas, high-capacity point-to-multipoint private-user fixed-link systems are appropriate to carry data, speech and video between customer buildings and the nearest network node. Mobile high-capacity systems may operate within public places (e.g. shopping areas and travel termini), domestic and office buildings (cordless telephones) and public transport. Radar systems have particular merits at EHF. Remote sensing of the surface and the atmosphere feature strongly in this part of the spectrum, both for research and operationally. Applications are considered fully in Chapter 19.

Atmospheric influences for submillimetric waves:	rain, hail, snow, cloud, mist, dust and smoke all cause very severe attenuation; localised refractive-index gradient (mirage); refractive-index fluctuations cause scintillation; absorption by atmospheric gases
Terrain influences for submillimetric waves:	screening by trees
Submillimetric wave system considerations:	mirror or lens antennas
Typical submillimetric wave services:	remote sensing and possibly short line-of-sight communications
Comments:	propagation restraints to communication are extreme, except for very short paths; equipment severely lacking

1.3.7 Infrared waves (3–430 THz) and optical waves (430–860 THz)

Atmospheric influences:	rain, hail, snow, cloud, mist, dust and smoke all cause very severe attenuation; localised refractive-index gradient (mirage); refractive-index fluctuations cause scintillation; absorption by atmospheric gases, notably carbon dioxide at higher (near-infrared) frequencies
Terrain influences:	screening by small objects
System considerations:	mirrors and lenses for antennas; lasers for near-infrared and optical
Typical services:	short-range and also indoor applications (e.g. to headsets in auditorium) for far-infrared; intruder alarms, smoke detectors, remote control systems (e.g. TV/video recorder etc.) and noninvasive spectrometry (e.g. beam passing through fruit etc.) for near-infrared; tellurometry and line-of-sight links for optical
Comments:	little communication use at present at far-infrared; no communication potential seen at present for near-infrared; these wavelengths are used a little for short line-of-sight communication in the atmosphere at optical wavelengths.

1.4 Radio services

The Radio Regulations of the ITU [1] list more than 30 service areas, and these make somewhat different demands on the radio spectrum according to the peculiarities of propagation and system considerations at different frequencies mentioned in Section 1.3. As described here, the services are grouped, and many have terrestrial and satellite aspects.

1.4.1 Fixed services

These services may be from *point-to-point*, or they may be from *point-to-multipoint* in a networking situation. Both may involve satellites or single terrestrial paths or a sequence of terrestrial relay links or a troposcatter path well beyond the horizon. In addition to these (tropospheric) path types, ionospheric and groundwave paths may exist for a range of distances, according to the frequency selected.

Frequency allocations for terrestrial *fixed* services extend in all regions of the spectrum. Although fixed-service communication at VLF may be worldwide, the information content is very low and data rate very slow. HF links still provide a background of long-distance communication, particularly inter-continental, whereas at SHF the paths may be essentially line-of-sight (or a succession of lines-of-sight), but the information content is very high. Such systems carry about 20% of the main trunk-public-telephone, data-link and video-circuit networks, some 1800 telephone channels (each of 4 kHz) occupying the bandwidth of one TV channel. Substantial use is also made of frequencies at the top of the UHF band for tropospheric-scatter (transhorizon) paths. These systems are used particularly for the transmission of an information bandwidth of a few hundred kilometres with high reliability where a series of line-of-sight paths is not practical. Various forms of local area network (LAN) are developing, as well as new short-range systems typically operating over less than 200 m. Multiservice systems are developing to carry many video and audio channels, and data, one- or two-way. *Fixed-satellite* services are long established at SHF (with allocations also at low EHF), having specific bands for Earth-to-space 'up' links and space-to-Earth 'down' links. However, these bands are generally shared with terrestrial fixed services, and determining the potential interference between such services forms a major part of Chapter 13.

1.4.2 Mobile services

Here again the radio paths may be via (one or more) satellites or line-of-sight terrestrial paths at SHF, within line-of-sight or slightly beyond at UHF (and further beyond line-of-sight at VHF) or via the ionosphere at lower frequencies. At VLF, the low information content referred to above may be acceptable in some communications with a submarine, where brief coded signals are sufficient.

Generally the upper VLF and lower LF *maritime-mobile* bands are used for telegraphy with coastal stations, with upper HF bands used for voice communication over long distances. MF is used for telephony over medium distances, and there are extensive VHF bands and some UHF bands used for near-line-of-sight ship-to-shore telephony, port operations service etc. The *aeronautical-mobile* bands extend from HF to SHF, again with a need for long, medium and short distances. (There is in addition the *aeronautical-fixed* service between fixed points provided primarily for the safety of air navigation and for the regular, efficient and economical operation of air transport.) *Land-mobile* applications extend from HF to EHF, the former including cordless telephones. Private mobile radio is at VHF and UHF, adjacent to public utilities (e.g. gas and electricity), paging, public radiophones, cellular phones etc. Diffraction round local obstacles and slightly beyond the radio horizon is effective at these frequencies, and small antennas may be used efficiently. The proliferation of hand-held radiotelephones has been most remarkable.

As with terrestrial-mobile, so satellite-mobile services comprise maritime-, aeronautical- and land-mobile services. Perhaps the most extensive of these is the *maritime-mobile-satellite* service, with now quite small terminals available. The ship-to-satellite and satellite-to-ship links are at high UHF, while the shore-to-satellite and satellite-to-shore links operate at SHF.

1.4.3 Radiodetermination

Radiodetermination (with or without the use of satellites) is the general term for systems which use radio waves to determine the position, velocity and/or other characteristics of an object, or to obtain information relating to these parameters. As a special case, *radio-navigation* systems (maritime or aeronautical, with or without satellites) are used for navigation (including obstruction warning), e.g. radars, radio beacons for direction finding etc. Radars mainly use UHF or SHF, the former for high-power long-range systems and the latter for higher-definition systems (and smaller systems for smaller ships and aircraft). Other navigational systems operate at VLF and LF for phase comparison systems such as Omega, Loran and Decca, where stability is required over long or moderately long distances, and at VHF, UHF and SHF for aeronautical instrument-landing systems and microwave landing systems. Satellite navigation transmissions are made at VHF/UHF, where the use of two frequencies allows correction of ionospheric distortion of the signal. Global-navigation-satellite systems (GNSSs) have revolutionised the art and science of navigation, giving 100 m accuracy for normal (uncoded) operation (using hand-held receivers) and centimetre precision over wide areas (e.g. for geodetic surveying) using coded operation.

1.4.4 Broadcasting

As with other services, there is a requirement for long-distance broadcasting (at LF, 'longwave'), and less distance (at MF, 'medium wave'), with several bands (at HF, 'shortwave') using the ionospheric (sky) wave, at VHF (for FM sound) and UHF for television. LF is only appropriate for those parts of the world where atmospheric noise is moderate to low. MF experiences severe mutual interference between transmissions at night. Broadcasting in the lower HF band is more attractive than in the MF band in the tropics because good coverage of a larger area may be achieved with relatively low atmospheric noise. The upper end of the HF band is used for intercontinental broadcasting. For localised broadcasting, VHF is more subject to transhorizon interference than is UHF, but the latter requires more fill-in transmitters to overcome diffraction losses in shadowed areas. *Satellite broadcasting* services have developed at SHF, where high definition will be possible. Reception may be by individual home receiver (direct broadcasting from satellites, DBS), or by centralised community systems. The up-link feeder systems form part of the fixed-satellite service. New terrestrial digital broadcasting services require new propagation studies at SHF.

1.4.5 Safety (emergency and distress) services

It could be claimed that these services have the greatest need of radio, and indeed the correct use of their allocated radio frequencies is a most serious matter. Again there is a wide range of frequencies used from HF to SHF for distress systems, which include the Emergency Position-Indicating Radio-Beacon (EPIRB) system designed to transmit from a small floating transmitter to a satellite at UHF. The fire, police and ambulance emergency mobile systems tend to be at VHF for vehicles and UHF for hand-held systems. Their fixed-link systems are generally at VHF. Essentially, existing systems will be integrated within the satellite-based Global Maritime Distress and Safety System which came into force in 1992 and will be worldwide in 1997.

1.4.6 Miscellaneous

Apart from the few main service areas outlined above, there remain many more. The *radioastronomy* service has allocations near many quantum-mechanical molecular-transition-line frequencies, and most of these are kept clear of other transmissions (though others have to operate on a secondary basis). The *Earth exploration-satellite* service is now well established as a means of examining resources and other characteristics of the Earth using active or passive sensors, and includes interrogation of instrumental airborne or Earth-based platforms. The *amateur* service also has allocations in the bands from MF to EHF, including amateur satellites. The *standard-frequency* and *time-signal* services transmit (with or without the use of satellites) to a high precision and are intended for general reception. The *meteorological aids* service involves radiosondes and hydrological observations as well as direct observation of weather patterns on the Earth's surface by optical and infrared sensors. Finally, the *space research* service, the original users of satellites, continue to have need of communication with low-orbit and geostationary orbit terminals as well as those in 'deep space' (greater than 2×10^6 km from Earth)

1.5   Frequency allocations and the ITU

The last Section indicated the many pressures for use of the electromagnetic spectrum, and it is helpful to see how specific frequencies are made available by international and national agreements. The general frequency-allocation plans are established at World Radiocommunication Conferences (WRCs) or sometimes Regional Radiocommunication Conferences (RRCs) convened by the ITU. Such conferences used to be called to examine specific problems, but they are now held every two years with agendas arranged two years earlier. The Conferences are empowered to make *allocations* of frequency bands for particular services to regions and/or administrations under special conditions. These are then listed in the Table of Frequency Allocations within the Radio Regulations. Subsequently, *assignments* are given by national administrations for radio stations to use specific frequencies (or radio-frequency channels) from specific locations under specified conditions. Any new or modified assignment has to be notified to the ITU Radiocommunications Bureau (BR) for inclusion in the Master International Frequency Register.

The ITU has some 175 member countries, and they are collectively responsible for formulating the Radio Regulations on the manner in which radio systems and services shall be operated. The ITU (founded in 1865) is the oldest intergovernmental organisation, and became the earliest of the Specialised Agencies of the United Nations. It now comprises the Telecommunications Standardisation Sector (ITU-T), the Radiocommunication Sector (ITU-R), the Telecommunications Development Sector and a General Secretariat. Each Sector has a Bureau (e.g. BR, above) forming a secretariat. The ITU-R now carries out its work in nine Study Groups, as follows: 1: Spectrum management; 2: Inter-service sharing and compatibility; 3: Radiowave propagation; 4: Fixed-satellite service; 7: Science services; 8: Mobile, radiodetermination, amateur and related satellite services; 9: Fixed service; 10: Broadcasting service (sound); 11: Broadcasting service (television). ITU-R Study Group 3 now combines the old CCIR Study Groups 5 and 6. Each of the Study Groups draws up Recommendations based on Questions for study on technical and operating matters relating to radiocommunications. Each allocated frequency band requires a set of agreed planning standards. Not only are these Recommendations essential to the Conference activity of the ITU, but those concerned with radiowave propagation have resulted from the participation of leading scientists and engineers and are now found quoted in the scientific literature as authoritative statements arrived at by consensus. Several will be referred to in other Chapters.

1.6   Conclusions

This introductory overview has looked at some of the propagation effects to be considered in this book, some of the services which require a knowledge of these propagation effects for their design, and finally the forum of the ITU-R where (among other things) international agreement is sought on modelling of the propagation processes to minimise interference between systems and maximise reliability.   In all the service areas (as considered in Section 1.4), there are continuing remarkable and rapid developments, constantly raising new questions about the effects of terrain and atmosphere on radiowave propagation.

Chapter 2

Electromagnetic-wave propagation

M.J. Mehler

2.1 Introduction

The objective of this Chapter is to introduce the principles of electromagnetic theory which are essential to an understanding of radiowave propagation. Where possible a practical engineering approach has been adopted. However, the material has been structured to provide those with a more mathematical background with a satisfactory account of the subject, together with references to further reading material.

Initially, propagation between two antennas in free space is considered by means of power flux concepts. This enables important ideas such as 'free-space path loss', 'antenna gain' and 'effective aperture' to be introduced. Maxwell's equations are presented and plane-wave solutions are derived as a means of introducing polarisation, wave impedance and electromagnetic power flow. Finally, radiation from a current distribution is examined and illustrated by deriving the fields of a dipole antenna.

2.2 Power budget

Radio-propagation engineers are often concerned with determining the power received by a distant antenna when the transmitted power is given. Such a calculation is the key to the design of radio systems and involves the determination of the power budget taking account of each element of the radio link. In practice, when both transmitter and receiver are in free space, this calculation can easily be made using a simple conservation-of-energy argument. Moreover, even when the antennas are not in free space the concept of a link budget, based on energy conservation, can be extended to take account of effects such as diffraction loss.

2.2.1 *Transmitting antennas*

Initially, consider a point source in free space radiating energy equally in all directions. If such an isotropic source radiates a total of P_t (W), then the power density F_0 at a distance d will be given by

$$F_0 = \frac{P_t}{4\pi d^2} \tag{2.1}$$

Note that such a source could not be physically realised since the presence of polarisation cannot be neglected. However, this concept is useful for specifying a reference against which antenna gain may be defined. In practice an antenna will possess a radiation pattern, as illustrated in Figure 2.1.

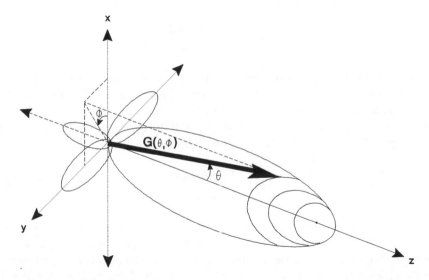

Figure 2.1   An antenna radiation pattern

If an antenna is fed with a power P_t, the power density $F_d(\theta,\phi)$ may be specified at a distance d in a given direction θ, ϕ. The gain of the antenna in the θ, ϕ direction may be defined with respect to an isotropic power density by taking the ratio of $F_d(\theta,\phi)$ with the flux density produced by an isotropic radiator at the same distance and fed with the same power P_t. Thus, the antenna gain is given by

$$G(\theta,\phi) = \frac{F_d(\theta,\phi)}{F_0} \tag{2.2}$$

The nature of the gain function $G(\theta,\phi)$ will depend on the type of antenna. Normally, antennas are used to concentrate the power flux in some specific direction and along this 'boresight' direction the antenna gain will be a maximum. Typically, wire antennas of the 'Yagi' type operating in the UHF bands yield gains of approximately 15 dBi, whereas large microwave earth-station antennas of about 3 m diameter can produce gains in the order of 50 dBi. Here the i denotes the gain with respect to that of an isotropic antenna.

2.2.2 Receiving antennas

The power density at a distance d in a given direction from a transmitting antenna is $F_d(\theta,\phi)$. A useful expression for this quantity can be found by combining eqns. 2.1 and 2.2, to yield

$$F_d(\theta,\phi) = \frac{P_t G_t(\theta,\phi)}{4\pi d^2} \tag{2.3}$$

where G_t is the gain of the transmitting antenna. A receiving antenna placed in this power flux will deliver a power P_r into a load connected to the antenna terminals or waveguide port. To facilitate the calculation of P_r we may consider the antenna to be presenting an effective collecting-aperture area A_e, to the incoming field. Hence, the received power becomes

$$P_r = \frac{P_t G_t(\theta,\phi) A_e}{4\pi d^2} \tag{2.4}$$

The concept of an effective aperture area is easily interpreted for reflector antennas, where a paraboloid of diameter D presents a projected area $\pi D^2/4$ to a normally incident wave. Even if ohmic losses in the antenna are neglected, some of the incident field will be reflected and hence not all of the incident energy will be absorbed in the antenna load. If the paraboloid is rotated, the incident wave will no longer be normal to the aperture. Hence, additional incident energy will be reflected and thus the received power in the load will be further reduced in a manner consistent with the antenna pattern. These considerations illustrate that the effective aperture area of an antenna is not necessarily equal to the physical area $A = \pi D^2/4$. Moreover, the effective area will depend on the direction of the incident field. It is possible to relate the effective aperture to the antenna gain through the relation [75]

$$A_e(\theta,\phi) = G(\theta,\phi)\frac{\lambda^2}{4\pi} \tag{2.5}$$

The ratio between the effective aperture and the physical area, in the direction of maximum gain, is a measure of antenna efficiency η, so that $\eta = A_e/A$. Combining this definition with eqn. 2.5 yields a useful expression for the maximum antenna gain, namely

$$G = \left(\frac{\pi D}{\lambda}\right)^2 \eta \qquad (2.6)$$

Although the above discussion has been concerned with reflector antennas, the concept of effective aperture area is also commonly applied to wire antennas.

2.2.3 Free-space transmission loss

Combining eqn. 2.4 with eqn. 2.5 permits the received power to be written in terms of the transmitting and receiving antenna gains G_t and G_r, respectively. The resulting expression takes the form

$$P_r = P_t G_t G_r \left(\frac{\lambda}{4\pi D}\right)^2 \qquad (2.7)$$

To interpret this result, note that, if both transmitting and receiving antennas possess a gain of unity, then in free space the ratio P_r/P_t depends on $(\lambda/4\pi d)^2$, which is simply the free-space loss.

2.2.4 Link power budget

Radio engineers usually express eqn. 2.7 in decibels, so that we may write

$$P_r(\text{dBw}) = P_t(\text{dBw}) + G_t(\text{dBi}) + G_r(\text{dBi}) - L_{BF}(\text{dB}) \qquad (2.8)$$

where

$$L_{BF} = 20\{\log_{10}(4\pi d) - \log_{10}(\lambda)\}$$

noting that $\lambda = c/f$ (c is the velocity of light in a vacuum $= 3 \times 10^8$ ms^{-1}) and taking f to be in MHz and d in km, the following standard result is obtained:

$$L_{BF} = 32.44 + 20\log_{10} f + 20\log_{10} d \qquad (2.9)$$

(Note that dBw is with respect to 1 W; dBi is the gain with respect to an isotropic antenna.)

Equation 2.8 can be extended to treat propagation between antennas which are not in free space by making allowance for additional sources of path loss. For example, propagation over a path with terrain features introduces loss in addition to that of free space. This additional diffraction loss, L_d, can easily be included into the link budget, so that eqn. 2.8 becomes

$$P_r = P_t + G_t + G_r - L_{BF} - L_d \qquad (2.10)$$

Clearly, this approach can be extended to include other loss mechanisms, such as antenna-feeder losses. Equation 2.10 forms the basis of many radio-system-link budget calculations.

2.3   Basic electromagnetic theory

The above discussion has only been concerned with the use of conservation-of-energy arguments where the electromagnetic field has been treated as a scalar power flux. In reality this model is inadequate and better agreement with experimental observation is achieved by treating electromagnetic effects in terms of the interaction of charged particles with a vector field. For engineering applications it is often sufficient to describe these electromagnetic phenomena by means of a classical field theory, namely Maxwell's equations. These equations relate the electric field E (V/m) and the magnetic field H (A/m) to the current density J (A/m^2) and the charge density ρ (C/m^3).

They may be written in the form [122]

$$\nabla \times E = -\mu \frac{\partial H}{\partial t} \qquad (2.11a)$$

$$\nabla \times H = J + \varepsilon \frac{\partial E}{\partial t} \qquad (2.11b)$$

$$\nabla .(\varepsilon E) = \rho \qquad (2.11c)$$

$$\nabla .(\mu H) = 0 \qquad (2.11d)$$

where ε and μ are the permittivity and permeability of the medium. They are often written as $\varepsilon = \varepsilon_0 \varepsilon_r$, and $\mu = \mu_0 \mu_r$, where $\mu_r = \varepsilon_r = 1$ in free space, and $\varepsilon_0 = 8.854 \times 10^{-12}$ F/m, $\mu_0 = 4\pi \times 10^{-7}$ H/m.

Significant simplification of Maxwell's equations can be achieved in free space where both J and ρ are zero. Taking the curl of eqn. 2.11a and using eqn. 2.11b to eliminate H gives

$$\nabla \times \nabla \times \boldsymbol{E} = -\mu_0 \varepsilon_0 \frac{\partial^2 \boldsymbol{E}}{\partial t^2} \qquad (2.12)$$

This may be further simplified by means of a standard vector identity to give

$$-\nabla^2 \boldsymbol{E} + \nabla(\nabla.\boldsymbol{E}) = -\mu_0 \varepsilon_0 \frac{\partial^2 \boldsymbol{E}}{\partial t^2} \qquad (2.13)$$

Since in eqn. 2.11c $\rho = 0$ and $\varepsilon = \varepsilon_0$ (constant), eqn. 2.13 may be further reduced to

$$\nabla^2 \boldsymbol{E} - \mu_0 \varepsilon_0 \frac{\partial^2 \boldsymbol{E}}{\partial t^2} = 0 \qquad (2.14)$$

This is the well known free-space wave equation for $\boldsymbol{E}$.

2.3.1 Plane-wave solutions

In general, solutions to eqn. 2.14 are difficult to obtain for many important engineering problems. However, a number of very useful properties of the electromagnetic field can be deduced by considering simple plane-wave solutions. The equation for a plane wave which is travelling in the positive z direction, with polarisation in the x direction, may be written as

$$\boldsymbol{E} = E_0 \cos(\omega t - kz)\boldsymbol{x} \qquad (2.15)$$

where E_0 is the amplitude of the wave, ω is the angular frequency $2\pi f$, and k the wave number $2\pi/\lambda$. Here, f (Hz) and λ (m) are the frequency and wavelength, respectively.

If we select a point along the wave of arbitrary constant phase, satisfying $\omega t - kz = \text{constant}$, then the velocity c of this point on the wave is $dz/dt = \omega/k = \lambda f$. It is easy to check that eqn. 2.15 is a solution of eqn. 2.14 provided that

$$k^2 = \omega^2 \mu_0 \varepsilon_0 \qquad (2.16)$$

Therefore, the wave velocity is given by

$$c = \frac{\omega}{k} = \frac{1}{\sqrt{\mu_0 \varepsilon_0}}$$

and it may be verified that $c = 3 \times 10^8$ m/s in free space.

To find the magnetic field corresponding to this plane-wave solution we substitute eqn. 2.15 into eqn. 2.11a to obtain

$$-\mu_0 \frac{\partial H}{\partial t} = \nabla \times \mathbf{E} = \frac{\partial}{\partial z} \{E_0 \cos(\omega t - kz)\} \hat{y}$$

$$-\mu_0 H = \hat{y}. E_0 k \sin(\omega t - kz) dt$$

so that

$$\mathbf{H} = H_0(\omega t - kz)\hat{y} \tag{2.17}$$

where

$$H_0 = \frac{E_0 k}{\omega \mu_0}$$

It can be seen that, for this plane-wave solution, the E and H fields are orthogonal, as illustrated in Figure 2.2.

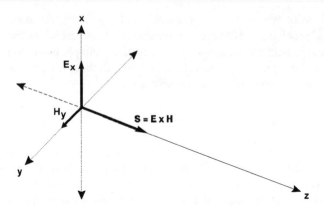

Figure 2.2   Relationship between the E and H fields

2.3.2 Wave impedance

It is evident that the ratio of the E- and H-field amplitudes in eqn. 2.15 and eqn. 2.17 is a constant, namely

$$\frac{|E|}{|H|} = \frac{E_x}{H_y} = \frac{\omega \mu_0}{k} = \sqrt{\frac{\mu_0}{\varepsilon_0}} \tag{2.18}$$

This ratio is called the free-space wave impedance since it has units of ohms, and is usually denoted by Z_0, which for this free-space case has an approximate value of 377 Ω. It is analogous to the ratio of voltage and current in an electric circuit.

2.3.3 Power flow and Poynting's theorem

In the same way that the product of voltage and current may be used to calculate power in an electric circuit, a vector product of E and H gives the power density in the electromagnetic field. For the plane-wave solution, the power flow along the Z direction is $(|E|\,|H|)/2$ W/m^2. Here the factor 1/2 in this expression has been introduced because E_0 and H_0 are peak values, whereas the power density is in root-mean-square units. Using eqns. 2.15 and 2.17, this expression reduces to

$$\frac{E_0^2}{2Z_0} = \frac{Z_0}{2} H_0^2$$

In general the E and H fields may not be orthogonal and in this case the above result is not valid. However, it is possible to show that the power flow in an electromagnetic field may be represented by the Poynting vector S [122], which points in the direction of power flow and has a magnitude equal to the power density, and is defined by

$$S = E \times H \qquad (2.19)$$

2.3.4 Exponential notation

Plane-wave solutions occur commonly in engineering problems. Indeed, it is possible to represent arbitrary field distributions by a sum of plane waves of varying amplitude, phase and direction. It is also possible to simplify the manipulation of plane-wave expressions by introducing exponential notation, namely

$$\text{Re}\left[\exp\{j(\omega t - kz)\}\right] = \cos(\omega t - kz) \qquad (2.20)$$

Hence, eqn. 2.15 is written as

$$E = E_0 \,\text{Re}\left\{\exp(j\omega t)\exp(-jkz)\right\}\hat{x} \qquad (2.21)$$

The $\exp(j\omega t)$ term is present because the field varies harmonically with time. Such steady-state solutions arise commonly in radio-engineering problems. In

Maxwell's equations the adoption of the harmonic time dependence of eqn. 2.21 implies that differentiation with respect to time is replaced by $j\omega$, so that eqns. 2.11a–d reduce to

$$\nabla \times E = -j\omega\mu H \tag{2.22a}$$

$$\nabla \times H = J + j\omega\varepsilon E \tag{2.22b}$$

$$\nabla.\varepsilon E = \rho \tag{2.22c}$$

$$\nabla.\mu H = 0 \tag{2.22d}$$

The wave equation corresponding to eqn. 2.14 follows by taking the curl of eqn. 2.22a and using eqn. 2.22b to eliminate H, and is given by

$$\nabla^2 E + k^2 E = 0 \tag{2.23}$$

It is easily shown that eqn. 2.21 is a solution of eqn. 2.23. Thus, using this shorthand notation, eqn. 2.21 becomes

$$E = E_0 \exp(-jkz)\hat{x} \tag{2.24}$$

When interpreting this equation, or any solution of eqn. 2.23, the need to multiply by $\exp(-jkz)$ and take the real part should be borne in mind.

2.4   Radiation from current distributions

So far the plane-wave solutions considered have been obtained without any reference to the current distribution which excites them. This is manifest by the arbitrary nature of E_0 in eqn. 2.15. In reality, E_0 would be determined by the source currents. More generally, if the field were represented by a superposition of plane waves, known as a plane-wave spectrum, the amplitude, phase and direction of each plane wave would be determined by the source-current distribution.

When calculating the field radiated by a current distribution, such as that supported on an antenna structure, it is often more convenient to work with a vector potential, namely A. This potential may be introduced through the standard result $\nabla.\nabla \times A = 0$, and it may be deduced from eqn. 2.11d that

$$\mu H = \nabla \times A \tag{2.25}$$

Substituting eqn. 2.25 into eqns. 2.11*a–d* yields a wave equation for *A* of the form

$$\nabla^2 A + k^2 A = -\mu_0 J \tag{2.26}$$

If the current distribution is bounded by a volume v_0, the field *A* in the free-space region outside the volume v_0 can be written explicitly in the form

$$A(r) = \frac{\mu}{4\pi} \int J(r_0) \frac{\exp(-jk|r - r_0|)}{|r - r_0|} dv_0 \tag{2.27}$$

Equation 2.27 is therefore a general solution to eqn. 2.26 for any current distribution and, as such, is a very useful result. The integration in eqn. 2.27 is taken over the source co-ordinates which are denoted by the subscript 0. The details of the derivation of eqns. 2.26 and 2.27 are beyond the scope of this Chapter, but may be found in Reference 122. The field quantities *E* and *H* can be derived once *A* is known, first by application of eqn. 2.25 to find *H* and then, from eqn. 2.22*b*, *E* is deduced, noting that in free space *J* = 0.

2.4.1 Radiation from a short current element

Consider the short current element of length *dl* as shown in Figure 2.3, which is assumed to carry a constant current *I*. Since the element is *z* directed, we may write down the expression for *A* in the form

$$A_z = \frac{\mu_0}{4\pi} \int_{-dl/2}^{+dl/2} I \frac{\exp(-jkr)}{r} dz, \qquad A_x = A_y = 0 \tag{2.28}$$

Equation 2.28 follows directly from eqn. 2.27 by using the approximation *dl* ≪ *r*, which implies that the observation distance is much greater than the length of the current element. The integral in eqn. 2.28 is easily evaluated to yield

$$A_z = \frac{\mu_0}{4\pi} I.dl \frac{\exp(-jkr)}{r} \tag{2.29}$$

The fields of antennas are usually presented in spherical co-ordinates; hence we convert eqn. 2.29 as follows:

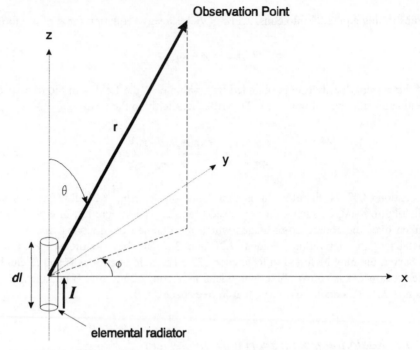

Figure 2.3   Radiation from a short current element

$$A = \frac{\mu_0 Idl}{4\pi} \frac{\exp(-jkr)}{r} (\hat{r}\cos\theta - \hat{\theta}\sin\theta)$$

To find H, eqn. 2.25 is applied to give

$$H = \frac{1}{\mu_0} \nabla \times A = \frac{Idl\sin\theta}{4\pi}\left(\frac{jk}{r} + \frac{1}{r^2}\right)\exp(-jkr)\hat{\phi} \tag{2.30}$$

Equation 2.30 shows that the magnetic field is arranged in loops around the z axis of the current element. Furthermore, the field appears to fall off with distance at two rates, as controlled by the terms $(1/r)$ and $(1/r^2)$. A more complex expression results when eqn. 2.22b is used to calculate E, namely

$$E = \frac{jz_0 Idl\cos\theta}{2\pi k_0}\left(\frac{jk}{r^2} + \frac{1}{r^3}\right)\exp(-jkr)\hat{r} - \frac{jz_0 Idl\sin\theta}{4\pi k_0}\left(-\frac{k_0^2}{r} + \frac{jk}{r^2} + \frac{1}{r^3}\right)\exp(-jkr)\hat{\theta}$$

$$\tag{2.31}$$

It is evident that the E field possesses both transverse and longitudinal components, together with a distance dependence which extends up to $(1/r^3)$ terms.

When r is large relative to the wavelength, the only important terms are those that vary as $(1/r)$. In this radiation, or far-field, zone the E and H fields reduce to

$$E = jZ_0 Idlk_0 \sin\theta \, \frac{\exp(-jkr)}{4\pi r} \, \hat{\theta} \qquad (2.32a)$$

$$H = jIdlk_0 \sin\theta \, \frac{\exp(-jkr)}{4\pi r} \, \hat{\phi} \qquad (2.32b)$$

When $r < \lambda$, the terms $(1/r^2)$ are important and cannot be neglected in the calculation. These near-field components are reactive and equate to the storage of magnetic- and electric-field energy. Hence the antenna may be visualised as being surrounded by reactive stored energy for $r < \lambda$, with radiation fields which become more dominant as r increases. As would be expected, the radiation field possesses a Poynting vector which points radially outward from the current element. This is evident if one calculates the radiated power S_r (RMS) from Poynting's theorem applied to eqn. 2.32, namely

$$S = \frac{E \times H}{2} = \frac{|E||H|}{2}(\hat{\theta} \times \hat{\phi}) = \frac{|E||H|}{2} \hat{r} \qquad (2.33)$$

$$S_r = \frac{I^2 dl^2 Z_0 k^2 \sin^2\theta}{32\pi^2 r^2} \hat{r} \qquad (2.34)$$

This result shows that the power falls off as an inverse-square law and flows radially away from the current element, as expected.

To calculate the total radiated power P_r it is necessary to integrate eqn. 2.34 over a complete sphere enclosing the current element, which yields the surface integral

$$P_r = \frac{I^2 dl^2 Z_0 k^2}{32\pi^2} \int_0^{2\pi} \int_0^{\pi} \frac{\sin^2\theta}{r^2} (r^2 \sin\theta) d\theta d\phi$$

Integration over ϕ is straightforward, but to facilitate the integration over θ the substitution $u = \cos\theta$ is made, to yield

$$P_r = \frac{I^2 dl^2 Z_0 k^2}{16\pi} \int_{-1}^{+1} (1 - u^2) du$$

which reduces to

$$P_r = \frac{I^2 dl^2 Z_0 k^2}{12\pi} \tag{2.35}$$

2.4.2 Radiation resistance

If a source is connected to the antenna terminals and supplies a given power P_R, we may regard the power as being supplied to an effective radiation resistance R_{rad}. Although this is not a physical resistance, it absorbs a power equal to that being radiated in to the far field. Since I is a peak value, equating the power supplied to R_{rad} to that which is radiated gives

$$\frac{1}{2} I^2 R_{rad} = \frac{I^2 dl^2 Z_0 k^2}{12\pi}$$

Therefore

$$R_{rad} = \frac{Z_0 (kdl)^2}{6\pi} = 80\pi^2 \left(\frac{dl}{\lambda}\right)^2 \tag{2.36}$$

As an example, at a frequency of 1 MHz (λ = 300 m) and with an antenna of length dl = 1 m, R_{rad} = 0.0084 Ω. In practice it would be difficult to match an antenna with such a small radiation resistance to a source and achieve efficient power transfer. As a result, practical wire antennas are usually of the order of one wavelength in length.

2.4.3 The halfwave dipole

The starting point for calculating the fields radiated by the short current element was an assumed current distribution. Since the short current element is assumed to be small compared with the wavelength, taking the current to be uniform along the element is a good approximation. However, for the halfwave dipole a physically realisable distribution would take the form of a cosinusoidal distribution falling to zero at the wire ends.

Such a distribution can be justified as a good approximation theoretically and may be written as

$$I = I_0 \cos\left(\frac{2\pi}{\lambda} Z\right) \qquad -\frac{\lambda}{4} \le Z \le \frac{\lambda}{4} \tag{2.37}$$

Again, the antenna is assumed to be located along the z axis, as illustrated in Figure 2.3.

Although the algebra is more complex with the current distribution of eqn. 2.37, a calculation analogous to that used for the short current element may be repeated to derive the halfwave dipole fields. For brevity, the results are merely stated here and in the far field are given by [75]

$$\mathbf{E} = \frac{jI_0 Z_0}{2\pi r} \exp(-jkr) \frac{\cos\left(\frac{\pi}{2}\cos\theta\right)}{\sin\theta} \hat{\theta} \tag{2.38a}$$

$$\mathbf{H} = \frac{jI_0}{2\pi r} \exp(-jkr) \frac{\cos\left(\frac{\pi}{2}\cos\theta\right)}{\sin\theta} \hat{\phi} \tag{2.38b}$$

The radiation resistance of the halfwave dipole can be calculated by following the approach adopted for the short current element, and may be shown to be approximately 73 Ω.

Chapter 3

Basic-radio system parameters

L.W. Barclay

3.1 Introduction

This Chapter introduces a number of topics which should be useful for the succeeding Chapters. Antenna gain, radiated power and transmission loss are commonly used terms when describing systems, but the precision given by the internationally agreed definitions of these terms is necessary if ambiguity is to be avoided. System performance is governed not only by the transmission loss, under some stated conditions, but also by the variability of the signal in time or space, which can then be described in statistical terms, and by the level of background signals — either broadband noise or interfering transmissions. The statistical probability distributions in common use are introduced, and the benefits of diversity reception are outlined. The types of radio noise are described together with the ways in which noise power from a number of sources may be combined for use in performance prediction.

3.2 Propagation in free space

As described in Chapter 2, a transmitter with power P_t in free space which radiates isotropically in all directions gives a power flux density s at distance d of

$$s = \frac{P_t}{4\pi d^2} \tag{3.1}$$

Using logarithmic ratios,

$$S = -41 + P_t - 20\log d \tag{3.2}$$

where S is the power flux density in decibels relative to 1 Wm^{-2}, P_t is the power in decibels relative to 1 kW and d is in km.

The corresponding field strength e is given by

$$e = \sqrt{120\pi s} = \frac{\sqrt{30P_t}}{d} \tag{3.3}$$

This relationship applies when the power is radiated isotropically.

A $\lambda/2$ dipole has a gain in its equatorial plane of 1.64 times (see Section 3.3) and in this case the field strength is

$$e \approx \frac{7\sqrt{P_t}}{d} \tag{3.4}$$

The power available, P_r, in a load which is conjugately matched to the impedance of a receiving antenna is

$$P_r = sa_e \tag{3.5}$$

where a_e is the effective aperture of the antenna which is given by $\lambda^2/4\pi$ for an isotropic radiator.

Thus, again for an isotropic radiator in free space,

$$P_r = \frac{P_t}{4\pi d^2}\frac{\lambda^2}{4\pi} = P_t\left(\frac{\lambda}{4\pi d}\right)^2 \tag{3.6}$$

and the free-space basic transmission loss is the ratio P_t/P_r (see Section 3.6).

3.3   Antenna gain

The ITU Radio Regulations [1] formally define the gain of an antenna as 'The ratio, usually expressed in decibels, of the power required at the input of a loss-free reference antenna to the power supplied to the input of the given antenna to produce, in a given direction, the same field strength or the same power flux density at the same distance'. When not specified otherwise, the gain refers to the direction of maximum radiation. The gain may be considered for a specified polarisation. Gain greater than unity (positive in terms of decibels) will increase the power radiated in a given direction and will also increase the effective aperture of a receiving antenna.

Depending on the choice of the reference antenna a distinction is made between:

(*a*) absolute or isotropic gain (G_i), when the reference antenna is an isotropic antenna isolated in space. (Note that isotropic radiation relates to an equal intensity in all directions. The term 'omnidirectional radiation' is often used for an antenna which radiates equally at all azimuths in the horizontal plane; such an antenna may radiate with a different intensity for other elevation angles);

(*b*) gain relative to a halfwave dipole (G_d), when the reference antenna is a halfwave dipole isolated in space whose equatorial plane contains the given direction;

(*c*) gain relative to a short vertical antenna conductor (G_v) much shorter than one-quarter of the wavelength, normal to the surface of a perfectly conducting plane which contains the given direction.

An isotropic radiator is often adopted as the reference at microwaves and at HF, while a halfwave dipole is often adopted at VHF and UHF, where this type of antenna is convenient for practical implementation. A short vertical antenna over a conducting ground is an appropriate reference at MF and lower frequencies where ground-wave propagation is involved and this usage extends to sky-wave propagation at MF and, in older texts, at HF.

The comparative gains of these reference antennas, and of some other antenna types, are given in Table 3.1.

Table 3.1 Gain of typical reference antennas

Reference antenna	g_i	G_i * (dB)	Cymomotive force (for a radiated power of 1 kW) (V)
Isotropic in free space	1	0	173
Hertzian dipole in free space	1.5	1.75	212
Halfwave dipole in free space	1.65	2.15	222
Hertzian dipole, or short vertical monopole, on a perfectly conducting ground [†]	3	4.8	300
Quarter-wave monopole on a perfectly conducting ground	3.3	5.2	314

* G_i = 10 log g_i
[†] For the Hertzian dipole, it is assumed that the antenna is just above a perfectly conducting ground.

3.4   Effective radiated power

The Radio Regulations also provide definitions for effective or equivalent radiated power, again in relation to the three reference antennas:

(i)   equivalent isotropically radiated power (EIRP): the product of the power supplied to the antenna and the antenna gain in a given direction relative to an isotropic antenna (absolute or isotropic gain). Specification of an EIRP in decibels may be made using the symbol dBi;

(ii)   effective radiated power (ERP) (in a given direction): the product of the power supplied to the antenna and its gain relative to a halfwave dipole in a given direction;

(iii)   effective monopole radiated power (EMRP) (in a given direction): the product of the power supplied to the antenna and its gain relative to a short vertical antenna in a given direction.

Note that ERP, which is often used as a general term for radiated power, strictly only applies when the reference antenna is a halfwave dipole.

An alternative way of indicating the intensity of radiation, which is sometimes used at the lower frequencies, is in terms of the 'cymomotive force', expressed in volts. The cymomotive force is given by the product of the field strength and the distance, assuming loss-free radiation. The values of cymomotive force when 1 kW is radiated from the reference antennas are also given in Table 3.1.

3.5   The effect of the ground

The proximity of the imperfectly conducting ground will affect the performance of an antenna. In some cases, where the antenna is located several wavelengths above the ground, it may be convenient to consider signals directly from (or to) the antenna and those which are reflected from the ground as separate signal ray paths. When the antenna is close to, or on, the ground it is no longer appropriate to consider separate rays and then the effect may be taken into account by assuming a modified directivity pattern for the antenna, including the ground reflection; by modifying the effective aperture of the antenna; or by taking account of the change in radiation resistance etc. A discussion of this for the ground-wave case, where the problem is most difficult, is contained in Annex II of ITU-R Recommendation P.341.

Information concerning the electrical characteristics of the surface of the Earth are contained in further ITU-R texts: Recommendations P.527 and P.832.

3.6   Transmission loss

The concept of free-space basic transmission loss was introduced in eqn. 3.6 as the ratio P_t/P_r. Transmission losses are almost always expressed in logarithmic terms, in decibels, and as a positive value of attenuation; i.e.

$$L_{bf} = 10\log\left(\frac{P_t}{P_r}\right) = P_t - P_r = 20\log\left(\frac{4\pi d}{\lambda}\right) \qquad (3.7)$$

or

$$L_{bf} = 32.44 + 20\log f + 20\log d \qquad (3.8)$$

where f is in MHz and d is in km.

The concept of transmission loss may be extended to include the effects of the propagation medium, and of the antennas and the radio system actually in use:

Free-space basic transmission loss L_{bf} relates to isotropic antennas and loss-free propagation;

Basic transmission loss L_b includes the effect of the propagation medium, e.g.
(*a*) absorption loss (ionospheric, atmospheric gases or precipitation);
(*b*) diffraction loss;
(*c*) effective reflection or scattering loss, in the ionospheric case including the results of any focusing or defocusing due to curvature of a reflecting layer;
(*d*) polarisation coupling loss; this can arise from any polarisation mismatch between the antennas for the particular ray path considered;
(*e*) aperture-to-medium coupling loss or antenna gain degradation, which may be due to the presence of substantial scatter phenomena on the path;
(*f*) effect of wave interference between the direct ray and rays reflected from the ground, other obstacles or atmospheric layers.

Transmission loss L includes the directivity of the actual transmitting antennas, disregarding antenna circuit losses;

System loss L_s is obtained from the powers at the antenna terminals; and

Total loss L_t is the ratio determined at convenient, specified, points within the transmitter and receiver systems.

The relationships between these loss ratios are illustrated in Figure 3.1. It is important to be precise when using the terms, and the full definitions are given in ITU-R Recommendation P.341.

3.7 System performance

Using the concepts of transmission loss discussed in Section 3.6, and including the appropriate loss in the propagation medium, to be dealt with in later Chapters, the available signal power at the receiver may be determined. This signal power will give a satisfactory system performance if the background level in the channel is sufficiently low; i.e. if the signal-to-noise ratio or the signal-to-interference ratio reaches defined standards. Required signal-to-noise ratios are given, for example, in ITU-R Recommendation F.339-7 [3], while the range of signal-to-interference ratios used as a basis for the determination of frequency re-use are given in various texts relating to specific services.

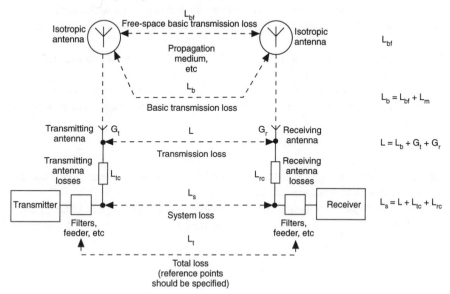

Figure 3.1   Graphical depiction of terms for transmission loss

Note that, for many systems, particularly newer systems involving digital techniques with high bit rates, it is not sufficient just to specify the signal-to-background-noise or signal-to-interference ratio. The spread of the signal in time (multipath) and in frequency (the Doppler spread) are of great importance and these factors may be indicated in various ways, e.g. as dispersion or by referring to impulse response, correlation bandwidth etc. These aspects, of key importance in the performance of digital systems, will be discussed in later Chapters.

3.8   Radio noise

There is a number of types of radio noise that must be considered in any design; though, in general, one type will pre-dominate in particular circumstances and that should be taken into account in the system design. Broadly, the noise can be divided into two types: noise internal to the receiving system and noise external to the receiving antenna.

The internal noise is due to losses in the antenna circuits or in the transmission line, or is generated in the receiver itself. It has the characteristics of thermal noise (i.e. white Gaussian noise). External noise arriving at the receiving antenna may be due to:

(i)   atmospheric noise generated by lightning discharges, or resulting from absorption by atmospheric gases (sky noise);
(ii)   the cosmic background, primarily from the Galaxy, or from the sun;
(iii) broadband man-made noise generated by machinery, power systems etc.

In some cases, e.g. noise emanating from computer systems, such noise may have considerable frequency variability.
   The noise power due to external sources P_n can conveniently be expressed as a noise factor f_a, the ratio of the noise power to the corresponding thermal noise, or as a noise temperature t_a:

$$f_a = \frac{P_n}{kt_0 b} = \frac{t_a}{t_0} \qquad (3.9)$$

where k is Boltzmann's constant $= 1.38 \times 10^{-23}$, t_0 is the reference temperature, taken as $288\,\mathrm{K}$, and b is the noise bandwidth of the receiving system in Hz.
   Note there is some confusion in the currently used terminology but here f_a is the numerical noise factor, and the term 'noise figure' F_a is used for the logarithmic ratio, so that

$$F_a = 10 \log f_a \qquad (3.10)$$

The available noise power in decibels above 1 W is given by

$$P_n = F_a + B - 204 \qquad\qquad \mathrm{dBW} \qquad (3.11)$$

where $B = 10 \log b$.
   When measured with a short vertical grounded monopole the corresponding vertical component of the RMS field strength is given by

$$E_n = F_a + 20 \log f_{MHz} + B - 95.5 \qquad\qquad \mathrm{dB\ (\mu V/m)} \qquad (3.12)$$

   Minimum, and some maximum, values for the external-noise figures are shown in Figures 3.2 and 3.3 (taken from ITU Recommendation P.372-6 [9]). Generally, one type of noise will predominate, but where the contributions of more than one type of noise are comparable, the noise factors (not the figures in decibels) should be added. Atmospheric noise due to lightning varies with location on the Earth, season, time of day and frequency. Maps and frequency correction charts are given in Recommendation P.372-6. Man-made noise varies with the extent of man-made activity and on the use of machinery, electrical equipment etc. The relationship for a range of environments is also given in the Recommendation. Note that some care may be needed to ensure that an appropriate curve is selected since the noise generated, for example in a 'business' area, may differ from country to country.

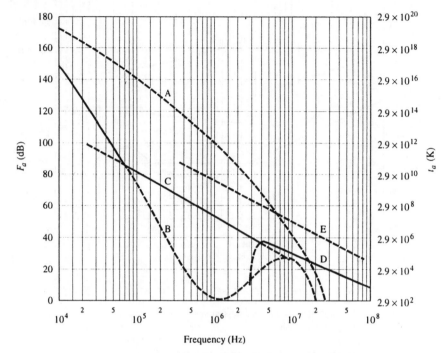

Figure 3.2   Noise figure F_a against frequency: 10 kHz to 100 MHz

A: atmospheric noise, value exceeded 0.5% of time
B: atmospheric noise, value exceeded 99.5% of time
C: man-made noise, quiet receiving site
D: galactic noise
E: median business-area man-made noise
———— minimum noise level expected

Both atmospheric and man-made noise are impulsive in character and an assessment based wholly on the noise power is likely to be inadequate. In some cases the dominant feature which determines system performance will be a parameter derived from the amplitude probability distribution of the noise. The Recommendation includes this information, together with examples of the prediction of system performance. However, in other cases, such as for digital systems, the characteristic duration and repetition rate of noise impulses may be important.

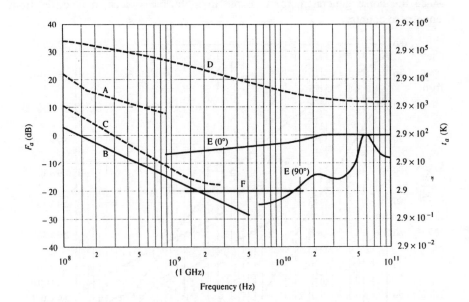

Figure 3.3　　Noise figure F_a against frequency: 100 MHz to 100 GHz

        A:　estimated median business-area man-made noise
        B:　galactic noise
        C:　galactic noise (toward galactic centre with infinitely narrow beamwidth)
        D:　quiet Sun (1/2° beamwidth directed at Sun)
        E:　sky noise due to oxygen and water vapour (very narrow-beam antenna);
            upper curve, 0° elevation angle; lower curve, 90° elevation angle
        F:　black body (cosmic background), 2.7 K
        ──────　minimum noise level expected

    When noise power is the appropriate parameter, the limit to system performance depends on the overall operating noise factor *f* from the combination of external and internal (receiver and antenna) sources. When the temperature of the antenna circuits etc. is t_0,

$$f = f_a + f_c f_t f_r - 1 \qquad (3.13)$$

where f_c and f_t are the losses (available input power/available output power) for the antenna circuit and transmission line, respectively; f_r is the noise factor of the receiver and f_a is the noise factor due to the external noise sources.

3.9   Fading and variability

Both signals and noise are subject to variations in time and with location. These changes in intensity arise from the nature of a random process, from multipath propagation, from movements of the system terminals or the reflecting medium, from changes in transmission loss etc.   A knowledge of the statistical characteristics of a received signal may be required in the assessing of the performance of modulation systems etc.

Statistics of the signal variability are also required for spectrum planning and for predicting the performance of systems. For these purposes it is important to know, for example:

(*a*) the signal level exceeded for large percentages of time or location (e.g. for the determination of quality of the wanted service or of the service area);
(*b*) the signal level which occurs for small percentages of time (e.g. to determine the significance of potential interference or the feasibility of frequency reuse).

In some cases signals are subject to rapid or closely spaced variations, superimposed on a slower variability. In such cases it may be possible to treat the phenomena separately, say by using a long receiver integration time or by 'averaging' the level of the signal (e.g. with AGC) so that the time interval adopted encompasses many individual short term or closely spaced fluctuations.

3.9.1 Normal (Gaussian) distribution

When the value of a parameter results from the cumulative effect of many processes, each of which has the same central tendency, the probability density $p(x)$ has a bell-shaped distribution peaking, at a central, mean, value $\bar{x}$. For n discrete values of a variable x, measured at regular intervals of time or location etc.,

$$\bar{x} = \frac{\sum x_n}{n} \tag{3.14}$$

Where the distribution is symmetrical, the values of the mean $\bar{x}$, the mode m, which is the most frequently recurring value, and the median x_{50}, the middle value when the individual values are listed in order, are all identical.

The shape of such a symmetrical, centrally peaking distribution is often that of the normal, Gaussian, distribution:

$$p(x) = \frac{1}{\sigma\sqrt{2\pi}} \exp\left\{-\frac{1}{2}\left(\frac{x-\bar{x}}{\sigma}\right)^2\right\} \tag{3.15}$$

where σ is a normalising parameter, the standard deviation; σ^2 is also called the variance.

$$\sigma = \sqrt{\frac{\sum(x_n - \bar{x})^2}{n}} \tag{3.16}$$

The cumulative probability function $F(x)$, for this distribution is given by

$$F(x) = \frac{1}{\sigma\sqrt{2\pi}} \int_{-\infty}^{x} \exp\left\{-\frac{1}{2}\left(\frac{t-\bar{x}}{\sigma}\right)^2\right\} dt \tag{3.17}$$

Statistical tables giving values for both $p(x)$ and $F(x)$ are readily available, as is graph paper with a graticule such that the normal cumulative-probability function appears as a straight line.

An approximation for the half of the distribution where $x < \bar{x}$ is given by

$$F(x) = \frac{\exp\left(-y^2/2\right)}{\sqrt{2\pi}\left\{0.661y + 0.339\sqrt{y^2 + 5.51}\right\}} \tag{3.18}$$

$$y = \frac{\bar{x} - x}{\sigma}$$

The upper half of the distribution may be obtained by using the above equation with $y = (x - \bar{x})/\sigma$, in this case giving $1 - F(x)$.

Table 3.2 gives some examples taken from the distribution.

Table 3.2 The normal distribution

Occurrence	
68 %	within 1 σ
95.5 %	within 2 σ
90 %	less than + 1.28 σ
99 %	less than + 2.33 σ
99.9 %	less than + 3.09 σ

In fact, in radiowave propagation, a normal distribution of signal power etc. only occurs when there are small fluctuations about a mean level, as might be the case when studying scintillation. Predominantly, it is the normal distribution of the logarithms of the variable which gives useful information: the log-normal distribution discussed below.

3.9.2 Log-normal distribution

In the case of a log-normal distribution, each parameter (the values of the variable itself, the 'mean', the standard deviation etc.) is expressed in decibels and the equations in Section 3.9.1 then apply. The log-normal distribution is appropriate for very many of the time series encountered in propagation studies, and also for the variations with location, for example within a small area of the coverage of a mobile system. Note that, when a function is log-normally distributed, the mean and median of the function itself are not the same: the centre of the distribution, when $F(x) = 0.5$, is still the median, whereas the mean of the numerical values is given by $\bar{x} + \sigma^2 / 2$.

Figure 3.4 is an example of normal probability paper, where a normal distribution is plotted as a straight line with a slope dependent on the standard distribution.

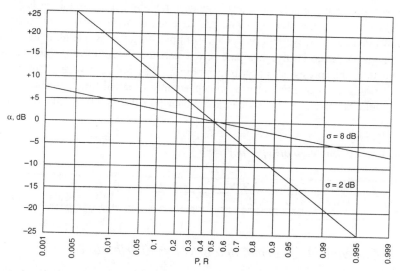

Figure 3.4   *Log-normal probability, showing two different distributions with standard deviations of 2 and 8 dB*

3.9.3 Rayleigh distribution

The combination of a number (say, more than three) of component signal vectors with arbitrary phase and comparable amplitude leads to the Rayleigh distribution. Thus, this is appropriate for situations where the signal results from multipath or scatter. In this case

$$p(x) = \frac{2x}{b^2} \exp\left(-\frac{x^2}{b^2}\right) \qquad (3.19)$$

and

$$F(x) = 1 - \exp\left(-\frac{x^2}{b^2}\right) \qquad (3.20)$$

where b is the root mean square value. (Note that x and b are numerical amplitude values, not decibels.)

For this distribution the mean is $0.886b$, the median is $0.833b$, the mode is $0.707b$ and the standard deviation is $0.463b$.

It is useful to note that, for small values of $F(x)$, $F(x) \approx x^2 / b^2$, so that when x is a voltage amplitude its power decreases by 10 dB for each decade of probability. However, this is not a sufficient test to determine whether a variable is Rayleigh distributed, since some other distributions have the same property. This property is shown in Table 3.3, which gives some examples from the Rayleigh distribution. Special graph paper is also available on which the distribution is plotted as a straight line, this presentation is used in Figure 3.5; note, however, that such a presentation greatly overemphasises the appearance of small time percentages and care should be taken that this does not mislead in the interpretation of plotted results.

Table 3.3   Rayleigh distribution

$F(x)$	0.999	0.99	0.9	0.5	0.1	0.01	0.001	0.0001
$20 \log(x)$	+10 dB	+8.2	+5.2	0	-8.2	-18.4	-28.4	-38.4

3.9.4 Combined log-normal and Rayleigh distribution

In a number of cases there will be a composite variation in the signal, in which rapid or closely spaced fluctuations, which may be due to multipath or scatter, follow a Rayleigh distribution but the mean of these, measured over a longer period of time or a longer distance, is itself subject to log-normal distribution.

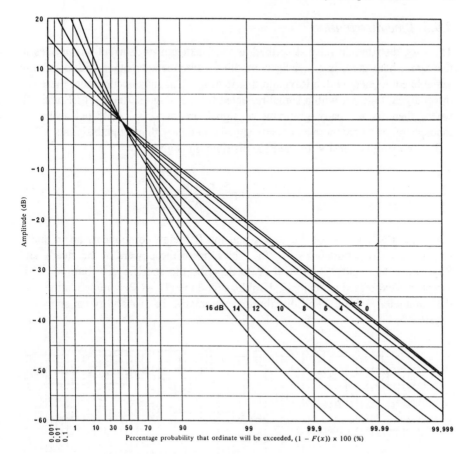

Figure 3.5   *Combined log-normal and Rayleigh distribution (with σ of log-normal distribution as parameter)*

This distribution is given by Boithias [58]:

$$1 - F(x) = \frac{1}{\sqrt{2\pi}} \int_{-\infty}^{\infty} \exp\left\{-x^2 \exp(-0.23\sigma u)\, \frac{u^2}{2}\right\} du \qquad (3.21)$$

where σ is the standard deviation (in decibels) of the log-normal distribution. This combined distribution is given in Figure 3.5. An alternative representation, normalised at the 50% probability, is given by Picquenard [147] and reproduced in ITU-R Report 266. This combination of distributions has also been studied by Suzuki, and his proposed formulation has been evaluated by Lorenz [125].

3.9.5 Rice distribution

The Rice distribution (also described as the Nakagami-*n* distribution) applies to the case where there is a steady, nonfading, component together with a random variable component with a Rayleigh distribution. This may occur where there is a direct signal together with a signal reflected from a rough surface; at LF and MF where there is a steady groundwave signal and a signal reflected from the ionosphere; or where there is a steady signal together with multipath signals.

The probability density for the Rice distribution is given by

$$p(r) = \frac{2r}{b^2} \exp\left(-\frac{r^2 + a^2}{b^2}\right) \tag{3.22}$$

where the RMS values of the steady and the Rayleigh components are *a* and *b* respectively. A parameter $K = a^2/b^2$, the ratio of the powers of the steady and Rayleigh components, is often used to describe the specific distribution. *K* is sometimes expressed in decibels. [Boithias (and the ITU-R Recommendation) use the parameter $= 1/K$]. The distribution, parametric in probability, is shown in Figure 3.6.

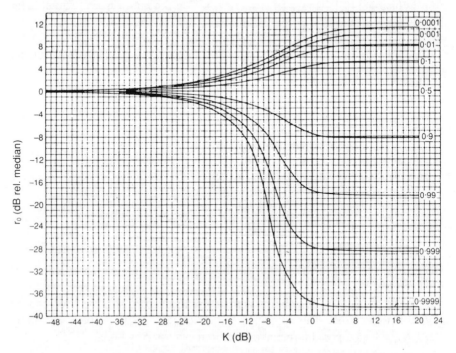

Figure 3.6   Nakagami–Rice distribution function F(r) [values of F(r) are shown on the curves]

In most cases, the power in the fading component will add to the power of the steady signal, where, for example, the multipath brings additional signal modes to the receiver. In some other cases, the total power will be constant where the random component originates from the steady signal.

3.9.6 Other distributions

Many further asymmetrical distributions have been studied and utilised in propagation studies. Griffiths and McGeehan [97] have compared some of these such as the exponential, Gamma, Wiebull, chi-squared, Stacy and Nakagami-*m* distributions. It may be appropriate to include such distributions in models of particular propagation behaviour. For example, Lorenz [125] has suggested that the Suzuki distribution is appropriate for VHF and UHF mobile communication in built-up areas and forests, while the Wiebull distribution is appropriate for area-coverage statistics where line of sight paths occur frequently.

However, before embarking on the use of an unfamiliar and complicated distribution, the user should be sure that the uncertainty and spread in the observations is small enough that the use of the distribution will result in a significant improvement in the accuracy of the model. The difference between various distributions for values between, say, 10% and 90% occurrence will often be small, and it is only in the tails of the distribution, where observations may be sparse, that a distinction could be made.

For applications concerned with the quality and performance of a wanted signal, or with the interference effects of an unwanted signal, it is seldom necessary to consider both tails of a distribution at the same time. In some cases half-log-normal distributions, applying a different value of the standard deviation on each side of the median, will be adequate. The mathematical elegance of a complete distribution should be weighed against the practical convenience of using the appropriate half of a more common distribution for the occurrence percentages of interest.

3.10 Fading allowances

For service planning, the specified signal-to-noise ratio for the required grade of service will probably include an allowance for the rapid fading which will affect the intelligibility or the bit-error ratio of the system. It may still be necessary to allow for other variations (hour-to-hour, day-to-day, location-to-location) of both signal-to-noise, which are likely to be log-normally distributed, but uncorrelated. It is appropriate to do this by first determining the overall median signal-to-noise ratio expected and then to apply a log-normal distribution to this where the variance σ^2 is obtained by adding the variances of each contributing distribution; i.e.

$$\sigma^2 = \sigma_1^2 + \sigma_2^2 + \dots \tag{3.23}$$

An example of this procedure is given in ITU-R Report 266. The procedure may, if necessary, be extended still further to include the probable error of the prediction due to the sampling involved in establishing the method etc. Where no allowance for this is included, the prediction has a confidence level of 50%, since one-half of the specific cases encountered are likely in practice to be below the predicted level. An assessment may be made of the probable error and, by applying a normal distribution, an allowance may be made for any other desired confidence level. This has been described by Barclay [51].

3.11 Diversity

The fading allowance necessary to achieve a good grade of service may demand economically prohibitive transmitter powers and antenna gains. In any case, the use of excessive radiated power conflicts with the need for good spectrum utilisation. Techniques for overcoming this problem include coding and diversity. Particularly for circuits with rapid fading, such as those where Rayleigh or Rician fading dominates, copies of the signal with the same characteristics are available displaced in time, position, frequency or, for some types of propagation, with angle or orthogonal polarisation. For example, a signal message may be repeated later in time if, when first transmitted, a fade had reduced the signal-to-noise ratio. For digital systems this process may be automated by the use of an error-detecting code and using a method of automatic repeat requests (ARQ) if errors are detected in the received signal. More modern techniques use sophisticated error-correction and error-detection codes to combat the effects of fading, and these may take account of the expected patterns of error occurrence. Spread-spectrum signals, both direct-sequence and frequency-hopping, employ techniques to take advantage of the frequency-selective nature of fading, and this is discussed in later Chapters.

Diversity techniques utilise two or more samples of the signal obtained from separated antennas, or sometimes from duplicated transmissions on several frequencies. These signals are then combined in the receiver to produce an output with a smaller fading variability. Signals may be combined by techniques such as:

(i)   selection of the stronger or strongest;
(ii)   combining the output of channels with equal gain;
(iii) weighting the combination according to the signal-to-noise ratio of the channel (maximal ratio combining).

Figure 3.7 shows the distributions for two-element diversity, where the signals are uncorrelated and each have a Rayleigh distribution, for various methods of combination. The corresponding distribution for four-channel diversity (as might be employed for tropospheric scatter systems) is shown in Figure 3.8. In fact, a substantial advantage is still obtained if the signals are partially correlated. Figure 3.9 shows, for two-element-selection diversity, the effect of varying the correlation coefficient.

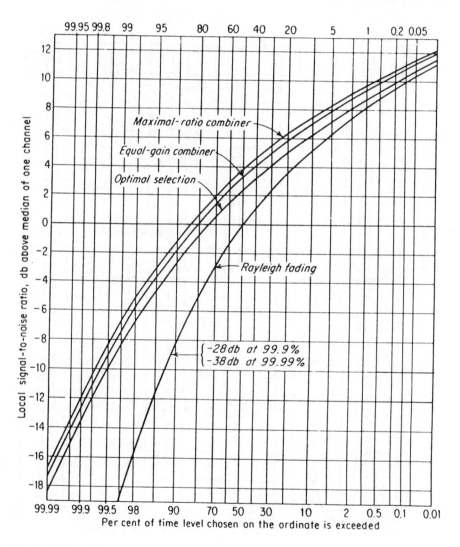

Figure 3.7 Two-element diversity

The advantage given to system performance by the application of diversity may be expressed in two ways, either:

(*a*) as diversity improvement — the ratio of the time percentages, with and without diversity, for which the signal fade depth exceeds a specified level, see eqn. 8.12; or
(*b*) as diversity gain — the increase in the signal level exceeded when using diversity for a specified time percentage.

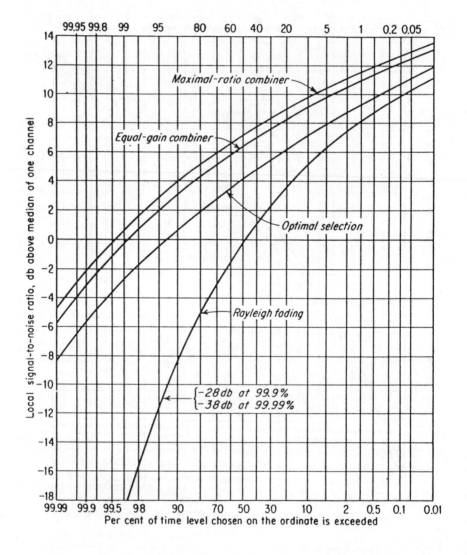

Figure 3.8   Four-element diversity

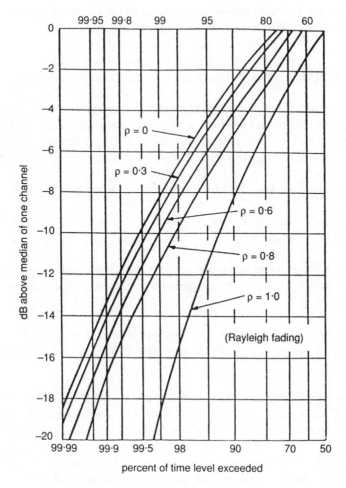

Figure 3.9   Two-element diversity, Rayleigh fading for various degrees of correlation

3.11.1 Correlation coefficient

The correlation coefficient is obtained as

$$r = \frac{\Sigma\{(x - \bar{x})(y - \bar{y})\}}{\left\{\sum(x - \bar{x})^2 \sum(y - \bar{y})^2\right\}^{1/2}} = \frac{\sum(xy) - n\bar{x}\,\bar{y}}{n\sigma_x \sigma_y} \qquad (3.24)$$

where x and y refer to simultaneous, or appropriately time-shifted, pairs of values from the two distributions and n is the number of pairs.

3.12   Reliability

Using the techniques above, the reliability of a system may be specified. Although precise definitions have not, so far, been universally adopted, the set of terms given in ITU-R Recommendation P.842-1 [31] has been applied, for example, to the assessment of HF broadcasting. The Recommendation includes:

Reliability: probability that a specified performance is achieved;

Circuit reliability: probability for a circuit that a specified performance is achieved at a single frequency;

Reception reliability: probability for a circuit that a specified performance is achieved by taking into account all transmitted frequencies associated with the desired signal; and

Service reliability: probability that a specified performance will be achieved at a specified percentage of the service area by taking into account all transmitted frequencies.

These definitions distinguish between 'basic' and 'overall' reliability where the limiting background is, respectively, noise, or the combination of noise and interference. For a given radio service, the definitions may need to be adapted to the requirements of that service. For example, for broadcasting applications, the term service reliability is replaced by the term broadcast reliability, and is calculated for a specified number of test points within the nominal service area.

3.13   ISDN performance requirements

For system planning it is often appropriate to consider a worst case. At MF and HF, where ionospheric modes are involved it may, for example, be useful to make predictions for sunspot minimum conditions and perhaps for the season which has the highest noise level, or the lowest MUF etc. At UHF and higher frequencies, where climatic effects are very important, it is valuable to know the expected performance in terms of a specified parameter in the worst month in a period of 12 consecutive months. The fraction of time during which a preselected threshold is exceeded in the worst month of a year is referred to as 'the annual worst-month time fraction of excess', and the appropriate statistic would be the long term average of that fraction. The average annual worst-month time fraction of excess p_w is calculated from the annual value p by use of a conversion factor Q, the ratio of two fractions. Typical values of Q are given in ITU-R Recommendation P.841 [29]. Definitions are in Recommendation P.581, and the topic is further discussed in the old ITU (CCIR) Report 723.

Chapter 4

Introduction to diffraction, reflection and scattering

D.F. Bacon

4.1   Diffraction basics

People do not have a natural feel for diffraction. With visible light it is normally too small to see. With radio it occurs at much larger scales, but we cannot see it.

A simple view of both light and radio is that it travels in straight lines (unless something specific causes reflection or refraction). The 'ray' is a familiar idea. In radio so is the concept of an expanding wavefront, which also suggests power radiating straight outwards. Figure 4.1 illustrates how a plane wavefront would advance if propagation consisted simply of energy flowing in straight lines. The power flowing through each point of a wavefront would simply carry on in a set of parallel rays to make the new wavefront position. On this basis an obstruction would cast a sharp shadow.

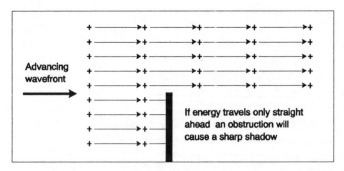

Figure 4.1   Over-simple straight-ahead model

In fact, radio shadows do not have sharp edges. Moreover, the variation of radio intensity at the edge of a shadow has an interesting and unexpected shape.

In Figure 4.2 uniform radio energy from the left falls on a totally absorbing obstruction cutting off the lower half of the radiation. On the right the resulting illumination is represented by a graph with a horizontal intensity scale plotted against height in relation to the obstacle. Without the obstacle this would be a uniform value shown as a broken line. With the obstacle present, the illumination changes to the solid line. Above the obstacle the illumination shows ripples around the original value. Below the obstacle the illumination decreases continuously. At the optical-cutoff point the illumination has fallen to –6 dB, or a quarter of the previous power. Somewhat above this point, the largest ripple gives a stronger signal than before by over 1 dB.

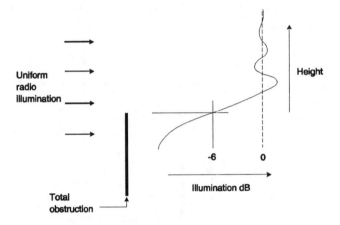

Figure 4.2   The shape of a radio shadow

The first reason for this shape is that radio energy does not simply travel in a straight-ahead manner. An early description of what actually happens came from the Danish scientist Christiaan Huygens, who devised a method for finding subsequent positions of an advancing wavefront. This is illustrated in Figure 4.3, in which we again have a plane wavefront advancing from the left. The energy from each point radiates in all directions to form many small spherical wavefronts, which Huygens called wavelets. It is the envelope of these wavelets which forms the new wavefront.

Note that the process of building up a new wavefront from the envelope of Huygens' wavelets involves a sideways interchange of energy. Some energy will contribute to the new wavefront after travelling at an angle to the general direction of propagation. This will tend to fill in the shadow behind an obstruction, as illustrated in Figure 4.4.

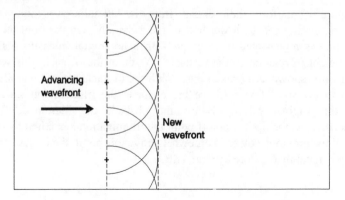

Figure 4.3   Huygens' wavelets

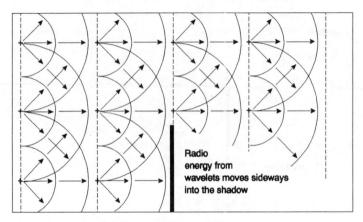

Figure 4.4   Huygens' wavelets filling a radio shadow

This shows that energy can leak into a radio shadow, but it does not explain the ripples in intensity above the obstruction in Figure 4.2. The exact calculation of diffraction is complicated, but a qualitative explanation can be derived from a construction known as the Cornu spiral.

Figure 4.5 shows a wavefront which has reached the line A–A'. We wish to consider the resulting signal level at point B, making allowance for the sideways redistribution of energy illustrated in Figure 4.4. We will take a set of rays from uniformly spaced points along the wavefront and allow them to converge onto B.

Radio signals form vector fields, and to add the various contributions at B we must take account of both amplitude and phase. We will do this graphically by adding arrows head-to-tail, representing, let us say, the addition of electric-field contributions.

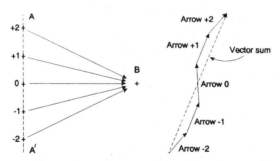

Figure 4.5   *Vector addition of contributions from a wavefront*

The ray which takes the shortest and most direct route to B gives the central, vertical, arrow in Figure 4.4. This is the longest, and we will call it 'arrow 0'. The adjacent rays above and below give 'arrow +1' and 'arrow –1', respectively. These are both slightly shorter than arrow 0, since they have had to travel further, and are coming out of the wavefront at an angle to the average direction of power flow. Owing to the longer path length, arrows +1 and –1 also have a phase difference relative to arrow 0, illustrated by an inclination to the right. In the same way, arrows +2 and –2 are shorter still, and leaning even more to the right.

In Figure 4.5 we have stopped at a total of five arrows. The amplitude and phase of their vector sum is represented by an arrow from the tail of arrow –2 to the head of arrow +2, indicated by the broken line. If the process of adding contributions is continued indefinitely, the sequences of arrows above and below 'arrow 0' will form two spirals. The Cornu spiral, which results from taking a large number of very small arrows, is reproduced in Figure 4.6. The complete double spiral, with each end circling inwards to the points X and Y, represents free-space propagation, with no obstruction to contributions arriving at B from anywhere on the advancing wavefront. The resultant in this case will be the vector from X to Y.

If we now imagine a rising obstruction progressively cutting off contributions to the spiral, starting at the bottom of the wavefront in Figure 4.5, the form of the curve in Figure 4.2 can now be visualised. Remember that the vector sum is the line joining one end of the spiral to the other. As the obstruction rises, arrows are progressively removed from the lower spiral, starting at X. Thus the lower end of the resultant will travel round an increasing spiral, while the upper end remains stationary at Y. This gives the ripples above the obstruction. At the optical-cutoff point we have just half the complete spiral, or a drop of 6 dB from free-space propagation. For increasing shadowing beyond this point the tail of the vector addition will spiral steadily inwards towards Y, giving continuously increasing attenuation.

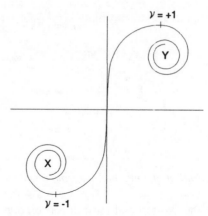

Figure 4.6   Cornu spiral

Figure 4.5 shows a two-dimensional slice through a wavefront. The principle, however, is the same for a real diffraction situation in three dimensions, as illustrated in Figure 4.7. Here circles are drawn on the advancing wavefront to represent equally spaced phase changes for contributions at point B. Although the geometry is more complicated, contributions from these zones will again form a Cornu spiral, and a wide variable-height obstruction will produce the same variation of illumination.

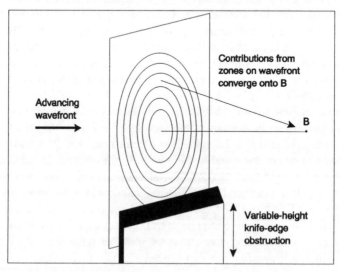

Figure 4.7   Three-dimensional diffraction

Figure 4.8 shows the actual variation of illumination for knife-edge diffraction. It is known as the Fresnel–Kirchoff curve, and is plotted as diffraction loss against a geometrical parameter v, which is explained in Section 4.3. Loss increases downwards so that the stronger signals are higher on the graph.

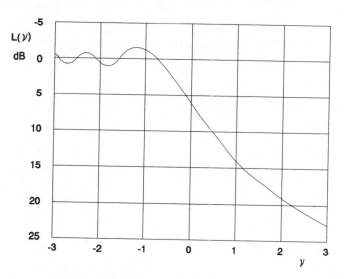

Figure 4.8   The Fresnel–Kirchoff knife-edge diffraction curve

The loss represented by the curve in Figure 4.8 is referred to as *diffraction loss*. It is understood to be in addition to any other losses which may occur along a propagation path, including free-space loss. For this reason it is sometimes called *excess diffraction loss*. However, the simple term 'diffraction loss' will nearly always mean 'the additional loss due to diffraction'.

4.2   Geometry related to diffracton

Before diffraction is considered in more detail, it will be useful to describe some geometrical methods which are frequently relevant.

4.2.1 Curved-Earth geometry

Many diffraction calculations concern the height of obstructions along a radio path. If this is more than a few kilometres long, the curvature of the Earth should be taken into account. For paths up to a few hundred kilometres in length, the following approximations are useful.

The 'bulge' b of the Earth's curvature above a true straight line between two points A and B on its surface, as shown in Figure 4.9a, is given by

$$b = \frac{d_1 d_2}{2r_e} \qquad (4.1)$$

The 'drop' h of the Earth's curvature below the tangent plane at a point P, as shown in Figure 4.9b, is given by

$$h = \frac{d^2}{2r_e} \qquad (4.2)$$

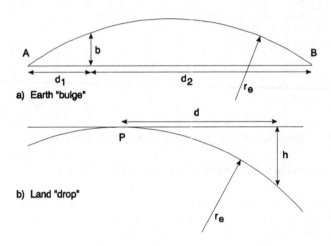

a) Earth "bulge"

b) Land "drop"

Figure 4.9   Curved-Earth geometries

The approximations used in curved-Earth geometry normally include ignoring the fact that vertical dimensions, such as b and h in Figure 4.9 are not strictly vertical in relation to the reference plane. Also, it is usually immaterial whether horizontal distances are measured along a true straight line or over the Earth's curved surface. It is, however, important that all dimensions, both horizontal and vertical, including the effective Earth radius r_e, should be in the same units.

4.2.2 *String analysis for locating obstacles*

Figure 4.10 shows the cross-section of a radio path containing hills. The terrain profile will consist of a set of co-ordinate pairs giving distance and height above sea level. The problem is to compute whether there is radio line of sight between the transmitter T and the receiver R and, if not, where the obstacles are.

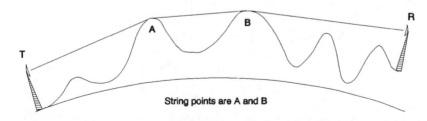

Figure 4.10   String analysis to locate obstructions

This is usually done by 'string analysis'. An imaginary string is stretched over the path from T to R. The profile points touched by the string are obstacles; if the string does not touch any point the path is line of sight.

One method for doing this is to start at T and compute the elevation angles, as seen by an imaginary observer, of all profile points, including R, and choose the highest. This is the first string point. If it is R then the path is line of sight. In Figure 4.10 the first string point will be A. Move the observer to A and repeat the process until R is the point with the highest elevation.

The geometry for computing the elevation angle of a profile point is shown in Figure 4.11. The curve AB is sea level between the observation point O and the profile point P at distance d. The line OC is parallel to the tangent plane AT. The various heights in Figure 4.11 are:

h_0 = observation point height above sea level;
h_p = profile point height above sea level;
h_d = drop below the tangent plane due to Earth curvature;

If θ is small, it is given in radians by PC/OC. Thus, noting that h_d is given by eqn. 4.2, we obtain:

$$\theta = \frac{h_p - h_0}{d} - \frac{d}{2r_e} \tag{4.3}$$

In eqn. 4.3 the first term gives θ for a flat Earth, and the second term accounts for Earth curvature.

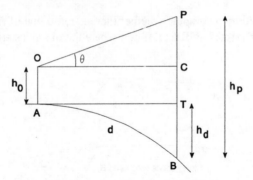

Figure 4.11   Elevation of point with curved Earth

4.3   Fresnel geometry

Figure 4.12 shows the essentials of Fresnel geometry. It represents a radio path diffracting over a knife edge obstacle of height h above the straight line joining transmitter T to receiver R. The obstacle height h can be either positive or negative, as illustrated in Figure 4.12a and 4.12b, respectively. The distances from T and R to the top of the obstacle are d_1 and d_2.

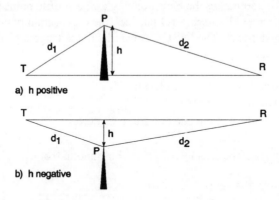

Figure 4.12   Fresnel geometry

Fresnel geometry is based on the path-length difference between a ray taking the direct path TR, and the indirect path TPR. Where the obstacle height h is much less than d_1 and d_2 the difference Δd in path length can be approximated by

$$\Delta d = \frac{h^2(d_1 + d_2)}{2d_1 d_2} \tag{4.4}$$

The loss due to diffraction over a knife-edge obstacle is normally computed from a parameter v which represents the geometry according to

$$v = h\sqrt{\frac{2(d_1 + d_2)}{\lambda d_1 d_2}} \tag{4.5}$$

where λ is the wavelength in the same units as d_1, d_2 and h.

Equation 4.5 can be arranged in other ways, and several forms are given in Recommendation 526-4 [14]. Note that in eqn. 4.5 v has the same sign as h.

Equations 4.4 and 4.5 show that v is a simple function of the path difference expressed in wavelengths:

$$v = 2\sqrt{\frac{\Delta d}{\lambda}} \tag{4.6}$$

This allows v to be mapped onto the Cornu spiral. For instance, setting Δd in eqn. 4.4 to a quarter wavelength gives $v = 1$. This corresponds to where the Cornu spiral has turned through 90° relative to its centre, as indicated in Figure 4.5.

The significance of the geometrical parameter v is apparent in the mathematical formulation of the Cornu spiral. Assuming $h \ll d_1$, $d_2 \gg \lambda$, the spiral is given to a good approximation by plotting $S(v)$ against $C(v)$ where:

$$C(v) = \int_0^v \cos\left(\frac{\pi s^2}{2}\right) ds \tag{4.7}$$

and

$$S(v) = \int_0^v \sin\left(\frac{\pi s^2}{2}\right) ds \tag{4.8}$$

In terms of numerical integration, these represent the summation of equi-length vector contributions from equally spaced points on the wavefront. Equations 4.7 and 4.8 also show that v equals the curved length of the Cornu spiral measured from the optical-cutoff point at $v = 0$. Further information on the complex Fresnel integral can be found in [113].

4.4   Fresnel clearance

The Cornu spiral shows that a radio path requires a certain amount of clearance around the central ray if the signal expected from free-space propagation is to be received. This is the principle of Fresnel clearance. It applies particularly to the design of point-to-point radio links, where communication is required along a single radio path. This is best done by using directional antennas and a line-of-sight path to use the lowest power necessary to overcome any fading mechanisms, thus causing minimum interference to other links using the same frequency.

Fresnel clearance is quoted in terms of Fresnel zones. The first Fresnel zone encloses all radio paths for which the additional path length does not exceed half of the wavelength (i.e. a phase change of $180°$). The second Fresnel zone encloses all paths for which the additional path does not exceed one wavelength, and so on.

The radius of a Fresnel zone at any point along a radio path can be obtained by replacing Δd in eqn. 4.4 by the required number of wavelengths, changing the obstruction height h to the Fresnel radius F, and solving for F.

Thus the radius of the first Fresnel zone F_1 is given by setting Δd to $\lambda/2$, which, for d_1 and d_2 both much more than F_1, gives

$$F_1 = 2\sqrt{\frac{\lambda d_1 d_2}{d_1 + d_2}} \quad = 17.3\sqrt{\frac{d_1 d_2}{fd}} \tag{4.9}$$

where f is measured in Ghz and d in km.

The shape of a Fresnel zone is shown in Figure 4.13. For a given Fresnel clearance no obstructions should exist in the three-dimensional volume produced by rotating the ellipse around the direct ray.

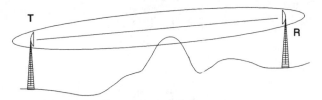

Figure 4.13   Fresnel zone around the direct ray

Equation 4.9 is an approximation which is invalid near the antennas. It gives zero Fresnel zone radius at each antenna, whereas the actual Fresnel zones extend behind the antennas, which are at the foci of the ellipse, as drawn in Figure 4.7. For most practical questions of path clearance this is unimportant.

Referring to the Cornu spiral in Figure 4.6, it can be seen that the first Fresnel zone is more than enough clearance. The edges of the first zone are where the individual vectors have turned through $180°$, well beyond the maximum amplitude of each spiral. In practice a clearance of 0.6 of the first Fresnel zone diameter is normally considered adequate for point-to-point links, corresponding to $v = -1.1$.

4.5   Diffraction loss

In some situations the radio planner cannot ensure Fresnel clearance. For instance, in broadcasting and mobile radio, propagation is frequently not line-of-sight, while for all types of radio service, when evaluating possible interference an unwanted signal may, usefully, be shadowed by an obstruction. In cases like these we need to calculate the loss due to diffraction over the obstacle.

The exact calculation of diffraction loss requires the evaluation of Fresnel integrals which, in effect, sum a very large number of small contributions travelling between successive wavefronts. This is usually difficult to do by pure mathematics. It is quite practicable with software, but requires large amounts of computing time. Fortunately, approximations to Fresnel integrals exist for a number of regular geometries, which with some ingenuity normally allows practical calculations to proceed.

4.5.1   Knife-edge diffraction loss

When an obstacle has a sharp top we have 'knife edge' diffraction, with the geometry shown in Figure 4.7. The loss in this case can be obtained by calculating the parameter v, as given by eqn. 4.5. Diffraction loss is then given directly by Figure 4.8, or an equivalent numerical model. A useful approximation to knife-edge diffraction loss in dB for v greater than about –0.7 is

$$J(v) = 6.9 + 20\log\left\{\sqrt{v^2 + 1} + v\right\} \qquad (4.10)$$

Equation 4.10 is a simplified form of the expression given in Recommendation 526-4 [14]. It covers shadowing by an obstacle which intrudes significantly into the first Fresnel zone, or which forms an obstruction to line-of-sight propagation. It does not, however, give the ripples in Figure 4.8 for an obstruction lying outside 0.6 Fresnel-zone clearance.

4.5.2   Cylinder diffraction loss

The diffraction loss over the curved surface of a cylinder can be computed from an extension of eqn. 4.10. The geometry is shown in Figure 4.14. The distances d_1 and d_2 are measured from T and R to the apex point where the cylinder horizon rays would cross above it. The height h of the obstruction is measured from the straight line joining T and R to this apex point. The radius of the cylinder is r.

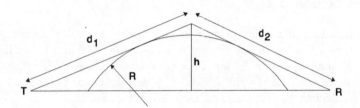

Figure 4.14   Geometry of diffraction over a cylinder

The diffraction loss L for this geometry is given by

$$L = J(v) + T \qquad (4.11)$$

where:

$J(v)$ = diffraction loss for a knife-edge obstruction extending up to the apex point, as given by eqn. 4.8;

T = additional loss due to diffraction over the cylinder surface between the points where the rays form tangents.

The correction T is formulated differently in recent revisions of Recommendation 526-4 [14]. The important point is that it is positive. In other words there is greater diffraction loss over a cylinder than over the equivalent knife edge, even though the crest of the cylinder is lower. This is because, in knife-edge diffraction, the process of eliminating parts of an advancing wavefront is considered to take place only once. For diffraction over a rounded surface it takes places for a succession of points along the surface. In general, objects with sharp corners, such as buildings, cause less diffraction loss than rounded objects such as hill tops.

4.6   Diffraction over irregular terrain

Most objects in the real world are not as tidy as knife edges or cylinders. A particular problem is diffraction over irregular terrain. There are two general approaches to this for calculations using approximations to Fresnel integrals. One models the terrain as a set of knife edges, the other as a series of horizontal cylinders.

4.6.1 Multiple-knife-edge methods

The proponents of this approach argue that hardly anything in the natural world looks smooth to radio waves, and thus a succession of knife edges is a suitable model. String analysis is used to find the obstruction points of a path profile, and each of these is treated as though a knife edge.

There is a variety of methods of this type; all of them sum the diffraction losses of individual knife edges. They differ in how the path geometry is interpreted in order to calculate the parameter v for each edge. The method of Epstein–Petersen [87] is illustrated in Figure 4.15, in which three hilltops are represented by knife edges at A, B and C. For each edge the loss is computed as though a diffraction path exists in isolation with the neighbouring points as terminals. Thus v is computed for TAB, ABC, and BCR, with the obstacle height in each case represented by the vertical line.

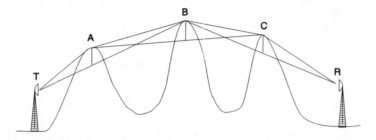

Figure 4.15   Epstein-Petersen construction

Other knife-edge methods include the 'Japanese' method [111], Deygout [81], and Giovaneli [92]. They tend to have systematic errors, and some use additional correction factors to overcome this. All tend to be inaccurate when two edges occur close together, and criteria are sometimes adopted for grouping multiple edges into a single equivalent obstruction.

4.6.2 Multiple-cylinder method

This is similar to the knife-edge method, but treats obstructions as horizontal cylinders. In other words, it requires the user to estimate the radius of curvature of obstruction tops. Thus in Figure 4.15 the geometry of Figure 4.14 is applied to each hilltop. This makes the interpretation of an irregular profile more complicated than the normal string analysis, and there is at present a lack of evidence as to how this is best done.

4.7   Reflection

In discussing diffraction we have assumed that obstacles completely absorb radio energy. This is rarely the case in practice. Radio can also be reflected, scattered or absorbed by the various objects it meets. The general situation is shown in Figure 4.16. The line AA' is a plane interface between media of different electrical properties. Radio energy meets this interface at an incidence angle θ_i. Owing to the different impedance presented by the new medium, part of the energy will be reflected, and by geometrical optics this will be at angle θ_r equal to θ_i. The remainder will penetrate the new material at angle θ_p, which in general will not equal θ_i. The change in direction of the penetrating ray is referred to as refraction.

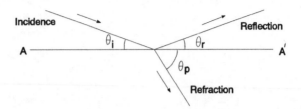

Figure 4.16   Reflection and refraction at a plane-medium boundary

The relative proportion of the energy reflected from or penetrating into the new material depends on the frequency, the angle of incidence and polarisation of the wave, and the electrical properties of the media. To simplify the discussion we will assume that reflection is caused by a horizontal plane, such as the ground, and that the upper medium is air. A point to note is that, in radio engineering, incidence and reflection angles are usually measured to the reflecting plane, whereas in optics they are measured to the normal. This modifies certain mathematical expressions.

4.7.1 Reflection by perfectly conducting ground

A perfect conductor reflects all the radio energy falling on it. Although real ground conducts less than perfectly, the ideal case is a useful starting point.

In Figure 4.17 a horizontal antenna of height h above perfectly conducting ground has a current element pointing into the paper. Horizontally polarised radiation at θ to the horizontal meets the plane at P and is wholly reflected at angle θ. At P we will assume that the incident E vector is also pointing into the paper. Since the tangential field at the surface of the perfect conductor must be zero, currents will be induced in it to produce an equal an opposite vector E'. These conditions are fulfilled everywhere on the conducting plane by an image antenna at h below it, with its current element exactly out of phase with the real antenna. Thus a phase change of 180° occurs when a horizontally polarised wave is reflected by a perfect conductor, and for analytic purposes the reflecting plane can be replaced by the image antenna.

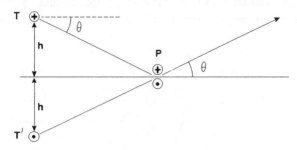

Figure 4.17 180° phase change for horizontal polarisation

In Figure 4.18 a short current element pointing upwards radiates downwards at angle θ, and this is reflected from the perfectly conducting plane. As before, the tangential electric field at the conducting plane must be zero, but the geometry is now quite different. The incident electric vector E can be resolved into horizontal and vertical components E_h and E_v respectively. Only E_h must be cancelled by currents induced in the plane. This condition is fulfilled by an image current element which must also point upwards. Because the vector E' produced by the image leans to the left at θ whereas E leans to the right at θ, their horizontal components E'_h and E_h cancel. Since their vertical components are in phase, no phase change occurs when a vertically polarised wave is reflected by a perfect conductor.

4.7.2 Reflection by imperfectly conducting ground

The reflection properties of real ground depend mainly on its conductivity and relative permittivity. It is characterised by a complex reflection coefficient ρ given for horizontal polarisation (see chapter 16 of [96]) by

$$\rho_h = \frac{\sin\theta - \sqrt{(\varepsilon_r - j\chi) - \cos^2\theta}}{\sin\theta + \sqrt{(\varepsilon_r - j\chi) - \cos^2\theta}} \tag{4.12}$$

and for vertical polarisation by

$$\rho_v = \frac{(\varepsilon_r - j\chi)\sin\theta - \sqrt{(\varepsilon_r - j\chi) - \cos^2\theta}}{(\varepsilon_r - j\chi)\sin\theta + \sqrt{(\varepsilon_r - j\chi) - \cos^2\theta}} \qquad (4.13)$$

where ε_r is the relative dielectric constant of the ground, χ is given by $18 \times 10^9 \sigma / f$, σ is the ground conductivity in S/m and f is the frequency in Hz.

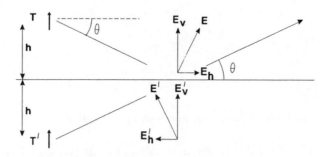

Figure 4.18   No phase change for vertical polarisation

Figure 4.19 shows the amplitude and phase angle of ρ_h and ρ_v plotted against incidence angle for $\sigma = 0.01$ S/m, $\varepsilon = 30$, and $f = 100$ MHz. For horizontal polarisation the behaviour is similar to that of the perfect-conductor model, with high amplitude and phase angle close to 180°. For vertical polarisation the situation is more complicated, and more variable with conditions. The amplitude passes through a minimum at what is known as the pseudo-Brewster angle. Above the Brewster angle the phase angle is close to zero, as predicted by the perfect conductor model. Below the Brewster angle the phase angle changes to 180°.

An important feature shown in Figure 4.19 is that, for grazing incidence, the reflection coefficient for both polarisations tends to –1, i.e. unity amplitude and the phase reversed. Since ground-reflection angles are normally small in radio propagation, this means that there tends to be little difference between the two polarisations as far as reflection from smooth ground is concerned. However, this does not apply to reflections from buildings, where large incidence angles are common. For the vertical side of a building, a horizontally polarised wave is effectively vertically polarised, and for a wide range of values will reflect less than a vertically polarised signal.

4.7.3 Flat-Earth reflection

The effect of reflection by smooth ground can be calculated simply if the Earth's surface is assumed to be flat, which is valid for relatively short paths. Figure 4.20

shows propagation by two rays between T and R separated by a horizontal distance d.

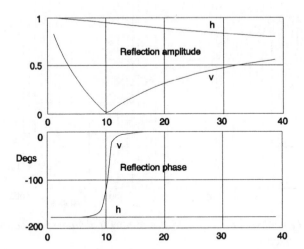

Figure 4.19   Amplitude and phase angle of ground-reflection coefficient

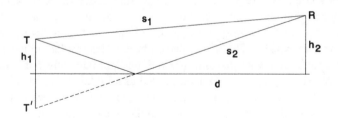

Figure 4.20   Two-ray propagation over flat Earth

The slope distances of the two rays are given by

$$s_1 = \sqrt{d^2 + (h_1 - h_2)^2} \tag{4.14}$$

and

$$s_2 = \sqrt{d^2 + (h_1 + h_2)^2} \tag{4.15}$$

If the antenna heights h_1 and h_2 are much less than d, we can assume grazing incidence and thus a reflection coefficient of -1. It then becomes simple to compute the vector sum of the field strengths due to the two rays, which is plotted against distance in Figure 4.21.

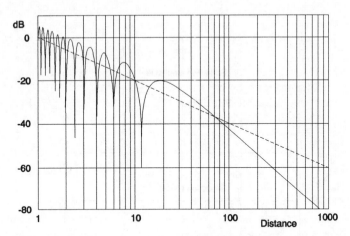

Figure 4.21   Typical variation of field strength with distance for ground reflection

The assumptions used to produce Figure 4.21 do not apply very frequently, although they could be true for a radio link passing above the surface of a lake. What is more important is the general form. As distance increases, the path-length difference for the two rays decreases, and thus the two contributions to the field strength pass periodically in and out of phase. This results in the successive maxima and minima. Note that the maxima are approximately 6 dB above the single-ray free-space propagation curve, plotted as a broken line. This is because the two vectors will have very similar amplitudes, and will thus result in twice the field strength when they are in phase. Beyond the last minimum the field strength falls at 40 dB per decade, which is inversely proportional to d^4.

For practical situations reflections are rarely as systematic as illustrated in Figure 4.21. However, the pattern of successive minima is extremely typical of propagation affected by reflections. This must be accounted for when making field-strength measurements. In nearly all real situations, a field strength will have a fine structure of variations due to reflections from the ground and other objects such as buildings. To obtain a representative value it is important to probe a volume of space and take some form of average of the field-strength pattern.

4.7.4 Fresnel clearance for reflection

The principle of Fresnel clearance applies to reflection. The first Fresnel zone on a reflecting surface encloses all reflection points for which the additional path length is less than half a wavelength. The calculation of reflection Fresnel zones is more complicated than for a direct radio path, and in general the zones will be ellipses rather than circles. If there is less than 0.6 of the first Fresnel zone there will be a corresponding loss in the reflected wave. This is important for situations where reflection is either wanted or undesirable. In the first case, many antennas in the LF, MF and HF bands rely on ground reflection to obtain the desired vertical radiation pattern. Good ground conductivity is thus required for most of the first Fresnel zone, and if necessary metallic wires or grids are placed on or under the ground to ensure this.

In the second case, where reflection might cause interference, its effect may be overestimated if the size of the reflecting surface is not taken into account.

4.8 Scattering

Scattering can be described as disorganised reflection from a rough surface. We are familiar with it from the behaviour of light. The surface of a table will normally show a reflected image, particularly if a well lit object is viewed in it at grazing incidence. One the other hand we can see the table from any angle, which means that it must be scattering the light that falls on it in all directions. Radio behaves in the same way, but the dimensions involved in 'smoothness' are larger owing to the longer wavelength.

4.8.1 Rayleigh roughness criterion

The Rayleigh roughness criterion provides a useful guide as to whether a surface should be considered smooth or rough as far as reflection is concerned. Figure 4.22 shows a reflecting surface with idealised roughness represented by rectangular steps of depth d. A ray incident at θ is may be reflected from either the upper or lower surface. The path difference is given by AC minus AB, from which the phase difference $\Delta\phi$ in radians is given by

$$\Delta\phi = \frac{4\pi d \sin\theta}{\lambda} \tag{4.16}$$

If the phase difference is small, e.g. less than 0.1 rad, the surface will support specular reflection. If it is large, e.g. more than 10 rad, the surface is rough and will produce scattering. For intermediate values there will be a mixture of scattering and more or less diffuse reflection. Practical surfaces do not usually consist of regular steps. For typical irregular surfaces the depth d in eqn. 4.16 can be replaced by a standard deviation representing the roughness. An additional factor is that real surfaces present different angles to incident radiation, which is not allowed for in Figure 4.22. Thus, the Rayleigh criterion is only a guide, but it is a useful one. Equation 4.16 also explains why most surfaces reflect better at small angles of incidence.

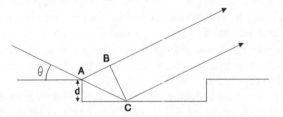

Figure 4.22   Geometry for Rayleigh roughness criterion

4.8.2 Accounting for scattering in propagation computations

When the effects of reflection are included in propagation computations, it is normally necessary to include the relative phases of direct and reflected signals. This is because true specular reflection produces a signal which is coherent with the incident wave. Accounting for phase in a scattering situation is usually not practicable, since reflections are then taking place from rough surfaces, or from a number of small objects such as raindrops. In this case it is only practicable to sum the scattered power.

Scattering can be accounted for by an extension of the free-space-loss equation. The power flux density (PFD) W_1 in W/m^2 at distance d_1 m from a transmitter radiating P_t W with antenna gain G_t is given by

$$W_1 = \frac{P_t G_t}{4\pi d_1^{\,2}} \qquad (4.17)$$

A scatterer at distance d_1 is assumed to scatter isotropically all incident energy which falls on an effective area A m^2, which accounts for the size, shape and scattering coefficient of the object. Thus the scattered PFD W_2 at distance d_2 m from the scatterer is given by

$$W_2 = W_1 \frac{A}{4\pi d_2^{\,2}} = \frac{P_t G_t A}{(4\pi d_1 d_2)^2} \qquad (4.18)$$

The power received by an antenna of gain G_r at distance d_2 from the scatterer is thus given by multiplying W_2 by the effective aperture of the receiving antenna:

$$P_r = W_2 \frac{G_r \lambda^2}{4\pi} = \frac{P_t G_t G_r A \lambda^2}{(4\pi)^3 (d_1 d_2)^2} \qquad (4.19)$$

Thus the transmission loss L_s over the scattering path is given by

$$L_s = \frac{P_r}{P_t} = \frac{G_t G_r A \lambda^2}{(4\pi)^3 (d_1 d_2)^2} \tag{4.20}$$

Equation 4.18 is known as the bistatic radar equation. For $G_t = G_r$ and $d_1 = d_2$ it becomes the monostatic radar equation, which applies to a primary radar system transmitting and receiving with the same antenna.

Chapter 5

Diffraction theory

P.A. Miller

5.1   Introduction

The aim of this Chapter is to extend the results presented in Chapter 2 to account
for the presence of obstacles in the space between a transmitter and receiver (see
Figure 5.1). This is an important class of problem, since it is not always possible
to ensure a free-space path. For instance, the path between a mobile telephone
and its transmitter may be blocked by hills, buildings or trees. The path loss along
such diffraction paths will usually be higher than the free-space value, and the
main problem we shall address is the determination of this diffraction loss.

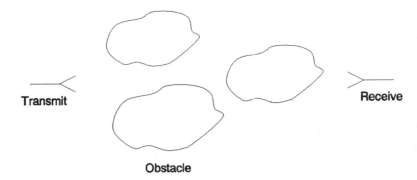

Transmit

Receive

Obstacle

Figure 5.1   Propagation path blocked by obstacles

To gain an understanding of the various scattering and diffraction mechanisms
involved, the problem will be formulated from classical electromagnetic theory.
However, in practice it is difficult to solve the resulting system of differential
equations for obstacles which are more than a few wavelengths in diameter, and it
is therefore necessary to adopt approximate or empirical techniques. In Section
5.2 the behaviour of electromagnetic fields at material boundaries is discussed,

and results derived for infinite plane waves. In Section 5.3 we look at the formulation of the general scattering problem, and discuss some approximations which can be applied to obtain solutions to practical problems, and in Section 5.4 we focus on approximations based on geometrical construction techniques.

5.2   Propagation across material boundaries

The fundamental cause of all scattering and diffraction phenomena is the conditions imposed by Maxwell's equations on electric and magnetic fields at material discontinuities. In this Section we shall derive these conditions and use them to account for the behaviour of plane waves at infinite plane boundaries.

5.2.1 Maxwell's equations in integral form

In Chapter 2 we were presented with Maxwell's equations in differential form. These relate the field quantities at a point in space to the current and charge densities at that point. These equations may also be written in integral form as follows:

$$\oint \boldsymbol{E}.d\boldsymbol{l} = -\frac{\partial}{\partial t} \iint_s \mu \, \boldsymbol{H}.d\boldsymbol{S} \tag{5.1a}$$

$$\oint \boldsymbol{H}.d\boldsymbol{l} = \iint_s \boldsymbol{J}.d\boldsymbol{S} + \frac{\partial}{\partial t} \iint_s \varepsilon \, \boldsymbol{E}.d\boldsymbol{S} \tag{5.1b}$$

$$\oiint \varepsilon \, \boldsymbol{E}.d\boldsymbol{S} = \iiint_v \rho \, dV \tag{5.1c}$$

$$\oiint \mu \, \boldsymbol{H}.d\boldsymbol{S} = 0 \tag{5.1d}$$

where $\boldsymbol{E}$ is the electric field (V/m), $\boldsymbol{H}$ the magnetic field (A/m), $\boldsymbol{J}$ the current density (A/m^2), ρ the charge density (C/m^3), ε the material permittivity (F/m) and μ the material permeability (H/m). It is easy to show these are equivalent to the point form of Maxwell's equations; for instance if we apply Stokes' theorem to the left-hand side of eqn. 5.1a and differentiate twice we recover the curl equation for $\boldsymbol{E}$.

5.2.2 Continuity conditions at material boundaries

Consider the boundary between two materials of permittivity ε_1, ε_2 and permeability μ_1, μ_2 (see Figure 5.2).

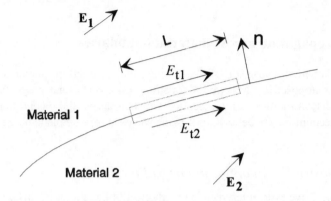

Figure 5.2 Continuity of tangential E field at a boundary

Let the electric field in material 1 be E_1 and that in material 2 be E_2. To find the relationship between these fields at the boundary, we apply eqn. 5.1a by evaluating the line integral around a loop of length ΔL which straddles the boundary, but is of negligible thickness. Since the loop has zero area, the right-hand side of eqn. 5.1a is zero, so

$$\left(E_{t1} - E_{t2}\right)\Delta L = 0 \tag{5.2}$$

where E_{t1} and E_{t2} are the tangential components of the electric field in the two materials at the boundary. This implies that the tangential components of the electric field are continuous across the boundary. If the normal to the surface at the boundary is n, then in vector notation this can be expressed as

$$\left(E_1 - E_2\right) \times n = 0 \tag{5.3a}$$

Similarly, if we consider the magnetic fields in the two materials, and apply eqn. 5.1b to a similar loop, we find that, provided that the conductivities of the materials are finite, the tangential component of the H field is also continuous across the boundary; therefore:

$$n \times \left(H_1 - H_2\right) = 0 \tag{5.3b}$$

The continuity conditions on the normal components of the field can be found by application of eqns. 5.1c and 5.1d to a small volume of area ΔS, and negligible thickness, straddling the boundary. It is then found that

$$(\varepsilon_1 E_1.n) - (\varepsilon_2 E_2.n) = \rho_s \qquad (5.3c)$$

where ρ_s is the charge density on the surface (C/m²), and

$$(\mu_1 H_1.n) - (\mu_2 H_2.n) = 0 \qquad (5.3d)$$

The derivation of eqns. 5.3c and 5.3d is left to the reader. For-time varying fields it is found that either the tangential or normal conditions are sufficient to solve all problems and the tangential components are normally used.

5.2.3 Fields incident on a perfect conductor

It is interesting to consider the continuity conditions when the second material is a perfect conductor. Although perfect conductors do not exist in reality, it is often mathematically convenient to consider good conductors to have infinite conductivity. The electric field in the conductor is related to the current density according to the expression $J = \sigma E$, which is Ohms Law expressed in field quantities, and σ is the conductivity (S/m). If the conductivity is infinite, for the current to remain finite the electric field in the perfect conductor must be zero. From eqn. 5.3a it is therefore apparent that the tangential component of the field in material 1 at the boundary is also zero. The magnetic field is likewise zero inside the conductor, but in general the tangential component just outside is not zero. This discontinuity can arise because of a surface current J_s (A/m) flowing on the conductor. The value of this current can be found by application of eqn. 5.1b to the loop in Figure 5.2 and is equal to the tangential component of the incident magnetic field. In summary, the tangential continuity conditions on the surface of a perfect conductor are

$$E_1 \times n = 0 \qquad (5.4a)$$

$$J_s = n \times H_1 \qquad (5.4b)$$

In real conductors with finite conductivity it is found that the fields and currents concentrate in a thin region or *skin* near the surface. The depth of the skin decreases as the conductivity increases.

5.2.4 Plane waves normally incident on an infinite dielectric boundary

One of the simplest boundary-value problems we can consider is a plane wave normally incident on an infinite dielectric boundary. An understanding of the mechanisms which arise in this case will lead to a qualitative understanding of what happens in the more general case of a three-dimensional wave incident on an arbitrarily shaped boundary.

You will recall that in Chapter 2 we learned that the E field in an homogenous source-free region is a solution of the homogenous Helmholtz equation

$$\nabla^2 E + k^2 E = 0 \qquad (5.5)$$

The general solution for a z-directed plane wave can be written in complex exponential notation [ignoring the $\exp(j\omega t)$ dependence] as

$$E = \{E_i \exp(-jkz) + E_r \exp(jkz)\}\hat{x} \qquad (5.6a)$$

where $k = 2\pi/\lambda$. The first term of eqn. 5.6a can be interpreted as a plane wave travelling in the $+z$ direction of amplitude E_i and the second term as a plane wave travelling in the $-z$ direction of amplitude E_r. The corresponding equation for the H field is

$$H = \{H_i \exp(-jkz) - H_r \exp(jkz)\}\hat{y} \qquad (5.6b)$$

from which it is apparent that the sign of the H field is reversed for the backward travelling wave.

We can now consider a plane wave normally incident on an infinite boundary between two regions (see Figure 5.3). Let the wave impedance in region 1 be η_1 $(=\sqrt{\mu_1/\varepsilon_1})$, and that in region 2 be η_2. Let the incident field in region 1 be $E_i\hat{x}, H_i\hat{y}$. At the boundary this will give rise to a backward-travelling or *reflected* wave in region 1, $E_r\hat{x}, H_r\hat{y}$, and a transmitted wave $E_t\hat{x}, H_t\hat{y}$ in region 2, and the problem is to establish the ratio of the reflected and transmitted fields to the incident field. Applying eqns. 5.3a and 5.3b we obtain

$$E_i + E_r = E_t \qquad (5.7a)$$

$$(H_i - H_r) = H_t = (E_i/\eta_1 - E_r/\eta_1) = E_t/\eta_2 \qquad (5.7b)$$

from which it is easy to show that

$$\frac{E_r}{E_i} = \frac{(\eta_2 - \eta_1)}{(\eta_2 + \eta_1)} \qquad (5.8a)$$

$$\frac{E_t}{E_i} = \frac{2\eta_2}{(\eta_2 + \eta_1)} \qquad (5.8b)$$

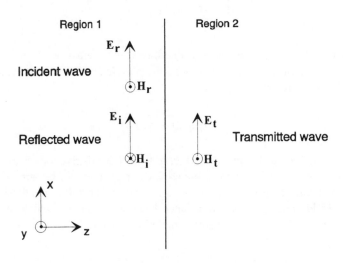

Figure 5.3   Plane waves normally incident on a planar boundary

It is clear from the above that, when $\eta_1 = \eta_2$, then $E_i = E_t$, and $E_r = 0$, as one would expect. It is also easy to show that eqns. 5.8a and 5.8b conserve energy, i.e. the incident power is equal to the sum of the reflected power and the transmitted power. If region 2 is a perfect conductor, we know that the transmitted fields must be zero. Application of eqns. 5.7a and 5.7b in this case yields

$$E_r = -E_i \qquad (5.9a)$$

$$H_r = H_i \qquad (5.9b)$$

5.2.5 Plane-wave propagation in a lossy medium

Most materials we will encounter in practical propagation problems, for instance brick or stone, will have finite nonzero conductivity, and we need to understand how this effects the properties of a plane wave and its behaviour at material boundaries. Maxwell's curl equation for the *H* field is

$$\nabla \times \boldsymbol{H} = j\omega\varepsilon \, \boldsymbol{E} + \boldsymbol{J} \tag{5.10}$$

which can be written as

$$\nabla \times \boldsymbol{H} = (j\omega\varepsilon + \sigma)\boldsymbol{E} \tag{5.11a}$$

or

$$\nabla \times \boldsymbol{H} = j\omega\left(\varepsilon + \frac{\sigma}{j\omega}\right)\boldsymbol{E} \tag{5.11b}$$

from which it should be clear that our previous results remain valid provided that we replace the permittivity by the complex number ($\varepsilon + \sigma/j\omega$). The value of k in our plane wave equation now becomes:

$$k = \omega\sqrt{\mu(\varepsilon + \sigma/j\omega)} \tag{5.12}$$

which can be written as complex number, $\alpha + j\beta$. For positive values of σ, α will be negative which implies that our plane-wave amplitude will be attenuated with distance. Not surprisingly, finite-conductivity materials result in the dissipation of energy. In addition, the wave impedance becomes complex, which means that waves may experience phase changes at material boundaries.

5.2.6 Oblique incidence on an infinite plane boundary

As a final part to our discussion on plane-wave behaviour at boundaries, we need to consider the more general problem of incidence at arbitrary angles. So far we have been concerned with a plane wave travelling parallel to the z axis of our co-ordinate system. To deal with oblique incidence, we need a more general expression for a plane wave travelling in some direction $\hat{\boldsymbol{k}}$ with direction cosines ($\cos\alpha$, $\cos\beta$, $\cos\gamma$). We can achieve this by a simple co-ordinate-rotation exercise, so that the z axis is rotated onto the required direction, from which we obtain

$$\boldsymbol{E} = E_0 \exp\{jk(\cos\alpha x + \cos\beta y + \cos\gamma z)\}\hat{\boldsymbol{x}} \tag{5.13}$$

assuming that the $\boldsymbol{E}$ field remains polarised in the x direction. A convenient general expression is thus

$$\boldsymbol{E} = E_0 \exp(j\boldsymbol{k}.\boldsymbol{r})\hat{\boldsymbol{x}} \tag{5.14}$$

where $\mathbf{k} = (2\pi / \lambda)(\cos\alpha \ \hat{x}, \cos\beta \ \hat{y}, \cos\gamma \ \hat{z})$ is the wave vector, and $\mathbf{r}$ is the position vector relative to our co-ordinate origin.

Returning now to the problem of an oblique plane wave incident on an infinite plane boundary (see Figure 5.4), where the incident wave vector is seen to have a component in the z direction normal to the boundary, and a component in the y direction tangential to the boundary. We now have four unknowns to find, the amplitudes of the reflected and transmitted waves, and their directions of propagation.

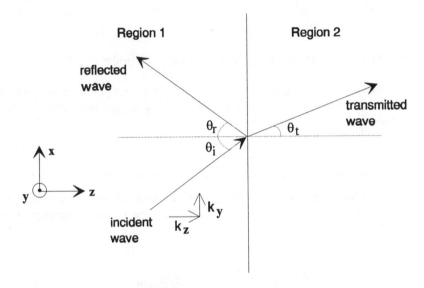

Figure 5.4 Oblique plane wave incident on a boundary

In region 1, the y-directed component of the wave is unaffected by the boundary, while the z-directed component has its direction reversed. It should be immediately apparent therefore that the angles of incidence and reflection, θ_i and θ_r, are equal, irrespective of the properties of region 2. This relationship will be no surprise to those familiar with ray optics. In region 2, continuity of the tangential E-field phase along the boundary can only be achieved if the y-directed component of the wave vector is equal to that in region 1. However, we know the magnitude of the wave vector in region 2, so it is now possible to derive the transmission angle. It is easily shown that

$$\frac{\sin\theta_t}{\sin\theta_i} = \frac{|k_1|}{|k_2|} \tag{5.15}$$

which is Snell's Law of refraction. It is comforting that light, which is known to be an electromagnetic phenomenon, exhibits behaviour which can be derived directly from Maxwell's equations.

The magnitudes of the reflected and transmitted fields can be found from application of the boundary conditions to the tangential components of the field, as in Section 5.2.3. The algebra is straightforward, but tedious and is not given here, a full development is given in [154]. Two sets of reflection and transmission coefficients can be derived for the *E* field polarised in the plane of incidence and normal to the plane of incidence, and in both cases the coefficients are dependent on the angle of incidence.

5.3   Wave solutions for general wave problems

So far, we have concerned ourselves with plane waves and planar boundaries. We now need to consider the more difficult problem of arbitrary wave functions incident on arbitrarily shaped obstacles.

5.3.1 Formulation of the general scattering problem

The general scattering problem consists of an arbitrarily shaped obstacle located in an otherwise homogenous region *R* which also contains some currents *J* (see Figure 5.5).

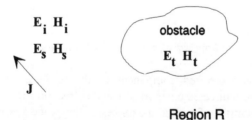

Figure 5.5   *General scattering problem*

In the absence of the obstacle we could calculate the fields from a knowledge of the currents using the techniques presented in Chapter 2; however, clearly the obstacle will modify the field in some way.

Before formulating the problem mathematically, it is useful to consider qualitatively what the obstacle will do to the field, by applying some of the mechanisms we observed in the plane-wave case. On the surface of the obstacle directly illuminated by the source, we might expect some of the field to be transmitted into the obstacle, and some of the field to be scattered back (reflected) into the source region. On the opposite side of the obstacle to the source, we might expect the obstacle to shadow the incident field, so that this area is illuminated by waves transmitted through the obstacle and leaking back out into region *R*. However, this is not the only source of illumination of this region. Some of the source field will *diffract* around the sides of the obstacle to illuminate this region, so that, even if the obstacle is perfectly conducting and the transmitted field is zero, the field in the shadow region will be nonzero owing to diffraction.

The general scattering problem is solved by enforcing the continuity conditions on the surface of the obstacle. Let the field in the region *R* without the obstacle be E_i, H_i, let the field scattered by the obstacle in this region be E_s, H_s, and let the field transmitted into the obstacle be E_t, H_t. The total field in the source region is thus $(E_i + E_s)$, $(H_i + H_s)$.

On the surface of the obstacle we require:

$$\left(E_i + E_s\right) \times n = E_t \times n \qquad (5.16a)$$

$$n \times \left(H_i + H_s\right) = n \times H_t \qquad (5.16b)$$

In principle, we can now solve the general scattering problem by obtaining solutions to the three-dimensional Helmholtz eqn. 5.5 in the two regions which satisfy eqns. 5.16a and 5.16b; however, in practice this can be an exceedingly difficult task. For the case when the source region is unbounded, rigorous analytical solutions can only be obtained for a very limited number of simple obstacle geometries. The most well known of these cases are the perfectly conducting cylinder, the perfectly conducting wedge and the perfectly conducting knife edge. The interested reader is referred to [110] for a full presentation of these results. The case of scattering from multiple obstacles can be formulated in a similar way, with a scattered and transmitted field arising from each obstacle. The scattered field from one obstacle will contribute to the incident field on the others, and the boundary conditions must be satisfied for all obstacles simultaneously. It is clear that the complexity of the problem increases rapidly as the number of scatterers increases.

Having shown that a direct solution of Maxwell's equations is impractical for most practical propagation problems, in the remainder of this Section we develop some approximations which can be applied to make the problem more tractable.

5.3.2 Equivalence theorem

The equivalence theorem allows us to replace the source currents of a field with alternative sources which produce the same field in some region of interest. This approach can simplify a broad range of scattering problems; however, we will need to introduce the concept of a magnetic current M (V/m^2). Magnetic currents have not been observed in reality; they are introduced purely as a mathematical convenience to replace some system of real sources. The M term is added to the curl equation for the E field to symmetrise the curl equations; thus:

$$\nabla \times E = -j\omega\mu\,H - M \qquad\qquad (5.17a)$$

$$\nabla \times H = j\omega\varepsilon\,E + J \qquad\qquad (5.17b)$$

We can also postulate the existence of a perfect magnetic conductor. When we apply the boundary conditions at the surface of this conductor we are forced to accept the existence of a surface magnetic current M_s (V/m), equal to

$$M_s = E \times n \qquad\qquad (5.18)$$

Now consider a region 2, containing real source current, surrounded by a source-free region 1 (see Figure 5.6a).

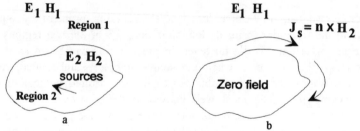

Figure 5.6   *Original and equivalent sources*
    a Original sources, b Equivalent sources

Let the field in region 1 be E_1, H_1, and let the field in region 2 be E_2, H_2. Continuity of field across the boundary requires that

$$E_1 \times n = E_2 \times n \qquad\qquad (5.19a)$$

$$n \times H_1 = n \times H_2 \qquad\qquad (5.19b)$$

The terms on the right-hand side of eqns. 5.19a and 5.19b can be regarded as sources giving rise to the fields in region 1. We can replace the right-hand terms by surface currents on the boundary of region 2, given by

$$M_s = E_2 \times n \qquad\qquad (5.20a)$$

$$\boldsymbol{J}_s = \boldsymbol{n} \times \boldsymbol{H}_2 \qquad\qquad (5.20b)$$

as shown in Figure 5.6*b*, and the field in region 2 is then zero. This is the equivalence theorem. Its importance lies in the idea that the fields over some closed surface can be regarded as the source for the radiated fields outside this region, by replacing the fields by their equivalent surface electric and magnetic currents.

5.3.3 Radiation from fields

Huygens' principle states that each part of a wavefront can be regarded as a new source of disturbance, sending out secondary waves which add to produce a new more distant wavefront. This principle was postulated long before Maxwell's equations, and has long been used to obtain qualitative answers to optical-diffraction problems. Here we shall consider a more quantitative discussion of this principle, using equivalent source currents.

Consider a rectangular element of area *dS* on a wavefront of a plane wave in free space (see Figure 5.7). The tangential components of the incident field are $\boldsymbol{E} = E_0 \,\hat{\boldsymbol{x}}$, $\boldsymbol{H} = H_0 \,\hat{\boldsymbol{y}}$, where $H_0 = E_0/\eta$. These may be replaced by equivalent source currents according to the equivalence principle: $\boldsymbol{M} = -E_0 \,\hat{\boldsymbol{y}}$, and $\boldsymbol{J} = -H_0 \,\hat{\boldsymbol{x}}$.

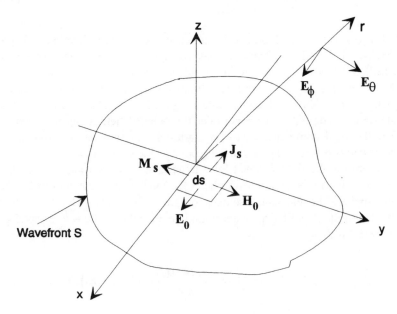

Figure 5.7   *Radiation from a wavefront*

In Chapter 2 we learned how to calculate the field radiated by an electric current using the magnetic vector potential A. To determine the fields radiated by a magnetic current, an electric vector potential F can be introduced in a similar way, defined by

$$\varepsilon E = -\nabla \times F \tag{5.21}$$

The F field in a free-space region, outside a volume v_0, can be determined from

$$F(r) = \frac{\varepsilon}{4\pi} \iiint\limits_{v_0} M(r) \frac{\exp(-jk|r - r_0|)}{|r - r_0|} dv \tag{5.22}$$

We are now in a position to calculate the total field radiated by an elemental area of wavefront. If we assume the observation distance $|r|$ is large in wavelengths, so that only components of the radiated field varying as $1/r$ need be considered, then the radiated E field, in spherical co-ordinates, can be shown to be [154]:

$$dE_\theta = \frac{jE_0 dS \exp(-jkr)}{2\lambda r}(1 + \cos\theta)\cos\phi \tag{5.23a}$$

$$dE_\phi = \frac{jE_0 dS \exp(-jkr)}{2\lambda r}(1 + \cos\theta)\sin\phi \tag{5.23b}$$

It is clear that the radiated field is maximum in the forward direction and zero in the backward direction. Thus, an electromagnetic wave in free space, once launched, continues to propagate in an outward direction. Note that an electric current sheet in isolation radiates equally on both sides, but crossed electric and magnetic current sheets of appropriate magnitude and phase can be made to radiate on one side only. If we consider the radiated field along the z axis, and convert the field components back to Cartesian co-ordinates, we find that the radiated field is polarised in the same direction as the source field.

To determine the field radiated by a large area of wavefront, we exploit the linearity of Maxwell's equations to obtain the total field as the sum of the fields radiated by all the elemental sources on the wavefront. The theta-directed component of the total field at an observation point r due to the whole wavefront S in Figure 5.7 is thus

$$E_\theta(r) = (1 + \cos\theta)\cos\phi \iint\limits_s \frac{jE_x^0(r_0)}{\lambda|r - r_0|} \exp(jk|r - r_0|)dS \tag{5.24}$$

Varying degrees of approximation can be applied to the evaluation of eqn. 5.24 depending on the size of our source wavefront, and how far away our point of observation is in wavelengths. In the Fraunhofer region (see Figure 5.8), the point of observation is sufficiently far away that individual field components from the elemental sources can be assumed to arrive parallel to one another.

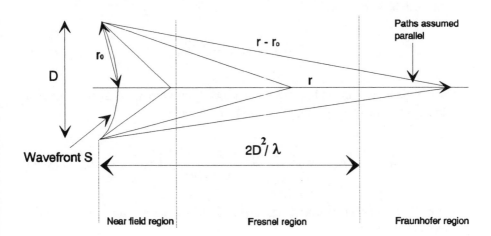

Figure 5.8   Definition of diffraction regions

This allows the distance $|r - r_0|$ in the exponential term in eqn. 5.24 to be approximated and the $1/|r - r_0|$ term is replaced by $1/|r|$. When these approximations are applied, we find that the radiated field is related to the source wavefront field by a two dimensional Fourier transform [154]. This has important consequences for the designers of directive antennas; if we want to contain the direction of radiation to a very narrow beamwidth, we need a very large-source wavefront or *aperture* field. The Fraunhofer region is usually considered to start at a distance of $2D^2/\lambda$ from the source wavefront, where D is the diameter of the wavefront. In the Fresnel region a more exact expression for the phase term in eqn. 5.24 has to be used, but amplitude variation due to the $1/|r - r_0|$ is still ignored. In the near-field region eqn. 5.24 is evaluated exactly.

This method of calculating the radiated field from a known source field is called *physical optics*. It is often applied to scattering from systems of near-perfect conductors, for instance large reflector antennas. In this case, the current induced on the conductor surface by the incident field is assumed to be the same as that due to an equal-magnitude plane wave incident on an infinite flat conductor, i.e. $J_s = 2n \times H$ (eqns. 5.4b and 5.16b). This is known as the *physical optics approximation*, and gives reliable results provided that the conductor size and radius of curvature are large in wavelengths.

5.3.4 *Local nature of high-frequency propagation*

It is interesting to apply the equivalence theorem to free space propagation between a source at O and an observation point P (see Figure 5.9). We can replace the source at O by an equivalent source field on an infinite plane U between O and P. The equivalent source field $E(u)$ on some elemental wavefront du at u can be written as

$$E(u) = \frac{K_1}{s(u)} \exp\{-jks(u)\} \tag{5.25}$$

where K_1 is a constant and $s(u)$ is the distance from O to u. If $E(u)$ produces a field dE_R at P, then the total field at P can be found from

$$E_R = K_1 K_2 \iint_U \frac{\exp\left[-jk\{s(u)+r(u)\}\right]}{s(u)+r(u)}\, du \tag{5.26}$$

where K_2 is another constant, and $r(u)$ is the distance from the elemental wavefront at u to P.

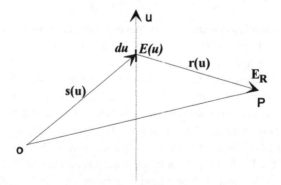

Figure 5.9   Equivalent field on a free-space path

Equation 5.26 is difficult to evaluate directly, but we can infer a number of conclusions by examining it qualitatively. The magnitude of the integrand in eqn. 5.26 is greatest when $s(u)$ and $r(u)$ are minima, i.e. when they lie on the line joining O to P, and the magnitude decreases as we move away from this line. The phase of the integrand changes by π radians every time $s(u) + r(u)$ increases by $\lambda/2$, and when these contributions are added in the integration they almost completely cancel, except in a small region close to the line joining O to P. The dominant contribution to E_R is thus from a small region of the plane U close to the direct path from O to P, and bounded by $s(u) + r(u) < OP + \lambda/2$. This result is useful for defining the necessary clearance volume for minimal impact on a free-space path, as was shown in an earlier Chapter.

This approach can also be applied to the case of scattering from a large object, and it is found that the dominant contribution to the field at some observation point arises from a small area close to the path which would be followed by a plane wave undergoing reflection from the object. These results demonstrate the local behaviour of high-frequency scattering. Provided that the scatterers are large in wavelengths (hence high-frequency), and the observation point is many wavelengths away from them, the field can be considered to propagate along local paths or *rays*, which are straight lines in an homogenous isotropic medium. This result allows us to use geometrical methods to calculate the scattered field, and these are discussed in Section 5.4.

The observant reader will have noticed we have dropped polarisation in the discussion and considered the electric field to be a scalar quantity. This is valid provided that there are no mechanisms present which can depolarise the signal; however, a scalar-field assumption is often employed in cases where depolarisation is possible.

5.3.5 *Diffraction from an absorbing knife edge*

So far in this Section we have discussed the formulation of wave solutions to scattering problems, and looked at various approximate techniques for finding solutions. Before leaving this Section it will be instructive to present a solution to a problem. We will address the example of diffraction from a perfectly absorbing knife edge (see Figure 5.10), and solve it by the method of physical optics.

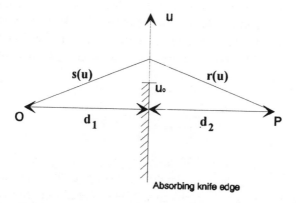

Figure 5.10   Diffraction from an absorbing knife edge

'Perfectly absorbing' means that all the incident field in absorbed, and there is no scattered or transmitted field. Again, this is a mathematical convenience; infinitely thin perfectly absorbing objects cannot be constructed. The perfectly conducting knife edge can be solved, but this is a substantially more difficult problem.

In Figure 5.10 the knife edge is located at a distance d_1 from the source O, and d_2 from the observation point P. Let the height of the knife edge be u_0. We can use eqn. 5.26 to calculate the field at the observation point E_R provided we restrict our integration to the area not blocked by the knife edge. Furthermore, we will assume that we are in the Fresnel region so that the amplitude dependence on distance can be ignored, and furthermore we will assume d_1, $d_2 \gg u_0$. We can then use the approximation

$$s(u) + r(u) = d_1 + d_2 + \frac{u^2}{2}\left(\frac{d_1 + d_2}{d_1 d_2}\right)$$ (5.27)

from which we obtain

$$E_R = K_3 \int_0^\infty \exp\left(-j\frac{\pi}{2}v^2\right)dv - K_3 \int_0^{v_0} \exp\left(-j\frac{\pi}{2}v^2\right)dv$$ (5.28)

where:

$$v = u\sqrt{\frac{2(d_1 + d_2)}{\lambda d_1 d_2}}$$ (5.29)

and K_3 is a constant. The integrals in eqn. 5.28 are Fresnel integrals, which occur frequently in diffraction problems, and are tabulated [47]. It is interesting to compare E_r with the field E_0 which would be obtained under free-space conditions without the knife edge. This can be achieved by setting v_0 to $-\infty$. The value of the Fresnel integrals over this range can be shown to be $(1 - j)$, from which we can obtain

$$\frac{E_r}{E_0} = \frac{1}{(1-j)}\{F(\infty) - F(v_0)\}$$ (5.30)

A plot of the magnitude of eqn. 5.30 against v_0 is given in Figure 5.11.

When $v_0 = 0$, i.e. the knife edge is at the same height as the source and observer, the field is 6 dB below its free-space value. If the knife edge is reduced below this height, the field is seen to ripple around its free-space value. When the knife edge is raised above this height, the observation point is shadowed, and the field strength falls off rapidly.

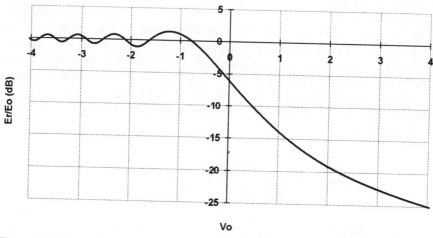

Figure 5.11   *Magnitude of the field diffracted by a perfectly absorbing knife edge relative to the free-space value, against the height of the knife edge*

5.4   Geometrical methods for scattering

5.4.1 *Geometrical optics*

Geometrical optics is the high-frequency limit of physical optics where the fields are considered to travel along rays and can be treated geometrically. At the boundary with a scattering obstacle, it is assumed that each part of the interface behaves as if it were locally part of an infinite planar interface and the incident field behaves as though it were locally part of a plane wave. The angular relationships derived in Section 5.2 can thus be used to determine the ray paths around a prescribed system of scatterers. The high-frequency assumptions under which the geometrical optics approximation is valid are:

(i)   all scattering obstacles are large in wavelengths;
(ii)   the radii of curvature of the obstacles is large in wavelengths;
(iii)   the RMS surface roughness is much smaller than a wavelength;
(iv)   the source and observation points are separated from each other and the closest scatter by distances large in wavelengths; and

(v)   the scatterers are separated from their closest neighbours by distances large in wavelengths.

Consider a plane wave from a source O, incident on a single wedge-shaped obstacle (see Figure 5.12). If we assume there is no field transmitted through the obstacle, then we can define three distinct observation regions. In region 1 we have both an incident ray and a reflected ray. In region 2 we only have an incident ray, while region 3 is shadowed, and the geometrical optics field here is zero. The boundary between regions 1 and 2 is called a reflection shadow boundary, and that between regions 2 and 3 an incident shadow boundary. Clearly, if we allow transmitted rays through the obstacle, then the field in region 3 will be nonzero.

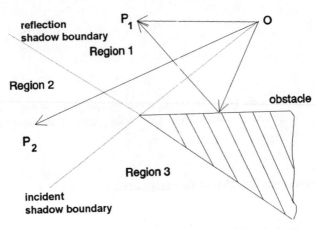

Figure 5.12   Geometrical optics construction for a wedge-shaped obstacle

Having established a method for determining the positions of rays, we need to determine the amplitude and phase of the wave propagating along each ray path. The phase can be determined using the plane-wave propagation constant $k = 2\pi/\lambda$ in each medium through which the ray passes. The amplitude variation along the ray can be determined from a conservation-of-energy argument, and is found to depend on the radii of curvature of the wavefront [110]. If the radii of curvature are both infinite, we have a plane wave, and the amplitude is constant along the ray. If only one radius is finite, we have a cylindrical wave, and the field amplitude varies as the reciprocal of the square root of the distance along the ray. If both the radii of curvature are finite and equal, we have a spherical wave, and the field amplitude exhibits a $1/r$ dependency along the ray, which is the result obtained for free-space loss. The complex reflection and transmission coefficients at boundaries are determined from the plane-wave case described in Section 5.2. For most practical propagation problems, we are interested in sources which are small compared with the ray lengths. These sources can thus be regarded as point sources, and the resulting wavefronts are spherical.

5.4.2 *Geometrical theory of diffraction*

A significant limitation of the geometrical-optics method, when applied to radio-frequency problems, is that it does not account for diffraction effects. Thus, in Figure 5.12 geometrical optics predicts zero field in region 3, whereas in practice a diffracted field, possibly of significant amplitude, may be present. To overcome this limitation, an extension to geometrical optics, called the geometrical theory of diffraction (GTD), was developed in the 1950s [110]. GTD was originally developed for the analysis of large antenna systems, but more recently has been applied successfully to propagation problems.

GTD is based on a small number of scattering problems which can be solved rigorously, the so-called canonical problems of GTD. These include the perfectly conducting infinitely long knife edge, the perfectly conducting infinitely long wedge and the perfectly conducting infinitely long cylinder. Considering the case of a perfectly conducting wedge illuminated by a plane wave, it is found that the rigorous solution can be written in terms of the geometrical-optics field plus a diffracted wave which appears to originate from a line source along the edge. Thus, for a wave incident normal to the edge, we have a circle of diffracted waves propagating out from the edge (see Figure 5.13*a*). For the more general case of oblique incidence, we have a cone of diffracted rays (see Figure 5.13*b*).

The amplitudes of the diffracted rays are related to the incident rays by a diffraction coefficient *D*, which is derived from the rigorous analysis. In the vicinity of the shadow boundaries, the equivalent-line-source argument becomes invalid and the *D* coefficient becomes ill conditioned; however, a number of extensions to correct for this behaviour have been proposed [110].

The rigorous analysis is applicable only to infinitely long conducting wedges illuminated by plane waves. The GTD method offers an approximate solution to the more general case of an arbitrary wavefront incident on a finite-length edge, by assuming that the edge behaves locally as if it were part of an infinite straight edge and the incident field behaves locally as a plane wave. More recently, approximations have been developed for dealing with a finite-conductivity wedge and a suitable expression for the diffraction coefficient is derived in [126], which is also stable over the boundary regions. It is somewhat involved, and so is not repeated here.

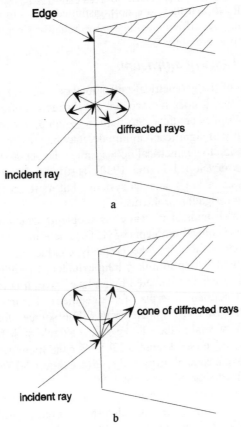

Figure 5.13   Wave incident normal to and at an oblique angle to an edge
   a Normal
   b Oblique

5.4.3 Predicting the signal strength around a building layout

With the addition of diffracted rays to our ray-tracing techniques, we are now in a position to determine the field distribution around a complicated scatterer configuration. As an example, we will consider the prediction of the signal strength around a prescribed building layout.

Figure 5.14 shows a view of the building layout and shows the position of the transmitter and an example receiver location. The buildings are modelled as rectangular boxes. We could add more detail such as roofs and wall features, but this will make our analysis more complicated, and we should take care not to violate the assumptions described in Section 5.4.1. To determine the reflection and diffraction coefficients, some values have to be assigned to the building-material conductivities and permittivities. In this analysis, the buildings are considered to be opaque, so that there are no transmitted rays. In real situations transmission coefficients are difficult to determine reliably since they strongly depend on the internal structure of the building.

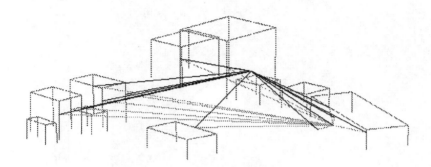

Figure 5.14   Building layout showing ray paths between transmitter and receiver

Figure 5.14 also shows ray paths between the transmitter and receiver. The Figure includes the line-of-sight path, and all the possible paths which involve one reflection or diffraction from a building, i.e. this is a first-order ray trace. The question of up to what order to trace rays in this type of problem is highly situation specific. As the order increases, the ray amplitude generally decreases since the reflection and diffraction coefficients are less than one, so beyond some limit the higher-order terms will have minimal impact on the received signal. The tracing of multiple diffracted rays where the first diffracted ray is close to a shadow boundary is dubious, since the GTD assumptions of an equivalent line source are invalid in this region.

If we move the receiver over an area, calculating the field at each point, we can build up a-signal strength or coverage-area plot (see Figure 5.15). This plot has been produced by tracing rays up to second order. The transmitter is located at the centre of the concentric circles, which are the result of reflections from the ground interfering with the line-of-sight signal. The shadowing effect of the buildings is clearly visible.

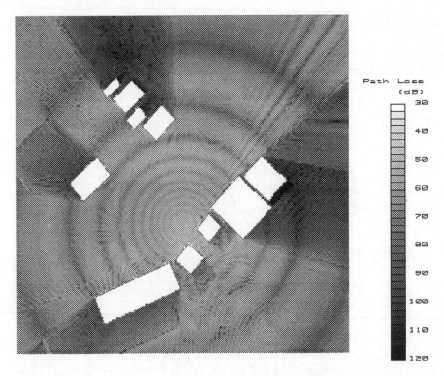

Figure 5.15   Coverage-area plot around a building layout from GTD analysis

Chapter 6

Clear-air characteristics of the troposphere

K.H. Craig

6.1 Introduction

This Chapter considers the effects of refractive-index variations on the propagation of radio waves in the troposphere, and in particular those mechanisms that lead to propagation beyond the normal line of sight. 'Clear air' implies that the effects of condensed water (clouds, rain etc.) are ignored, although gaseous absorption is included. The influence of terrain diffraction is covered in Chapter 4, but terrain reflections are discussed here insofar as they contribute to the clear-air space wave. The frequencies of interest are above about 100 MHz; below this frequency refractive-index variations are not strong enough to cause significant effects, and the ground wave and ionospheric mechanisms dominate at transhorizon ranges.

The emphasis is on the meteorological mechanisms that give rise to anomalous propagation, and the basic models that have been developed to predict the effects of refractive-index variations on radiowave propagation. Statistical procedures for the prediction of radio-link reliability are the subject of Chapter 8.

6.2 Causes and effects of refraction in the troposphere

6.2.1 Electromagnetic waves

Electromagnetic waves propagating in the troposphere are refracted and scattered by variations in the radio refractive index n. Recall that the electromagnetic field of a plane wave propagating in a medium of constant refractive index n has a space r and time t variation given by

$$E(r,t) = E_0 \exp\{i(nk_0 \cdot r - \omega t)\} \tag{6.1}$$

where $\omega = 2\pi \times$ frequency and $\boldsymbol{k}_0$ is a vector normal to the wavefront with a magnitude equal to the free-space wavenumber ($= 2\pi$ / wavelength).

In the troposphere, the refractive index is not constant. At microwave frequencies, however, it varies slowly on the scale of a wavelength. In this case it is still possible to write

$$E(\boldsymbol{r},t) \approx E_0 \exp\left[i\{n(\boldsymbol{r})\boldsymbol{k}_0.\boldsymbol{r} - \omega t\}\right] \tag{6.2}$$

although the magnitude of $E(\boldsymbol{r},t)$ will, in general, vary with position. The value and variations of $n(\boldsymbol{r})$ are fundamental to understanding the way in which electromagnetic waves propagate through the troposphere. For example, Snell's law of refraction and the Fresnel coefficients for reflection and transmission at an interface follow from eqn. 6.2 by applying appropriate boundary conditions across a boundary separating media of different refractive index. In this Chapter we are principally interested in refractive effects. We first consider the determination of n in the troposphere.

6.2.2 Radio refractive index

The radio refractive index of the troposphere is due to the molecular constituents of the air, principally nitrogen, oxygen, carbon dioxide and water vapour. The value of n deviates from unity because of (a) the polarisability of these molecules due to the incident electromagnetic field; and (b) quantum mechanical molecular resonances. The latter effect is limited to narrow frequency bands (for example, around 22 GHz and 60 GHz). We first discuss the former effect which is independent of frequency at the frequencies of interest (up to millimetre waves).

The deviation of n from unity is very small in absolute terms, a typical value being 1.0003 at the Earth's surface. Because of the closeness of n to unity, it is usual to work with the refractivity N defined by

$$N = (n-1) \times 10^6 \tag{6.3}$$

N is dimensionless, but for convenience is measured in 'N units'. N depends on the pressure P (mbar), the absolute temperature T (K) and the partial pressure of water vapour e (mbar):

$$N = 77.6\frac{P}{T} + 3.73 \times 10^5 \frac{e}{T^2} \tag{6.4}$$

This is derived from the Debye formula [80] for the polarisability of polar (i.e. with a strong permanent electric-dipole moment) and nonpolar molecules. The first (dry) term is due principally to the nonpolar nitrogen and oxygen molecules, while the second (wet) term is from the polar water-vapour molecules. The constants are empirically determined, based on experimental measurements [168]. An excellent discussion of eqn 6.4 is given by Bean and Dutton [55].

The variation of P, T and e can be considered at various scales:

(*a*)   on the largest (global) scale the troposphere is stratified in horizontal layers due to the effect of gravity;
(*b*)   on the medium scale (100 m–100 km) the ground and meteorology (local or mesoscale) can produce spatial and temporal variations; and
(*c*)   on the small scale (< 100 m) turbulent mixing causes scattering and scintillation.

We are primarily concerned with the large and medium scale. Turbulent scatter is discussed only briefly.

The macroscopic, large-scale structure of the troposphere varies much more rapidly vertically than horizontally. Figure 6.1 shows contours of potential refractivity (i.e. refractivity reduced to a standard pressure level) derived from aircraft measurements made over the English Channel. Bearing in mind the greatly exaggerated vertical scale, the variations are about two orders of magnitude greater in the vertical direction than in the horizontal. (Actually the inhomogeneities in Figure 6.1 are more severe than in many locations because of the influence of the coastal zones at each side of the Channel.) An assumption of horizontal stratification of the troposphere on this scale is therefore justified. In practice, the same stratification may persist over a horizontal region tens or hundreds of kilometres in extent. The inhomogeneities can have significant effects on radiowave propagation, however, as will be illustrated later.

In an atmosphere at rest with no heat sources, the pressure can be shown to decrease exponentially with height, dropping to a fraction $1/e$ of its value at the surface at a height of approximately 8 km (the *scale height*). In unsaturated air, the temperature falls linearly with height at about 1 °C per 100 m (the *dry adiabatic lapse rate*). The behaviour of water-vapour pressure is more complicated. Ignoring condensation, it would fall exponentially at the same rate as the pressure. However, air at a given temperature can hold only a limited amount of water vapour; the limit occurs at the *saturated water-vapour pressure*, e_s, which varies from 43 mbar at 30 °C to 6 mbar at 0 °C. Above this limit, water condenses to form water droplets (clouds). Since the saturation vapour pressure decreases as temperature decreases, and temperature decreases with height, condensation will occur above a certain height, reducing the water-vapour content of the air. Thus the water-vapour pressure decreases more rapidly with height than pressure and for practical purposes is negligible above 2 or 3 km. Above the condensation level, the temperature follows the *saturated adiabatic lapse rate* which is less than the dry adiabatic lapse rate because of the latent heat released by the condensation process.

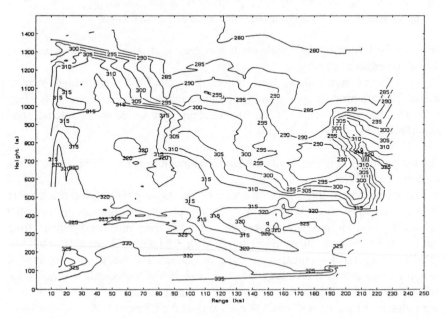

Figure 6.1   Contours of potential refractivity measured over the English Channel

For reference, note that several quantities other than water-vapour pressure are used to characterise the water-vapour content of the air: examples are *relative humidity* ($= e/e_s$, expressed as a percentage), *water-vapour density* ($= 216.7\ e/T$, g/m³) and *humidity mixing ratio* ($= 622\ e/P$, g/kg); it may also be specified in terms of the *dew-point* or *wet-bulb* temperatures.

The net effect of the variations in P, T and e is that N decreases with height. On average, N decreases exponentially in the troposphere:

$$N = N_s \exp(-z / H) \tag{6.5}$$

where N_s is the surface value of refractivity, z is the height above the surface and H is the scale height. Average midlatitude values are $N_s = 315$ and $H = 7.35\ \text{km}$ [13]. Maps showing the geographical and seasonal variation of N_0 (the value of N_s at sea level) are given by Bean and Dutton [55] and in ITU-R Recommendation P.453-5 [13].

6.2.3 *Effect of the refractive index on radiowaves*

If the refractive index were constant, radio waves would propagate in straight lines. For present purposes an adequate physical picture is to consider the radio waves as propagating out from a transmitter along 'ray' paths. Initially we assume stratification of the atmosphere and ignore variations in the horizontal direction. Changes in refractive index with height then determine the bending of the ray paths in a vertical plane. The amount of bending between level 1 and level 2 is determined by Snell's Law (i.e. $n_1 \sin\theta_1 = n_2 \sin\theta_2$, where $n_{1,2}$ and $\theta_{1,2}$ are, respectively, the refractive index and the angles that the rays make to the vertical in levels 1 and 2). Since n decreases with height, rays are bent downwards toward the Earth. An immediate consequence is that the radio horizon lies further away than the visible horizon (Figure 6.2).

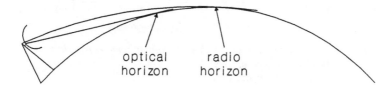

Figure 6.2   *Extension of radio horizon due to tropospheric bending (greatly exaggerated scale)*

For a radio path extending through the atmosphere, this refractive bending causes the elevation angle of a ray at the ground to be greater than if the atmosphere were not present. Figure 6.3 shows the computed relationship between the elevation-angle correction and the true elevation angle for a slant path through the atmosphere in a tropical and a polar climate [99]. The curve for the UK lies close to the mean of those shown. These elevation-angle offsets can be important, for example, in estimating target heights from radar returns.

For heights much less than the scale height, the exponential in eqn 6.5 can be approximated by the first term in its expansion, giving a linear decrease of refractivity with height at a rate of about $40\,N$ units per kilometre at midlatitudes. This approximation is excellent for terrestrial paths but is inadequate for airborne radar calculations and Earth–satellite paths at low elevation angles.

It can be shown that the radius of curvature C of a ray is very well approximated by

$$C = -\frac{dn}{dz} \tag{6.6}$$

at low elevation angles. The curvature of the Earth is $1/a$ where a is the Earth's radius (6378 km). Thus the curvature of the ray *relative* to the curvature of the

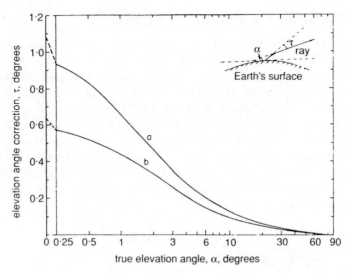

Figure 6.3   *Error in elevation angle due to tropospheric refraction*
         a Tropical maritime air (July)
         b Polar continental air (April)

Earth is $1/a - C$. Since we are often mainly interested in this relative curvature, it is useful to introduce the concept of an effective Earth radius $a_e = ka$, where k is known as the k factor. Then we have

$$C - \frac{1}{a} = K - \frac{1}{a_e} \tag{6.7}$$

where K is the effective ray curvature associated with the effective Earth radius a_e. We have already seen that $C = 40\times10^{-6}\,\mathrm{km}^{-1}$ in the average midlatitude atmosphere, while $1/a = 157\times10^{-6}\,\mathrm{km}^{-1}$ (note that the curvature of the Earth substantially exceeds the downward curvature of the ray). Straight-line $(K = 0)$ ray propagation relative to the effective Earth radius can then be arranged by setting $1/a_e = 117\times10^{-6}\,\mathrm{km}^{-1}$, corresponding to $k = 4/3$. This is a derivation of the well known 4/3-Earth's-radius construction so useful in engineering calculations: a ray propagating in a straight line over terrain based on a 4/3-effective-Earth radius is equivalent to a ray propagating in an atmosphere with the average lapse rate of 40 N/km over the actual terrain. Figure 6.4 shows different k factor representations of the same picture. Rays emanate from a transmitter on the left at a height of 25 m into a standard atmosphere ($- 40\,N$/km): note that the rays curve downwards slightly for $k = 1$ (when $K = 40\times10^{-6}\,\mathrm{km}^{-1}$) and are straight for $k = 4/3$ (when $K = 0$). For terrestrial radio links it is a simple matter to check for terrain clearance or obstruction by joining potential transmitter and receiver positions by straight lines on 4/3-Earth-radius graph paper.

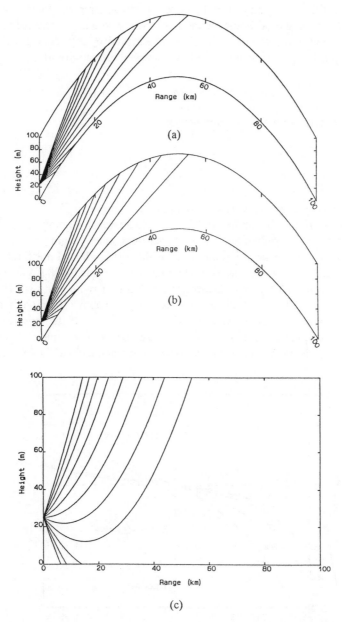

Figure 6.4   *k factor-representations of rays propagating in a standard*
            atmosphere
            a $k = 1$
            b $k = 4/3$
            c $k = \infty$ (flat Earth)

There is a third viewpoint, useful in ducting studies: replace the Earth with a flat Earth ($k = \infty$) and modify the curvature of the ray so that the relative curvature between ray and Earth is preserved. This is achieved by introducing a fictitious medium, replacing the refractivity N by the modified refractivity M:

$$M = N + 10^6 \times z / a \quad = \quad N + 157z \tag{6.8}$$

where the height z is given in kilometres, i.e.

$$\frac{\partial M}{\partial z} = \frac{\partial N}{\partial z} + 157 \tag{6.9}$$

Note that rays curve *upwards* relative to a flat Earth (Figure 6.4).

While N decreases by about $40\,N/\text{km}$ (M increases by about $117\,N/\text{km}$) in average conditions at midlatitudes in the lower troposphere, significant deviations from the average do occur. Figure 6.5 shows the distribution of mean refractivity gradient in the UK from the surface up to a range of heights [99]. If the lapse rate of N is less than $40\,N/\text{km}$, the downward curvature of radio rays will decrease, shortening the radio horizon and reducing the clearance above terrain on terrestrial paths; this is known as *subrefraction*. On the other hand if the lapse rate of N exceeds $40\,N/\text{km}$, the ray curvature will increase, extending the radio horizon and increasing path clearance; this is known as *super-refraction* (see Figure 6.6).

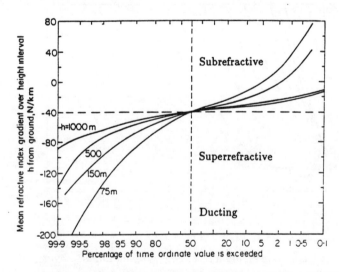

Figure 6.5   *Distribution of mean refractive-index gradient in the UK with height interval above ground level [99]*

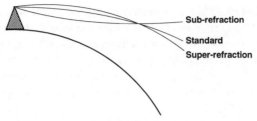

Figure 6.6   Classification of refractive conditions

When the lapse rate of N exceeds 157 N/km, i.e. $\partial N/\partial z < -157$, or equivalently, $\partial M/\partial z < 0$, the rays are bent towards the Earth more rapidly than the Earth's curvature. This is known as *ducting* and can cause rays to propagate to extremely long ranges beyond the normal horizon. The usual classification of propagation conditions in terms of refractivity gradients is given in Figure 6.7. The simple criterion for ducting in terms of modified refractivity gradients (the existence of a negative slope on the M/z graph, irrespective of the scales of the axes) is one reason why M is the most useful quantity for ducting studies. Ducting is discussed later.

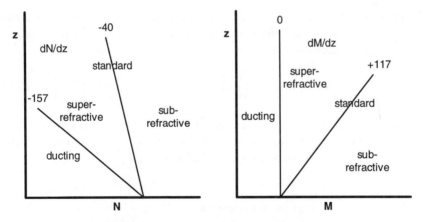

Figure 6.7   N/z and M/z plots of refractive condition classes
(Gradients are in N/km or M/km)

6.2.4 Gaseous absorption and complex refractive index

In Section 6.2.2 it was pointed out that molecular resonances make a significant contribution to the radio refractive index in certain frequency bands. Only oxygen and water vapour are relevant at frequencies below 350 GHz. These resonance lines can cause significant absorption of radio waves at frequencies near the lines.

The oxygen molecule has a permanent magnetic-dipole moment due to paired electron spins. Changes in orientation of the combined electron spin relative to the orientation of the rotational angular momentum give rise to a closely spaced group of 'spin-flip' or hyperfine transitions near 60 GHz, and a single line at 119 GHz. The water molecule has a permanent electric-dipole moment, and rotations of the molecule with quantised angular momentum give rise to spectral lines at 22, 183 and 325 GHz.

Figure 6.8 shows the attenuation rate per kilometre at ground level caused by oxygen and water vapour, calculated using the methods given by Liebe [123]. At low altitudes the lines are greatly widened by pressure (collision) broadening, and the complex of separate lines at 60 GHz cannot be resolved individually. At lower pressures near the top of the troposphere, the separate lines can be resolved. The absorption spectrum of water vapour has very intense lines in the far-infrared region; the low-frequency tails of these lines are seen as the sloping baseline of the water-vapour spectrum shown in Figure 6.8. Very significant absorption can occur, notably around the oxygen-line complex at 60 GHz. This limits path lengths to a few kilometres at these frequencies. Unlike oxygen, the amount of water vapour in the atmosphere is variable even at the ground, and consequently the attenuation near the water-vapour lines can vary significantly from place to place and day to day.

From the point of view of modelling, it is worth noting that gaseous absorption and refraction can be dealt with in a unified way. If the refractive index is regarded as a complex number $[n = \mathrm{Re}(n) + i\mathrm{Im}(n)]$, eqn. 6.2 can be written

$$E(\boldsymbol{r},t) \approx E_0 \exp\!\left(i\!\left[\mathrm{Re}\{n(\boldsymbol{r})\}\, \boldsymbol{k}_0 . \boldsymbol{r} - \omega t \right] \right) \exp\!\left[-\mathrm{Im}\{n(\boldsymbol{r})\}\, \boldsymbol{k}_0 . \boldsymbol{r} \right] \qquad (6.10)$$

The imaginary part of n causes the field to decay exponentially with range, characteristic of absorption. In fact the specific attenuation is just $0.182 \times 10^6\, f\,\mathrm{Im}(n)$ (dB/km) where f is the frequency (GHz).

6.2.5 Refractive-index measurements

The most widely available source of refractive-index data is the radiosonde. A radiosonde consists of an instrument package carried aloft by a freely ascending, gas-filled balloon. The instruments measure (directly or indirectly) pressure, temperature and humidity, from which height and refractive index along the ascent can be derived. The data are sampled and transmitted to a receiver on the ground. Radiosondes are launched twice daily at a large number of meteorological stations around the world for the purpose of weather forecasting. The vertical resolution of traditional radiosondes is poor in the first kilometre: thin ducting layers tend to be missed, or at best smoothed out, so underestimating the effect of ducting layers. The newer minisonde systems give better vertical resolution, are portable and can be launched by one person.

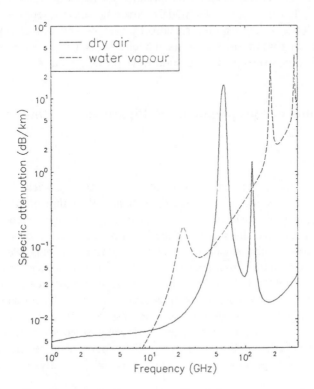

Figure 6.8   Specific attenuation at ground level due to gaseous absorption by oxygen and water vapour

(Water-vapour density = 7.5 gm^{-3}, temperature = 15 $^{\circ}$C)

An alternative *in situ* measurement system is the refractometer. This instrument is basically an open-ended resonant microwave cavity, whose resonant frequency is determined by the refractive index of the air within the cavity [188]. Advantages of refractometers over radiosondes include rapid response time (and hence high resolution) and direct measurement of refractivity. However, the bulk and cost of these systems rule them out as throwaway sensors for balloon ascents. They have principally been used as high-precision instruments mounted on aircraft or helicopters for detailed case studies.

Remote sensing of the troposphere is an active area of study. Techniques being investigated for refractivity sensing include Doppler radars, lidars (laser sounders) [133], sodars (acoustic sounders) and satellite-borne instruments. While useful qualitative results have been achieved (particularly on the spatial and temporal distribution of ducting layers), none of these methods has yet been developed into a practical, quantitative tool.

Numerical weather models hold out promise for the wide-scale forecasting of refractivity. Recent mesoscale models covering limited regions have good vertical resolution. Although not specifically designed for refractivity purposes, all the necessary parameters exist in the models. Much testing and refinement are required before these can be used for radiowave prediction purposes.

6.3   Anomalous propagation: multipath and ducting

6.3.1   Types of duct

We have seen that, if the vertical lapse rate of refractivity exceeds 157 N/km, the waves bend downwards with a curvature greater than that of the Earth, and ducting can occur. Radio energy can become trapped between a boundary or layer in the troposphere and the surface of the Earth or sea (*surface* duct) or between two boundaries in the troposphere (*elevated* duct). In this waveguide-like propagation, very high signal strengths can be obtained at very long range (far beyond line-of-sight). Indeed the signal strength may well exceed its free-space value. Equation 6.4 shows that two processes can cause the formation of high lapse rates: a rapid decrease in water-vapour pressure with height, and an increase in temperature with height (a temperature inversion); these mechanisms often occur together. The vertical pressure gradient never deviates much from its standard value—large-scale air movements (winds) rapidly restore pressure equilibrium.

The sensitivity of N to variations in the meteorological parameters can be found by differentiating eqn. 6.4. Assuming typical atmospheric conditions ($P = 1000$ mbar, $T = 293$ K, $e = 15$ mbar), the variation in N is given by

$$\delta N = 0.26\delta P + 4.3\delta e - 1.4\delta T \qquad (6.11)$$

Differences of a few degrees in T and a few millibars in e can occur between adjacent air masses in certain meteorological conditions. This can lead to changes of several tens of N units over a height interval of tens of metres, and the formation of a ducting layer.

There are three important types of layer/duct; they are illustrated in Figure 6.9 using actual radiosonde data. As discussed in Section 6.2.3, a *ducting layer* is immediately identifiable by a negative slope in the modified refractivity/height curve; the *duct* (marked by a vertical bar in Figure 6.9) extends from the top of the ducting layer down to the ground (for a surface duct) or down to the height at

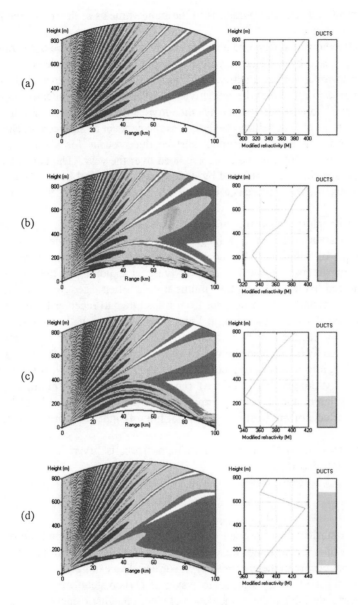

Figure 6.9 *Definition of duct types, and their effect on a 3 GHz transmitter at 20 m height*

The ducts are indicated by the vertical bars

a Standard atmosphere

b Surface layer, surface duct

c Elevated layer, surface duct

d Elevated layer, elevated duct

which the modified refractivity returns to the same value as at the top of the layer (for an elevated duct).

Figure 6.9*a* shows the standard atmosphere for reference. Figure 6.9*b* shows a surface duct caused by a surface layer. Figure 6.9*c* and *d* shows elevated layers: Figure 6.9*c* forms a surface duct, while Figure 6.9*d* forms an elevated duct. It should be borne in mind that the refractivity profiles can be complicated and that multiple layers do occur (such as the low surface duct in Figure 6.9*d* accompanying the elevated duct); nevertheless the concept of surface and elevated ducts is useful. In northern Europe, surface ducts occur for about 5%, and elevated ducts for 5–10%, of the time, averaged over the year. The percentage of the time that a radio link is affected by ducting will in general be less than these figures, because of the importance of the path geometry relative to the ducting layers.

The contour plots show the effect of the associated layers on a 3 GHz transmitter located on the left at a height of 20 m over the sea; the contours are given in terms of path loss. The common feature in all the plots is the lobing caused by interference between energy arriving at a point via the direct and ground-reflected paths. The term 'multipath' is often used to describe the situation where radio waves propagate from transmitter to receiver by more than one path, especially on a line-of-sight link. The secondary paths may occur by ground reflection (as here) or by refraction. The distinction between ducting and multipath caused by interference between two or more refractive paths through the atmosphere is more a descriptive convenience than a fundamental difference in mechanism. At longer (transhorizon) ranges, the number of multiple paths becomes very large, and it is more appropriate to use models based on ducting theory. The effects of multipath on terrestrial line-of-sight paths are discussed in Chapter 8.

The distance d (km) to the normal radio horizon, is given in terms of the transmitter height h (m) by

$$d = 4.12 \sqrt{h} \qquad (6.12)$$

Both types of surface duct are seen to propagate energy well beyond the normal radio horizon (at 18.4 km range for a 20 m high transmitter). However, note the 'skip zone' that occurs in the elevated-layer, surface-duct example.

A ducting layer will only trap radio waves if certain geometrical constraints apply. In particular, the angle of incidence of electromagnetic energy at the layer must be very small. A simple rule of thumb derived from the total-internal-reflection condition of geometrical-optics is that the maximum angle of incidence θ_{max} (degrees) is related to the change in refractivity ΔN (N units) across the layer by

$$\theta_{max} = 0.081 \sqrt{|\Delta N|} \qquad (6.13)$$

As ΔN rarely exceeds $50 N$ units, θ_{max} will be limited to $0.5-1°$. Simple geometrical considerations show that even energy launched horizontally will intercept an elevated layer at a nonzero angle owing to the Earth's curvature. It follows that ducting layers higher than about 1 km will not significantly affect terrestrial radio links. For example, although the elevated duct of Figure 6.9*d* distorts the interference lobes significantly, it is not strong enough to trap the energy completely.

The dependence of ducted field strength on frequency and path geometry is complicated. At long ranges from the transmitter, a simple single-mode propagation model of a surface duct (see Section 6.4.3) predicts that the field strength should decrease exponentially with distance (i.e. linear in decibels) because of mode attenuation. This is indeed reflected in the ITU-R method for the prediction of interference levels caused by ducting and layer reflection (see Chapter 13, eqn. 13.6, which is derived empirically from experimental data). However, the attenuation rate should theoretically become smaller at higher frequencies for a uniform duct; this is the opposite to what is observed, indicating that horizontal inhomogeneities and small-scale scatter begin to be important at microwave frequencies. The strongest coupling of energy into or out of a duct occurs when the transmitter or receiver itself lies within the duct: the ITU-R prediction method incorporates an antenna-height-dependent factor to account for coupling into coastal advection ducts.

If the change in refractivity between two air masses is very abrupt, it is more appropriate to consider the radio waves to be reflected by the layer, rather than refracted by it. A layer of thickness *t* can be considered to be a discontinuity if

$$t \quad < \quad 14 \lambda / \theta \qquad (6.14)$$

where λ is the wavelength (in the same units as *t*) and θ is the angle of incidence (degrees). In this case the strength of the field reflected and transmitted by the layer can be calculated by means of Fresnel's formulae for reflection and transmission coefficients. From eqn. 6.14 the distinction between refraction and reflection is frequency dependent, and is really a choice in the way the mechanism is modelled, rather than a difference in the mechanism itself. The layer-reflection model is most useful at lower (VHF) frequencies, while refraction/ducting is a better model at UHF and above.

The meteorological mechanisms causing duct formation are now discussed. More details can be found in [46] and [55].

6.3.2 Evaporation

A shallow surface-based duct, the evaporation duct, exists for most of the time over the sea (and other large bodies of water). It is caused by the very rapid decrease of water-vapour pressure with height in the lowest few metres above the sea surface; the mean duct thickness varies with geographical location. It ranges from 5–6 m in the North Sea through 13–14 m in the Mediterranean to over 20 m in the Gulf area. The evaporation duct is the dominant propagation mechanism for ship-based radar and communication systems. It can also have an important modifying effect at coastal sites in the presence of other surface ducts, such as the advection duct described below.

Because of its influence on naval-sensor performance, much effort has gone into understanding the evaporation duct. There are well established boundary-layer models [53,112,146,158] based on the similarity theory of turbulence; these enable the evaporation-duct height to be estimated from a bulk measurement of sea temperature and the air temperature, water-vapour pressure and wind velocity a few metres above the surface. In open ocean conditions where the sea temperature and meteorology vary relatively slowly with range, the evaporation duct can extend for hundreds of kilometres with almost constant duct height. In coastal regions, and in enclosed areas such as the Gulf area, significant variations in duct height can occur.

Figure 6.10 shows the effect of a 15 m evaporation duct on the coverage of a radar at a height of 25 m above the sea at 3 GHz (S band) and 18 GHz (Ku band). The contours here are given in terms of propagation factor (field strength relative to free space); this is a better quantity than path loss for comparing systems operating at different frequencies. Note that the effects are stronger at the higher frequency, and that the interference lobes are trapped in the duct even though the antenna lies above the duct.

6.3.3 Nocturnal radiation

Radiative heat loss from the ground during clear, still nights results in a temperature inversion. This can result in the formation of a duct over inland or coastal regions, depending on the humidity profile. If there is sufficient water vapour present, condensation can occur, forming a radiative fog; the temperature inversion will cause an increase in water-vapour pressure with height, leading to subrefraction. On the other hand if the air is dry, the temperature inversion causes super-refraction and ducting. In dry, hot climates (such as North Africa or the Middle East), radiation inversions cause severe problems to broadcast services [52].

The inversion layer weakens after sunrise, and is finally destroyed by solar heating. Hence periods of anomalous propagation associated with nocturnal radiation ducting are fairly short (one or two hours). In northern Europe this type of ducting tends to be localised, since it depends on the nature of the ground cover and of local topographical features. It is therefore less likely than other forms of ducting to cause interference over long distances. However, it can cause a severe multipath problem for low-lying terrestrial links, particularly when the ducting layers are in the process of forming and breaking up.

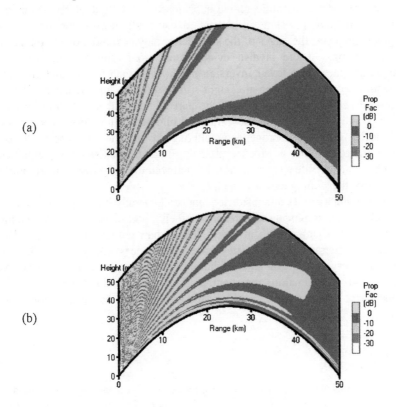

Figure 6.10   Effect of a 15 m evaporation duct on radar at a height of 25 m over the sea
a 3 GHz
b 18 GHz

6.3.4 Subsidence inversion

Elevated ducts can be formed by large scale atmospheric subsidence occurring during anticyclonic conditions. Large cooler air masses associated with high-pressure systems are heated by adiabatic compression as they descend to lower levels. A strong temperature inversion forms between the descending air and the well mixed air near the surface (the inversion layer is often visible at the ground owing to the trapping of atmospheric pollutants, or the formation of a layer of cirrus cloud at the inversion). The inversion may be accompanied by a sharp decrease in humidity.   Both mechanisms tend to cause the formation of an elevated duct.

In the early stages of anticyclonic subsidence, the height of the inversion is generally too great (1–2 km or higher) to cause significant ducting.   As the anticyclone evolves, the edges of the subsidence may descend close to the ground, causing ducting.   There is also evidence of strong diurnal variation in the inversion height, with a lower inversion height (less than 500 m) and anomalous propagation occurring during the night.   Anticyclonic subsidence can be very widespread (500–1000 km) although the part of the anticyclone contributing to ducting is unlikely to exceed a few hundred kilometres in extent.   The inversion layer can be relatively homogeneous, although it tends to slope down from the centre to the edge of the subsidence 'dome'.   Figure 6.11 shows the synoptic chart and radiosonde profiles for a period of anticyclonic subsidence over northern Europe; note the sloping elevated duct.

There are other causes of subsidence that can generally be neglected because of their limited geographical occurrence or their short duration.   For example, sporadic areas of subsidence at levels lower than normal may occur when a weather front interacts with an anticyclone.   This process can yield thin inversion layers with very strong gradients of humidity and temperature, and hence very severe periods of anomalous propagation.   This type of subsidence is short-lived and localised, and appears more like a perturbation of the general anticyclonic pattern.

6.3.5 Advection

Advection ducts are of considerable importance in coastal regions such as the North Sea or enclosed seas with adjoining hot, dry land areas such as the Mediterranean and the Gulf area.   Advection is the large-scale motion of air masses. A duct can be formed when a warm, well mixed air mass flows off a land surface over a cooler sea. In Northern Europe, advection is generally associated with anticyclonic subsidence: when a summer anticyclone is positioned over continental Europe, warm dry air is carried by advection from the continent out over the North Sea.   The interaction of the continental air with the underlying cooler, moister air just above the sea surface results in the formation of high humidity lapse rates and temperature inversions, producing marked refractivity gradients.   Advection ducts are usually surface based, typically less than 200 m thick (Figure 6.9b shows an example).

Whereas subsidence ducts are characterised by an almost trilinear shape, the typical advection duct profile is smoother. As a result, these ducts are weaker and leakier than subsidence ducts.   Nevertheless they constitute a major cause of anomalous propagation in coastal areas as the surface duct couples strongly to terrestrial links and advection ducts tend to persist for several days at a time. Like subsidence, advection shows some diurnal variation, with a peak of activity expected in the early evening. Advection ducts weaken away from the coast, but can still be significant several hundred kilometres from it.

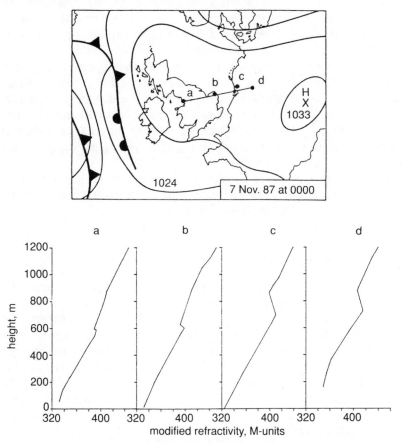

Figure 6.11   *Synoptic chart and radiosonde-derived modified refractivity profiles at the indicated locations showing a sloping elevated duct during anticyclonic subsidence*

6.4   Propagation models

6.4.1 Statistical and deterministic models

Two types of prediction model are required for radiowave propagation. Statistical models are often used for planning purposes to estimate the reliability of a system, or the level of interference to be expected to a service: the models are required to predict the path loss which is exceeded for a given percentage of the time of interest. Statistical models are often based on the underlying physics of the problem but, because of a shortage or absence of the necessary input meteorological and other environmental data, and the need for wide applicability and short computation times, these models tend to be semi-empirical, the model parameters being deduced from experimental data.

A different requirement arises when the operating environment of a system is well characterised, the expense of obtaining real-time meteorological measurements is justified and there is a requirement to predict the performance of a system in near real time; this may include the need for a propagation forecast a day or two ahead. The performance of military surveillance radars and communication systems comes into this category. Here a more complete model based on physical principles is required.

We now describe the most important deterministic methods currently used to model radiowave propagation in clear-air conditions. Statistical models of clear-air propagation are described in Chapters 8 and 13.

6.4.2 Geometrical optics

Probably the simplest conceptual model for radiowave propagation is geometrical optics. When the medium changes slowly on the scale of a wavelength, the electromagnetic field is a plane wave locally, and the energy propagates along rays that are trajectories orthogonal to the wavefronts. The rays are traced outwards from the transmitter, with the radius of curvature of the ray paths depending on the local refractive-index gradients. Differential equations describing the ray paths can be derived from repeated application of Snell's law. These can be integrated numerically in general, or, for one-dimensional vertical refractive-index profiles, can be integrated semi-analytically, yielding a computationally efficient algorithm.

Figure 6.12 shows an example of ray tracing for a transmitter at 25 m height in the presence of a simple elevated ducting layer. Ground-reflected rays have been included. Note that the lower turning points of the ducted rays occur above the ground, typical of an elevated duct. There are several limitations to this 'naive' ray trace:

(*a*)   when an individual ray encounters the layer it either penetrates it (such as the ground-reflected rays) or is turned round by it; there is no concept of a ray splitting into two components (leakage out of the duct, partial reflection);

(*b*)   the ray trace is frequency independent;

(*c*)   a single ray by itself carries no amplitude information.

The field strength can be derived from geometrical optics either by explicitly calculating the cross-section of a pencil of rays, or by integrating a further set of differential equations—the transport equations. To calculate the field strength at a point, it is necessary to categorise the rays carefully into separate families (each family containing rays which have followed similar trajectories), and to add coherently the contributions from all ray families passing through the point: this coherent addition requires the phase of the field, obtained from the optical path length of the ray. Two-dimensional field plots based on these methods have been generated [104].

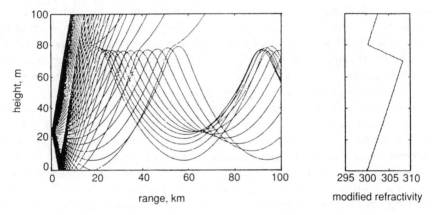

Figure 6.12   Ray trace for a transmitter at 25 m height in the elevated duct shown

Apart from the difficulty of classifying rays into families automatically, there is the problem that geometrical optics fails at ray caustics. A caustic is a locus of points defining the boundary of a ray family. On one side of the caustic, two rays pass through each point, while on the other side there are no rays of that family. A caustic occurs between 60 and 80 km at a height of 20 m in Figure 6.12, and there is a second caustic above this one at the boundary of the ray envelopes; the two caustics meet at a point called a cusp. The ray equations predict infinite field strength at caustics and cusps, with no energy reaching the region beyond a caustic from rays in the family giving rise to it. The problem is that geometrical optics ignores the effects of diffraction. Unfortunately, these situations are likely to arise in association with the anomalous refractive-index structures which give rise to ducting. At long ranges in the presence of ducts, many rays may link transmitter and receiver, and the number of caustics will build up with distance from the transmitter. In addition, the numerical accuracy required to calculate the optical path lengths of these long rays is difficult to achieve, even if it is assumed that the refractive index along the path is sufficiently well characterised for stable results, which it rarely is in practice. If surface-reflected rays are present, as in the case of a surface duct, the problem is compounded by the difficulties of modelling the reflection, particularly if the surface is rough.

While several sophisticated ray-tracing programs are available which give good *qualitative* results, and quantitative results in the shorter range, line-of-sight region, geometrical optics has severe difficulties for quantitative calculations in the transhorizon region.

6.4.3 Mode theory

At ranges beyond a transmitter's radio horizon, or in regions where the assumptions of ray theory break down, field-strength calculations can be made by 'full-wave' methods.   These attempt to find exact solutions (in principle) to Maxwell's equations for the given refractive-index profile and transmitter–receiver geometry.   For real situations, of course, such a solution is computationally intractable, and approximating assumptions are inevitable.

The most common approach is to impose strong constraints on the representation of the measured refractive-index profile.   For example, if the refractive index is assumed to vary only in the vertical direction, the wave equation can be separated and the imposition of matching conditions yields a one-dimensional equation, the mode equation.   Even this one-dimensional equation is difficult to solve in general; to simplify the problem, the profile is often approximated by simple shapes (such as linear segments) to which there are standard solutions.   Mode theory was first applied to tropospheric radiowave propagation in [59] and is fully described in [67]; the application to elevated ducts was first given in [183].

Mode theory in the troposphere is similar to mode propagation in a parallel-plate waveguide.   A duct of a given width will support a certain number of propagating (trapped) modes, the number being larger at higher frequencies. Higher-order modes are cutoff and are evanescent.   Each mode has an eigenvalue and eigenfunction: the eigenvalue defines the phase velocity and attenuation rate of the mode; the eigenfunction is a height/gain function which defines the vertical distribution of energy in the mode.   However, there is an important difference between tropospheric and waveguide modes: in the troposphere, at least one of the duct boundaries is 'soft', as it is defined by a change in the refractive-index gradient, rather than by a perfectly conducting plate.   The consequence is that *all* tropospheric modes are leaky to some extent (i.e. they are attenuated as they propagate) although the attenuation rate for the lowest-order modes may be very small.   A physical picture of tropospheric modes is given in [74]; in particular eqn. 9 in [74] shows that the inequality of eqn. 6.13 is the condition for a mode to be well trapped.   The field at a receiver is obtained by summing the field contributions from all significant modes.   Each contribution is a product of three terms: the height/gain function evaluated at the transmitter height, expressing the strength of coupling of the transmitter to the mode; the height/gain function evaluated at the receiver height; and a term expressing the attenuation of the mode with range.

Mode theory has been successfully applied to long-range propagation in the evaporation duct where there are a few, well-trapped modes, and the transmitter and receiver are both strongly coupled into the duct. Mode theory becomes intractable at short range near the line-of-sight, and at high frequencies (or for deep ducts) where the number of trapped modes becomes very large.

While a one-dimensional approximation may be adequate for evaporation ducts, it is inadequate for representing the meteorological structures which lead to elevated ducts. In general, horizontally homogeneous mode-theory models overestimate the path loss for terminals that are located outside a duct. Inhomogeneities can give rise to higher field strengths due to mode mixing between the strongly coupled but 'leaky' modes and the low-loss well trapped modes. Extensions of the theory to account for mode conversion in a stepped layer model have been made, but these are complex, and there seems little future for this approach in practical prediction models. Another disadvantage of mode theory is that there is no simple way of including the effects of terrain diffraction at the Earth's surface.

6.4.4 Parabolic equation

A recent alternative to mode theory is the parabolic-equation (PE) model [75,76,84,117]. This allows field-strength calculations to be made for essentially arbitrary two-dimensional refractive-index structures. The model is based on a parabolic approximation of the wave equation which admits an efficient numerical solution. The solution is matched in range in a rectangular domain, once the initial field and the boundary conditions at the top and bottom of the domain have been specified. The method works equally well at short and long ranges, thus avoiding the need for different models at these two extremes.

There are several numerical methods that can be used to find a solution of the PE. The computationally most efficient is the 'split-step' algorithm: fast-Fourier-transform methods make it possible to calculate radar coverage diagrams/field strength contours on a personal computer. The PE works equally well with real or complex refractive index. It is therefore very easy to include the effects of gaseous absorption (which can be variable from point to point) within the PE framework.

All the field-contour plots in this chapter were generated using the PE. Figure 6.13 shows the PE results at three frequencies (3 GHz, 9.5 GHz and 18 GHz) for the same refractive-index profile as the ray trace of Figure 6.12, together with the standard-atmosphere picture at 9.5 GHz for reference. In the radar's line-of-sight region the main feature is the interference lobing; the lobes are seen to be distorted significantly by the presence of the layer (note that the lobes becomes more numerous at higher frequencies; in fact the number of lobes in a sector is proportional to h/λ where h is the antenna height and λ is the wavelength). As the frequency increases, the pictures become more 'optical' and

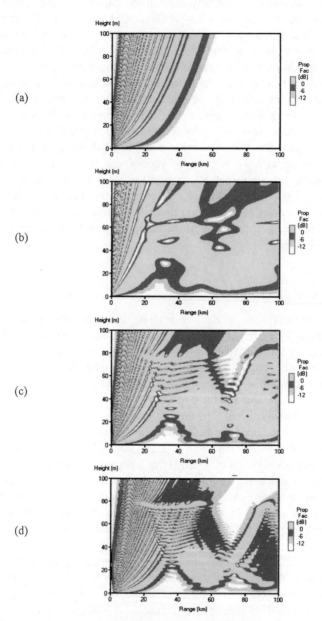

Figure 6.13   Parabolic equation derived contours for a radar at 25 m height over the sea
a 9.5 GHz radar in standard atmosphere
b–d the same elevated duct as in Figure 6.12
b 3 GHz; *c* 9.5 GHz; *d* 18 GHz

begin to resemble the ray trace as one would expect (compare Figures 6.13*d* and 6.12); note, however, that the full-wave-PE calculation contains much information missing from the ray trace: the PE results are quantitative, and are frequency dependent; leakage through the top of the layer is visible at all frequencies — at 3 GHz the field is much more diffuse than the ray trace would suggest; and the PE solution is valid at caustics and cusps, giving finite results. One of the principal advantages of the PE method is its ability to handle two-dimensional refractive-index data. Figure 6.14 shows the picture for the same case as Figure 6.13*c*, except that the ducting layer has been assumed to weaken with range, returning to a standard atmosphere at 100 km (such as could occur in coastal advection). The second-layer 'bounce' at 90 km no longer occurs.

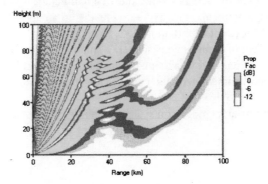

Figure 6.14   *The same situation as Figure 6.13c except that the ducting layer*
            weakens with range from the radar

Radiowave PE methods were first developed for propagation over the sea. Recent developments have extended their applicability to terrain diffraction and three-dimensional problems. PE and hybrid ray-PE models are now incorporated in operational-radar performance tools.

6.5   Turbulent scatter

In the troposphere small-scale irregularities in the propagation medium are caused by small deviations of the temperature or humidity from the background value. The process of creating these irregularities begins with the shear forces between two moving air masses; this force creates large turbulent eddies at the outer scale of turbulence (typically 100 m). These large eddies spawn smaller and smaller eddies, transferring energy from the 'input' range to smaller-scale structures in the 'transformation' range. In fully developed turbulence, this mixing continues right down to the inner scale of turbulence (about 1 mm) where the energy is dissipated;

in this 'dissipation' range, the eddies are dominated by the influence of viscosity and diffusion and cannot sustain turbulent activity. The theory of propagation in a turbulent medium is given in [172].

There are two ways in which turbulence affects radiowave propagation. First, within the radio beam, rapid amplitude and phase fluctuations or 'scintillation' occurs. This arises because the irregularities within the beam focus and defocus energy. Fading occurs with a characteristic frequency spectrum owing to the movement of the irregularities and angle-of-arrival variations arise which can cause increased fading when narrow-beam antennas are used. The largest amplitude fluctuations observed in the aperture plane of an antenna are produced by eddies with scale sizes of the order of the Fresnel zone size. For antenna apertures larger than the Fresnel zone, the fluctuations are spatially averaged by the antenna: this antenna aperture averaging results in an apparent loss of antenna gain, reducing the level of scintillation.

The second effect of turbulence is scatter: a small portion of the incident energy is redirected and provides a weak signal outside the incident beam. Troposcatter is always present to some extent and can provide communications on beyond-the-horizon paths.

The effects of turbulent scatter on system reliability are discussed in Chapter 8.

Chapter 7

Nature of precipitation and cloud

J.W.F. Goddard

7.1 Introduction

Water appears in the atmosphere in a variety of forms, usually referred to by the term 'hydrometeor', which includes particles as diverse as cloud, raindrops, snowflakes, ice crystals, hail and graupel. Of these, rain, hail, graupel and snow are generally recognised as 'precipitation'. The effects that hydrometeors have on communications systems are dependent both on the system frequency and the type of particle present. At any given instant, of course, more than one type of particle will effect a given link. For example, an Earth–space link may often encounter rain over the lower part of its path, and snow at greater heights.

In the following discussion, the various forms of hydrometeor will be considered first, together with their relevance to radiowave propagation. A brief theoretical framework will then be presented as a background to the model-development process, and, finally, specific effects of hydrometeors on systems will be considered.

7.2 Hydrometeors

7.2.1 Raindrop-size distributions

In modelling the effects of rain on radio waves propagating through it, or scattered from it, the usual problem is to numerically relate two physically distinct quantities. For example, we might wish to relate a microwave specific attenuation to a rain rate or to a radar reflectivity, a rain rate to a reflectivity, or a differential attenuation between two polarisations to an absolute attenuation in one etc. All these quantities are obtained as integrals over the drop-size-distribution function $N(D)$, defined such that $N(D)dD$ is the number of drops per cubic metre with drop diameters D between D and $D + dD$. (As all but the smallest drops are nonspherical, D is usually defined as that of an equivolumic sphere.) Nearly always the contributions to the two quantities to be related have a different D dependence, and some assumptions about the shape of the function $N(D)$ must be made.

Satisfactory results are generally obtained by using an ideal mathematical form for $N(D)$, described by a small number of free parameters. Large departures from the ideal forms undoubtedly occur in real rain, but the situation is eased by the fact that relationships between quantities are, in practice, often only required to hold on a mean statistical basis. Also fine structure in $N(D)$ tends to disappear in the integration.

Much modelling to date has used an exponential form for $N(D)$ of the form

$$N(D) = N_0 \exp(-\Lambda D) \tag{7.1}$$

Early work [128] indicated that Λ tended to increase with rain rate R, and that these distributions could be reduced to a one-parameter family by treating the relation as deterministic, with

$$\Lambda = 4.1 R^{-0.21} \qquad \left(\Lambda \text{ in mm}^{-1}, R \text{ in mmh}^{-1}\right) \tag{7.2}$$

Once Λ is determined from R, so is N_0 by the requirement that the integral of $N(D)D^3 V(D)$, where V is the drop terminal velocity, actually gives R. Typical values of N_0 were found to be around $8000\,\text{mm}^{-1}\,\text{m}^{-3}$, with this form of distribution becoming known as the Marshall–Palmer distribution. The parameter Λ can be shown to be related to the median volume drop diameter of the distribution D_0 through the relation

$$\Lambda = 3.67 / D_0 \tag{7.3}$$

The Laws–Parsons distribution [120], which predates the Marshall–Palmer one and is also much used, is an empirically measured form for $N(D)$ which is tabulated numerically rather than expressed mathematically. It is similar to the Marshall–Palmer form except that it has slightly fewer small drops.

Subsequent work which followed the development of disdrometers (instruments capable of automatically recording the size distribution of drops), showed that N_0 could, in fact, vary quite considerably, both from one event to another and even within a rain event. In addition, the shape of the distribution was also found to vary significantly, which led to a more general form of distribution, the gamma-type distribution, being proposed [175]. This takes the form

$$N(D) = N_0 D^m \exp\left\{-(3.67 + m)D / D_0\right\} \tag{7.4}$$

Positive values of m reduce the numbers of drops at both the large and small ends of the size spectrum, compared with the exponential distribution obtained when $m = 0$. In contrast, negative values of m increase the numbers of drops at each end of the spectrum. *Figure* 7.1 shows this effect for three different distributions, all with the same rainfall rate (5 mm h^{-1}) and the same D_0 (1 mm).

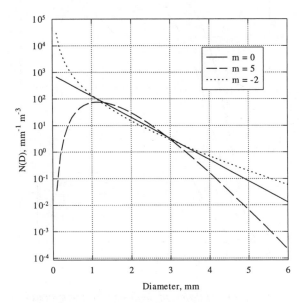

Figure 7.1   *Examples of three different drop-size distributions, with a rainfall rate of 5 mm h^{-1}, and a D_0 value of 1 mm*

Note from Figure 7.1 that the three distributions have very different numbers of small and large drops, but the same integrated rainfall rate. The reason for this can be seen in Figure 7.2, which indicates the contribution from each drop size to the total rainfall rate. For example, drops between 1 and 2 mm diameter contribute 60% of the total rainfall rate of 5 mm h^{-1}.

It must be stressed that there is at present no single model for drop-size distributions which is generally accepted as representing physical reality, even as a statistical mean over many rain events. Observations of quantities which are very sensitive to large drops, such as radar differential reflectivity, and certain crosspolarisation measurements, have tended to indicate the presence of fewer very large drops than in an exponential distribution. They have been well fitted using a gamma distribution with m between 3 and 5. This model, incidentally, reduces the assumed density of small drops (below 1 mm diameter) even though the experimental data may have little to say about them. Other measurements, particularly of attenuation at 30–300 GHz, are more sensitive to small drops and make no strong implication about the presumed large-drop cutoff, while work in the UK has suggested that a log-normal form of distribution would be appropriate for millimetric attenuation prediction [100]. Deductions about small drops have sometimes been contradictory [100,175]

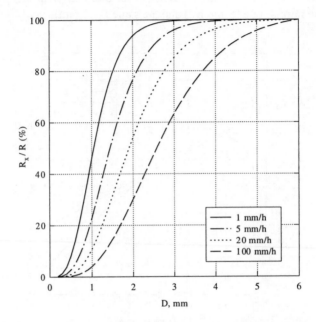

Figure 7.2   Contributions of different drop sizes to integrated rainfall rate, for various exponential drop-size distributions with $N_0=8000\,mm^{-1}m^{-3}$

It has generally been found that the Marshall–Palmer distribution is satisfactory for statistical predictions of attenuation in the 10–30 GHz range, if not on an event-by-event basis, and probably remains a good guess for statistical work at slightly higher frequencies. Fortunately, for particular modelling purposes it is not essential that the assumed distribution represents physical truth at all drop sizes.

7.2.2 Liquid-water clouds and fog

Nonprecipitating clouds containing only liquid water are not very significant for frequencies below about 100 GHz. The liquid-water content is too low to cause much absorption of energy; the droplets are too small to cause much scattering of energy and, being virtually spherical, do not cause measurable crosspolarisation. The most useful parameters to note here are:

(i) *Liquid-water content*

typical in small cumulus clouds:	$0.5\,\mathrm{gm^{-3}}$;
stratiform clouds:	$0.05\text{–}0.25\,\mathrm{gm^{-3}}$;
cumulus congestus and cumulonimbus:	$0.5\text{–}2.0\,\mathrm{gm^{-3}}$ (extreme);

(ii) *Droplet sizes*

small cumulus: many particles 3–10 μm radius, few above 35 μm; those in range 15–30 μm contribute about half the total volume;

cumulus congestus and cumulonimbus: large numbers of particles in the range 10–20 μm, with few above 80 μm. Particles in the range 20–45 μm contribute about half the total volume.

The liquid-water content typically peaks about 2 km above the cloud base and then decreases towards the top of the cloud which may be several kilometres higher. For much greater detail, see [129].

Fog can be considered to have similar physical properties to cloud, except that it occurs nearer the ground.

7.2.3 Ice hydrometeors

There is an enormous variety of forms of atmospheric ice particles, and the physics of their formation is extremely complex. However, some simple classifications may be made for radio purposes. Harden et al. [100] give a much more detailed discussion.

7.2.3.1 Ice crystals

From the radio-propagation point of view, single crystals at high altitudes appear to be the most important form of atmospheric ice, as they give rise to crosspolarisation effects (Section 7.6.5). A very crude picture, which nevertheless appears adequate to interpret radio effects (particularly crosspolarisation), is to divide them broadly into 'plate' and 'prism' (or needle) forms. The first group includes simple hexagonal plates with typical diameters up to 0.5 mm and thicknesses perhaps one-tenth of the diameter, and the classical dendritic form with complex hexagonal symmetry. The latter may reach 5 mm in diameter, with a thickness probably less than 0.05 mm. The second group includes a variety of long thin shapes including both needles and hexagonal prisms; typical lengths are around 0.5 mm, and the ratio of length to breadth ranges from one to five. For modelling radio scatter when the crystals are small compared with the wavelength, a start can be made by treating them as either very oblate or very prolate spheroids.

The type of crystal depends in a complex way on the temperature at which it formed, and hence on the height. In the range –8 to –27 °C, plate types are typical, with dendrites dominant in the subrange −14 to −17 °C. Below −27 °C prisms are formed. Both classes can form in the 0 to −8 °C range, but they are probably less important for propagation, except where differential phase effects are of concern.

7.2.3.2   Snow
This consists of aggregated ice crystals, with large flakes forming only at temperatures just below freezing, when the crystal surfaces become particularly 'sticky'. In stratiform rain most of the ice is present as large flakes down to a few hundred metres above the melting level, while at greater heights single crystals are mainly present. Dry snow does not appear to be very important for propagation of radio waves, at least below 30 GHz. This is partly because of its low-density structure (generally around 0.1 gm^{-3}), giving it a permittivity close to unity, and (for polarisation effects) partly because the flakes tend to tumble without a preferred orientation.

7.2.3.3   Hail and graupel
These particles are mainly formed by the accretion of supercooled cloud droplets in convective storms. Hail particles have densities close to water, and are often of roughly spherical shape, although a wide variety of shapes have been recorded. Graupel has a density intermediate between hail and snow and is usually of conical shape. Although of a higher apparent permittivity than snow, dry hail and graupel are only weakly attenuating below 30 GHz and do not appear to be very important for wave-propagation effects. However, when they begin to melt, they scatter like very large raindrops.   Noticeable polarisation effects might be expected but these seem to be of little significance on a statistical basis, at least in a climate such as that of the UK.

7.2.3.4   Melting layer
In stratiform rain, partially melted snowflakes exist within a height interval of about 500 m around the 0 °C isotherm. The melting particles combine a large size with a large apparent permittivity, and have a high spatial density because of their relatively small fall velocity.   The melting layer thus produces intense radio-scattering effects. An example of this is shown in Figure 7.3, where the 3 GHz backscattered-radar reflectivity is shown as a function of height during a typical stratiform rain event.

Also shown in the Figure is the difference in scattered power between horizontal and vertical polarisation, where it can be seen that the large melting snowflakes scatter the horizontal polarisation much more strongly than the vertical, because of their much larger horizontal dimensions. The melting phase has been modelled in some depth; see, for example [115], where it is predicted that the strongly enhanced scatter apparent at frequencies below about 15 GHz rapidly disappears above that frequency, owing to non-Rayleigh scatter from the large particles present.

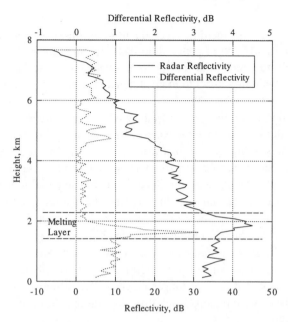

Figure 7.3   *Radar reflectivity and differential reflectivity (as functions of height) obtained with the Chilbolton radar*

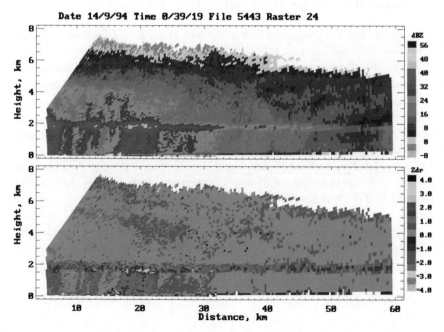

Figure 7.4   *Vertical radar section through rain obtained on 14 September 1994*
Top: reflectivity   Bottom: differential reflectivity

Although effects of the melting layer on most Earth–space paths are small because of the limited length of path within the layer, in climates such as that of the UK it is likely that terrestrial-link paths frequently traverse large lengths of melting layer and could be subject to more attenuation than expected from the rain rate seen at the ground (see Section 7.5.2).

Above the melting layer, the lower permittivity of snow reduces the magnitude of differential scatter to close to zero, except where, as near 5 km height, there are significant quantities of highly asymmetrical ice plates. Below the melting layer, oblate raindrops produce significant differential scatter, in this case of the order of 1 dB. In this example, the changes in reflectivity and differential reflectivity with respect to height within the rain are due to wind-shear effects, rather than, for example, evaporation. These shear effects can be clearly seen in the full radar scan of the same event, shown in Figure 7.4. This Figure also gives an indication of the variability of rain on a horizontal scale, even in stratiform conditions, with moderately heavy precipitation of around 10 mmh^{-1} at 20 km range, but much weaker rain of around 1 mmh^{-1} at longer ranges. The difference in mean drop sizes between these two regions can also be inferred from the differential reflectivity signatures.

7.3   Refractive indices of ice and water

For understanding the interactions of radio waves with hydrometeors, a vital factor is obviously the complex refractive-index n (or equivalently the permittivity ε_r, equal to n^2) of the water or ice forming the particle. Figure 7.5 shows the dependence of n on frequency for water. The high values for water arise from the polar nature of the water molecule, and are consistent with the well known value of about 80 for the relative permittivity of water at low frequencies. The peak in Im(n) around 20 GHz can be regarded as the 22 GHz line of water vapour, greatly broadened in the liquid state.

Although not shown on the Figure, Re(n) for ice is very close to 1.78 over the entire frequency range shown, while at 0 °C, Im(n) falls from 0.025 at 300 MHz to 0.0002 at 300 GHz (corresponding values for –20 °C: 0.006 to 0.00004).

A plane wave travelling in a medium of refractive-index n has its field intensity reduced by exp$\{-2\pi$ Im(n)$\}$ in travelling through one free-space wavelength. Thus in water at 300 GHz and 0 °C, waves are attenuated in voltage by about exp(-2π 0.5) = 0.04 in travelling through 1 mm. This shows that waves penetrating raindrops at 300 GHz and above are well attenuated within the diameter of the drop.

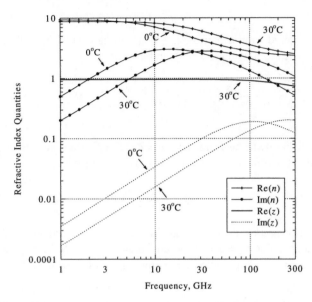

Figure 7.5   *Real and imaginary components of refractive index of water (n)*

Further insight can be obtained by considering the quantity $z = (\varepsilon_r - 1)/(\varepsilon_r + 2)$. A plot for water is included in Figure 7.5. It is important to note that, although Im(n) peaks at a certain frequency (dependent on temperature), Im(z) rises almost linearly with frequency throughout most of the frequency range. This is because Re(n) is falling with frequency. Re(z) is close to 1 (> 0.94) up to 30 GHz, then falls to 0.79 (0.91) at 100 GHz and 0.63 (0.74) at 300 GHz, for 0 °C (30 °C). For ice Re(z) is always close to 0.42 while Im(z) is about 0.0009 at 0 °C and 10 GHz, and falls with frequency. Note that it is very much smaller than the value for water.

Except for Re(n) and Re(z) for ice, and Re(z) for water below 100 GHz, all the quantities have a noticeable temperature dependence. Behaviour outside the frequency range shown is complex and further details can be found in [156]. Lower frequencies are almost unaffected by hydrometeors, while higher frequencies must be regarded as outside the 'radio'-frequency range.

7.4   Hydrometeor scattering theory

7.4.1 General

The far field scattered from a particle is commonly described by a dimensionless function S of the scattering angles [176], and is defined by

$$E_{scat} = E_{inc} S\left\{\frac{(\theta,\phi)}{jkr}\right\} \exp(-jkr + j\omega t) \qquad (k = 2\pi / \lambda) \qquad (7.5)$$

where r is the radial distance from the particle. For forward scatter, S is written $S(0)$. The total extinction cross-section of the particle is given by

$$C_{ext} = \lambda^2 / \pi \ \mathrm{Re}[S(0)] \qquad (7.6)$$

A plane wave propagating through a medium containing N randomly distributed particles per unit volume experiences an attenuation of $NC/2$ nepers per unit distance. (1 neper is a voltage ratio of e, a power ratio of e², or 8.7 dB.)

$$\alpha = 8.7\left(\frac{N\lambda^2}{2\pi}\right) \mathrm{Re}[S(0)] \qquad \text{dB per unit distance} \qquad (7.7)$$

The corresponding specific phase shift is

$$\beta = \left(\frac{N\lambda^2}{2\pi}\right) \mathrm{Im}[S(0)] \qquad \text{radians per unit distance} \qquad (7.8)$$

For rain with a distribution of drop sizes, the term NS can be rewritten as an integral over D of $N(D)S(0,D)$.

7.4.2 Rayleigh-scattering region

A scattering particle of radius a and refractive-index n is in the Rayleigh-scattering region when it is both electrically small ($2\pi a/\lambda \ll 1$) and phase shifts across it are small ($2\pi na/\lambda \ll 1$). In this condition Rayleigh's approximation can be used, which assumes that:

(*a*)   the scattered field is that of a dipole; and
(*b*)   the dipole moment induced in the particle is related to the incident electric field in the same way as for electrostatic fields.

At radio frequencies, the approximation is especially useful for cloud droplets and atmospheric ice crystals, and it gives partial insight into rain scatter.

If P is the induced dipole moment per unit incident field, $S(0)$ is given by

$$S(0) = jP\omega^3\mu_0 / 4\pi c \qquad (7.9)$$

then P can be written as

$$P = \varepsilon_0(\varepsilon_r - 1)\xi v = \varepsilon_0 U v \qquad (7.10)$$

where v is the particle's volume and ξ is the ratio of the internal to the external field. This ratio is given by $\xi = 3/(\varepsilon_r + 2)$ for a sphere. For a very oblate spheroid ('flat plate'), ξ is 1 when the field is applied along the longest axis, and is $1/\varepsilon_r$ when applied along the shortest axis. For a very prolate spheroid ('thin needle') it is 1 when the field is parallel to the long axis, and is $2/(\varepsilon + 1)$ when at right angles to it. Values of U for water and ice spheres and ice needles and plates are shown in Table 7.1. Imaginary parts are not shown as they are very small for ice.

Table 7.1   Values of U for different hydrometeors

	Water Sphere	Ice Sphere	Ice Needle	Ice Plate
$U_{parallel}$	$3z$	1.26	2.32	2.32
$U_{perpendicular}$	$3z$	1.26	1.04	0.68

Specific attenuation and phase shift in a cloud of the particles are given by

$$\alpha = (\pi V / \lambda)\,\mathrm{Im}(U) \qquad \text{nepers per unit distance} \qquad (7.11)$$

$$\beta = (\pi V / \lambda)\,\mathrm{Re}(U) \qquad \text{radians per unit distance} \qquad (7.12)$$

Here $V = Nv$ is the fractional volume of space occupied by the particles. Important features of these two expressions are their direct proportionality to frequency (if n is frequency-independent) and to total particle volume, regardless of how distributed between sizes. The first equation includes only extinction due to absorption, and the Rayleigh approximation can be improved by adding to C_{ext} a scatter cross-section equal to the total power reradiated by the induced dipole, which is $P^2/3\varepsilon_0^2\lambda^4$. This term varies as $D^6 f^4$ and becomes noticeable for raindrops at electrical sizes which are approaching breakdown of the Rayleigh approximation.

7.4.3 Optical- and resonance-scattering regions

The optical-scattering region is that in which the incident wavelength is much less than the diameter of the scattering particle. The larger raindrops are approaching this condition at 100 GHz and above. The scattering may be described by a ray model. Provided that the rays which enter the particle are well attenuated within the diameter, which is true in the high radio range but not of course at truly optical frequencies, the extinction cross-section approaches a value of twice the scatterer's geometric area. The factor of 2 is known as the 'extinction paradox', see [176].

Between the Rayleigh and optical regions lies the resonance region in which no simple approximation for the scattered fields is available. Mie's theory gives exact results for a sphere but requires considerable computation. Several numerical methods have been developed for nonspherical particles, for example [106]. Figure 7.6 shows extinction cross-sections for raindrops as functions of drop size and frequency, and several features of the three scattering regions can be observed.

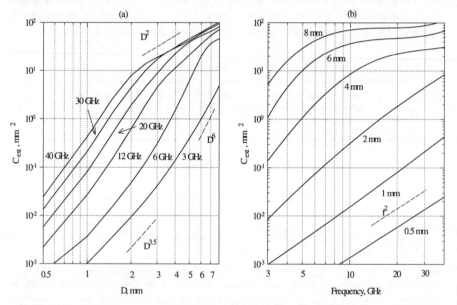

Figure 7.6   *Extinction cross-sections against drop diameter and frequency*
a Diameter
b Frequency

Taking Figures 7.6a and b together, we can see cross-sections increasing roughly as $D^3 f^2$ in the Rayleigh region. Equation 7.7 predicts this kind of dependence, with absorption increasing in proportion to particle volume and to $f\text{Im}(z)$, and it was already noted that $\text{Im}(z)$ increases roughly as f over most of the frequency range. In Figure 7.6a, a transition to D^2 behaviour can be seen for large drops, the transition occurring at a smaller size as the frequency is raised. Both graphs show fine structure due to resonance. In Figure 7.6*a*, it is seen that the resonances become increasingly damped at the higher frequencies; at the lower frequencies, as *D* is increased towards the first strong resonance, the dependence becomes a good deal steeper than D^3. This is particularly clear at 6 GHz. Similarly, Figure 7.6*b* shows a much steeper dependence than f^2 at intermediate sizes and lower frequencies and the relative independence of frequency for larger particles at the higher frequencies is also clear.

7.5　Attenuation effects

7.5.1　Rain and cloud

The total attenuation due to rain is due to the sum of the contributions of each individual drop. It is therefore necessary to assume a form of drop-size distribution $N(D)$, as described in Section 7.2. Then the total attenuation A can be computed from

$$A = 4.34 \, L \int_0^\infty C_{ext} \, N(D) \, dD \qquad\qquad \text{decibels} \qquad\qquad (7.13)$$

Figure 7.7 shows the results for specific attenuation, using eqn. 7.12 over a 1 km path together with the extinction cross-sections in Figure 7.6 and assuming an exponential distribution with $N_0 = 8000 \text{ mm}^{-1} \text{ m}^{-3}$. The behaviour can be well understood from the previous discussion of cross-section dependencies on size and frequency, remembering that fine structure due to resonance is smoothed out by the variety of drop sizes.

To illustrate the effects of cloud or fog, it is instructive to use eqn. 7.11 to calculate attenuation and phase shift in a water cloud, where the droplets are very small. Results for a cloud at $0\,^\circ\text{C}$, 1 km thick and containing 1 gm^{-3} of water are shown in Table 7.2. As expected, attenuation increases approximately with frequency squared, and phase shift roughly linearly, because $\text{Im}(z)$ increases and $\text{Re}(z)$ is fairly constant.

Table 7.2　*Cloud attenuation and phase shift*

Frequency, GHz	3 0	10	30	100	300
Attenuation, dB	0.009	0.09	0.77	5.5	10.2
Phase shift, degrees	1.7	5.5	16	45	108

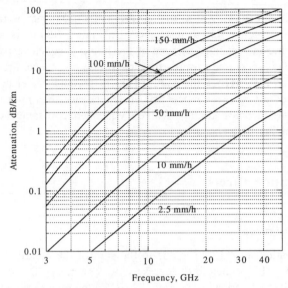

Figure 7.7   Specific attenuation due to rain

An interesting point may be seen by comparing the U values for spheres and crystals. If the water in the cloud froze into spheres, it would produce negligible attenuation but the phase shift would only be reduced to about one-half. If the water in the cloud froze into flat-plate crystals, the cloud could produce a differential phase shift between two linear polarisations equal to about 0.55 of the total phase shift in the water cloud. This is the reason why clouds of ice crystals can be strongly crosspolarising, yet cause very little attenuation. A differential phase shift of 3° would be feasible at 10 GHz, and it would increase linearly with frequency.

7.5.2 Melting layer

It was noted in Section 7.2.3.4 that attenuation in the region of melting snowflakes can be significantly larger than in the rain below. An example of this effect is shown in the model calculations plotted in Figure 7.8 [106]. The frequency is 20 GHz and a Marshall–Palmer dropsize distribution has been assumed. The extra attenuation in the melting layer is also predicted to depend on the initial density of the melting snowflakes. In Figure 7.8, densities of 0.2 and $0.3 \, \text{gm}^{-3}$ have been assumed, and it can be seen that the lower density almost doubles the attenuation.

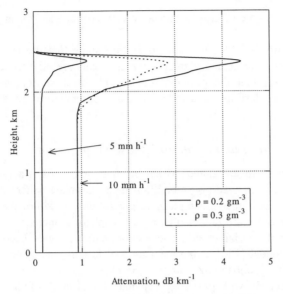

Figure 7.8   Attenuation in rain and melting layer at 20 GHz

7.6   Polarisation dependence

Because hydrometeors are not spherical, if a wave is propagated through them along a line-of-sight path it will usually change its polarisation as it travels. A 'crosspolarised' component in the orthogonal polarisation may be generated, which may be a problem for communication systems using polarisation orthogonality to maintain isolation between channels.

7.6.1 Concept of 'principal planes'

For any line-of-sight path traversing a particular assemblage of hydrometeors, there always exist two polarisations which will propagate over that link without changing. These polarisations may be frequency dependent and time varying as the rain, ice, etc. change. They can be referred to as 'principal planes' of the link but need not theoretically be linear polarisation. (Principal planes can also be defined, not for the whole link, but as a local property of the hydrometeors at any point along it.) Although waves transmitted in the principal planes will arrive unchanged, because of the nonspherical particles they will have experienced different attenuations and phase shifts. Consequently, any transmitted polarisation which is not one of the link's principal planes will be crosspolarised on reception. A powerful way of describing the effect on a given link is to specify the *principal planes*, the *differential attenuation* and the *differential phase shift*. With this information, the degree of crosspolarisation can be predicted for any transmitted polarisation.

A great simplification for rain effects is to assume that the raindrops are all rotationally symmetric, and that their symmetry axes are all vertical. It is then physically obvious that the principal planes are linear vertical and horizontal polarisations. This description is a very good approximation for most purposes.

7.6.2 *Differential attenuation and phase shift in rain*

Raindrops have a mean shape which is flattened. Satisfactory scattering calculations have been made by modelling them as oblate spheroids, and a few calculations have been made using a more precise description of the drop shapes. The true shape has a more flattened base than a spheroid, increasingly as the drop size increases, and for diameters D above about 4 mm the base becomes concave. A widely used model [150] of the axial ratio of the drop as a function of D has minor/major axis = 0.92, 0.82 and 0.65 at D = 2, 4 and 6 mm, respectively, although more recent work [93] suggests small raindrops (< 3 mm diameter) are more spherical than this model suggests, while large raindrops (> 5 mm diameter) are more oblate.

Figure 7.9 shows differential specific attenuation and phase shift for rain with spheroidal drops. At low frequencies, it confirms the expectation from Rayleigh theory that attenuation and phase shift will be greater in horizontal polarisation. The reversal in sign of the differential phase at high frequency is purely a resonance phenonemon in which large drops produce negative differential phase outweighing a positive contribution from the smaller ones. Note that, while differential attenuation, and phase below 18 GHz, increase with f for a given rain event, they decrease for a given fade depth. This is partly because less deformed smaller drops make a greater relative contribution to the total attenuation as frequency is raised.

7.6.3 *Attenuation/crosspolar-discrimination relations*

If a given microwave link experiences a given fade depth during rain, this could be due to a variety of drop-size distributions. Even approximating the drop-size distributions as a one-parameter family, with Λ dependent on rain rate (see eqns. 7.1 and 7.2), that rain rate would depend on the path length over which the rain was operating to produce the given fade.

Despite this variability, calculations show that the crosspolarisation is closely related to total attenuation. The possible variations in drop-size distribution for a given fade tend to vary the differential attenuation and differential phase in opposite senses, tending to keep the crosspolar amplitude (but not its phase) fairly constant. Thus for system-design calculations it is useful and convenient to approximate the relation between crosspolarisation and attenuation as a deterministic one.

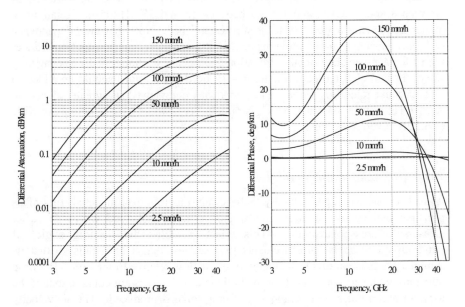

Figure 7.9   *Differential propagation (horizontal/vertical) in rain*

The crosspolar magnitude is usually expressed as crosspolar discrimination (XPD), defined as the decibel ratio, at the receiving point of a link, of the field intensity in the wanted or normal polarisation to that induced by the propagation medium in the orthogonal polarisation. XPD is worst (numerically lowest) when the transmitted polarisation is either circular or is a linear polarisation bisecting the principal planes of the medium.

As a first-order approximation, it can be assumed that the crosspolar field intensity (voltage) is proportional to the total amount of rain traversed and thus to the total fade. Reverting to the decibel XPD scale, this leads to the equation

$$XPD = U - 20 \log A + I(\theta) \tag{7.14}$$

where A is the fade depth of the copolarised signal in dB, U is a constant for a particular link frequency, and $I(\theta)$ is the improvement factor applicable when the transmitted polarisation is not one of the worst-case ones just mentioned. It is approximately given by $20 \log|\sin 2\theta|$, if the transmitted polarisation is linear and makes an angle of θ with the principal planes of the rain medium.

Some experimenters have derived a relationship in which the term $20 \log A$ is replaced by $V \log A$, V being a second frequency-dependent coefficient. This can make some allowance for the tendency of deeper fades to be associated with higher rain rates and thus larger raindrops.   Between 8 and 35 GHz, ITU-R recommends: $U = -30 \log f$, where f is the link frequency in GHz, with $V = 20$ between 8 and 15 GHz, and $V = 23$ between 15 and 35 GHz [16].   Relations at higher frequencies appear not to be well established.   Note that the f dependence of U is consistent with the previous comment that differential propagation reduces with f for a given fade depth.   Equation 7.2 assumes a horizontally propagating wave.   For an elevated path, the term $-40 \log(\cos$ elevation$)$ should be added to XPD because raindrops appear more spherical when viewed at a greater elevation. At 10 GHz, a fade of 15 dB, which is quite severe, would give a worst-case XPD of 6.5 dB at zero elevation.

7.6.4 Raindrop canting angles

It has been shown experimentally that the rain medium exhibits small tilts of its principal planes, so that horizontally and vertically polarised waves do suffer measurable crosspolarisation.   This effect is called the 'canting angle' of raindrops.   Crosspolar measurements [65] suggest that a drop falls so that its symmetry axis is parallel to the velocity of air flow relative to the drop. If a drop falls through a vertical wind shear, its horizontal velocity is not equal to that of the local air and a horizontal component of drag force is produced.   The air-velocity vector relative to the drop is thus tilted from vertical; hence the symmetry axis is also.

Calculations show that terrestrial links within about 40 m of the ground could see canting angles up to about 5°, and experimental values have been of this order. As wind shears reduce with height, the model implies equally strongly that tilts seen on Earth–satellite paths at elevations of more than a few degrees would be unlikely to see angles greater than 1°, which is negligible for practical purposes. This has not been conclusively demonstrated in experimental data, owing to measurement errors.

7.6.5 Ice-crystal principal planes

Of the two general types of ice-crystal shape thought to affect elevated Earth–satellite paths, it is known that 'plate' types generally fall with their flat faces in a horizontal plane.   For a medium of these crystals, the principal planes are clearly linear vertical/linear horizontal.

Needle crystal types are thought to fall with their symmetry axes (which are also their longest axes) in a horizontal plane.   It is also believed that the azimuths of these long axes may be systematically aligned in one direction by wind shear or sometimes by atmospheric electrostatic fields.   The evidence for this comes from observations of rapid changes in XPD coinciding with changes in field strength measured by nearby electric-field probes [103].

Experimental data suggest that all principal-plane angles occur occasionally, but show [44] that they spend about 80% of the time within $\pm15°$ of vertical/horizontal when the event is significant. This is consistent with the picture that needles are usually accompanied by plates at a lower altitude.

7.6.6 *Differential phase shift due to ice*

Statistics of ice differential phase shift on elevated paths are inferrable from measurements made at a few sites using satellite beacons. In Europe a typical value exceeded for 0.01% of the time at about $30°$ elevation would be about $10°$ at 12 GHz [44] corresponding to a worst-case XPD of 22 dB. Such statistics cannot yet be related to any features of the ground rainfall. The phase shift can, however, be scaled linearly to much higher frequencies still in the Rayleigh region. It is also very difficult from any ground-based radio or radar measurement to make any detailed inferences about ice-particle shapes and size distributions. Ice may exhibit resonant scatter in the millimetric region, which would be hard to model on the basis of existing data.

7.7   Bistatic scatter in rain

'Bistatic scatter' refers to a situation where a cloud of particles scatters a signal via an indirect path between two antennas which have intersecting beams but are not strongly coupled by a direct path. This is usually considered as a cause of interference, although it is also sometimes used intentionally in the form of bistatic radars. An example of an interference geometry is shown in Figure 7.10, where a terrestrial link is shown sharing the same frequency band as an Earth station.

When no rain is present, the only path for interference from the terrestrial link into the Earth station would be through the latter's far sidelobes, and even this route would normally be blocked by local terrain features. However, if rain occurred in the common volume of the two systems, energy could be scattered from the terrestrial transmitter into the main beam of the Earth station, with an increase in the possibility of interference. Figure 7.11 shows data from an experimental link operating between Chilbolton and Baldock in the UK, at a frequency of 11.2 GHz, together with some results from a prediction method developed within COST 210 [46] and subsequently adopted by the ITU-R [12]. In the absence of rain, no signal is received at the Baldock satellite-monitoring station from the transmitter at Chilbolton. However, when rain is present in the common volume of the transmitter and receiver, a clear signal is seen. The Figure shows the statistics of the transmission loss over a two-year period. The transmission loss is defined as the ratio of the transmitted power to the received power. The power received from a volume containing rain can be calculated using the bistatic radar equation of which an approximate form is

$$P_r = \frac{P_t \, G_t \, G_r \, \lambda^4 \, N \, V \, |S(\theta,\phi)|^2}{64 \, \pi^4 \, R_1^2 \, R_2^2} \tag{7.15}$$

Here R_1, R_2 are distances from the antennas to the beam intersection volume, t denotes transmission and r denotes reception, G denotes antenna gain and P power. V is the volume of the beam intersection and is assumed to contain N particles per unit volume with scattering function S at the appropriate angle. In using this equation, careful account must be taken of the two antenna polarisations. As it stands, the equation does not take account of attenuation. At the higher frequencies, attenuation on the paths to and from the scattering volume, and within it, is very significant and may lead to intense rain producing less scatter than moderate rain. Attenuation also limits the distance over which interference is likely to be significant at frequencies above 30 Ghz.

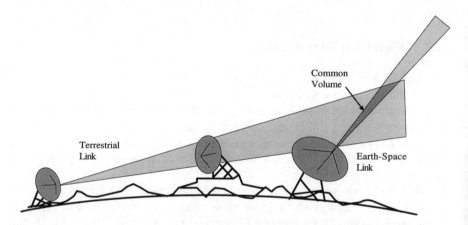

Figure 7.10   An illustrative rainscatter-interference geometry

Except in the Rayleigh and optical regions, the scattering functions $S(\theta,\varphi)$ of raindrops exhibit very complex angular variations and require detailed calculation. For example, Figure 7.12 shows the bistatic reflectivity factor as a function of scattering angle (θ) for a rainfall rate of 20 mm/h at a frequency of 11 GHz (for further details of this and other aspects of interference due to hydrometeor scatter, see [46]). It is obvious from this Figure that the scatter geometry can affect the scattered signal by tens of dBs when the plane of polarisation is parallel to the scatter plane.

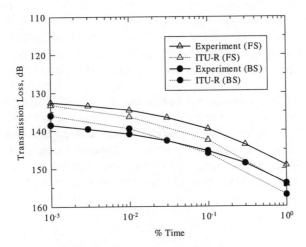

Figure 7.11   *Statistics of transmission loss at 11.2 GHz*
Transmitter–receiver distance 131 km
FS = forward scatter, BS = backward scatter

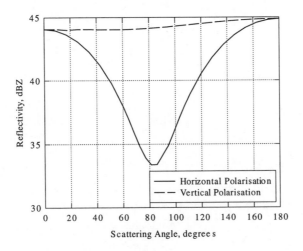

Figure 7.12   *Angular dependence of rain scatter at 11 GHz*
Rainfall rate of 20 mm/h

It is important to appreciate that a signal arriving via bistatic scatter may be very different in its time waveform and spectral shape from the transmitted signal, as it is essentially incoherent and randomly fluctuating. The transfer function of the path is strongly dispersive and frequency selective in both amplitude and phase, and varies randomly. The arriving signal may be described roughly as the sum of geometry, the spread of time delays between the individual paths may be in the order of microseconds, and the relative phasing between them changes as the particles move. A consequence of this is that it may be difficult to distinguish interference due to rain scatter from a decrease in signal-to-noise due to attenuation of the wanted signal, particularly as the two phenomena will usually occur together.

If a CW signal is transmitted, the bistatic signal is usually Rayleigh distributed and fluctuates on a time scale of milliseconds. It is believed that if a signal with a modulation bandwidth $\gg 1/T$ is transmitted, where T is the time-delay spread of the paths, the received signal will resemble Gaussian noise. The short-term spectrum of this noise will display intense frequency selectivity on a scale of $1/T$, with the structure itself changing randomly on the millisecond scale.

Chapter 8

Prediction of reliability when degraded by clear-air effects

K.H. Craig

8.1 Introduction

Chapter 6 described the clear-air refractive-index mechanisms which can affect the propagation of radio waves, and also described the basic (deterministic) propagation models which are used to predict field strength (or path loss) in the troposphere. System-performance requirements are normally specified in terms of the percentage of time for which some performance criterion is met (see Chapter 3). In this Chapter, statistical methods for predicting the cumulative distributions of transmission loss are described, based on the methods of the relevant ITU-R Recommendations. Terrestrial line-of-sight paths, terrestrial beyond-the-horizon paths and Earth-space paths are covered. Methods for protecting system performance against clear-air-propagation degradations are also discussed where appropriate. Each Section ends with a worked example to illustrate the prediction methods.

8.2 Terrestrial line-of-sight paths

8.2.1 Mechanisms

In the design of terrestrial line-of-sight links, two major causes of degradation must be considered:

(i) Attenuation due to rain and other hydrometeors: this topic is covered in Chapter 9 and is not discussed further here.
(ii) Fading and enhancements due to clear-air effects. The principal cause of these is multipath propagation, although certain refractive conditions can give rise

to 'flat' fading. While signal fading is the most obvious degradation, large enhancements (which occur under the same general conditions) also effectively reduce a system's performance: the operating point of the receiver must be chosen to avoid saturation, and this will reduce the dynamic range available for fading.

Before the prediction method for line-of-sight paths is presented, multipath propagation is discussed further.

8.2.2 Multipath

In Chapter 6 it was shown that, under certain meteorological conditions, signals emitted by a transmitter can travel to a receiver via multiple paths. On a well designed line-of-sight path, there will always be a direct (line-of-sight) path. Signals may also be received via one or more delayed paths as a result of reflection from the ground and/or refraction occurring along the path. The resultant signal amplitude at the receiving antenna will be the vector (phasor) sum of all the multipath components. Since each signal will arrive at the receiver with a different phase (due to different path lengths) and amplitude, the multipath components can interfere destructively, giving rise to fading, or can interfere constructively, giving rise to enhancement, compared with the normal situation where only the direct path is present. Multipath fading can be very deep. In the most common situation where only two paths exist, the cancellation would be complete (there would be a 'null' in the received signal) if the two signals had the same amplitude and 180° phase difference.

Figure 8.1 shows an example of multipath fading on a link operating in the lower 6 GHz band. It is apparent that the fading comprises two components: a rapidly fluctuating component that is caused by frequency selective fading, and a much more slowly varying component (giving rise to the mean depression of the signal in Figure 8.1) which is not frequency selective.

8.2.3 Flat and frequency-selective fades

Much confusion has arisen over the terms 'flat' and 'frequency-selective' fading. Loosely, they are used subjectively to describe the mean depression of signal strength and the rapid fading component seen in signal traces such as Figure 8.1. However, strictly speaking, the terms refer to the underlying mechanisms causing the fading, and whether these are frequency-selective or nonfrequency-selective.

In a broad sense, all propagation mechanisms are frequency selective; what is important is the amount of variation in the propagation characteristics over the operating bandwidth of the communications system under discussion. In this sense, all fading is 'flat' for a CW signal. However, the term 'flat fading' is normally used to describe fading caused by nonmultipath mechanisms, although it can occur in association with multipath propagation. Flat fading is due to a reduction in the strength of the direct ray path, giving rise to a mean depression of the signal level. This can occur for several reasons:

(*a*) The antennas on a line-of-sight link will normally be aligned during normal atmospheric conditions, assuming a 4/3 Earth radius. When subrefractive (or super-refractive) conditions occur, the ray that constitutes the direct path between the transmit and receive antennas has to be launched and received below (or above) the normal clear-sky path. This offset angle results in attenuation at both antennas owing to the reduced off-boresight antenna gain (Figure 8.2). If subrefraction is particularly severe, or the link has been badly planned so that insufficient ground clearance has been allowed (see Section 8.2.4.1), the direct-ray path may be very close to the ground and additional diffraction loss incurred.

(*b*) More complex refractive-index structures are sometimes encountered where a varying refractive-index gradient acts as a 'concave lens' and causes defocusing of the direct path. Defocusing of the direct path is particularly important when it occurs at the same time as a strong surface-reflected signal: the resultant (frequency-selective) multipath fading can be particularly severe.

(*c*) If the radio path has only a very small slope between the transmit and receive antennas, it is possible that atmospheric layering at an intermediate height prevents energy from arriving at the receiver (see Figure 8.3).

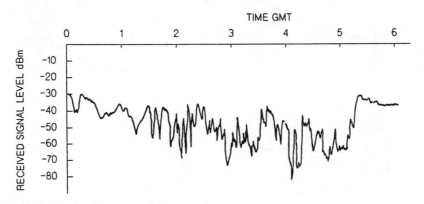

Figure 8.1   Typical multipath event

Compared with flat fading, frequency-selective fading is a genuine multipath phenomenon: it arises from the interference between the direct and delayed signal paths, and therefore there can be significant amplitude and phase variations over bandwidths relevant to digital-communications systems. The mechanism and the system impact of frequency-selective fading are discussed in Chapters 10 and 11.

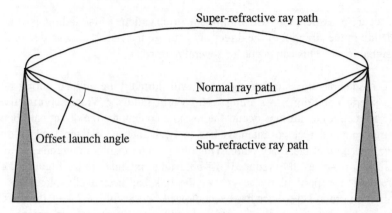

Figure 8.2   Launch-angle offsets under sub- and super-refractive conditions

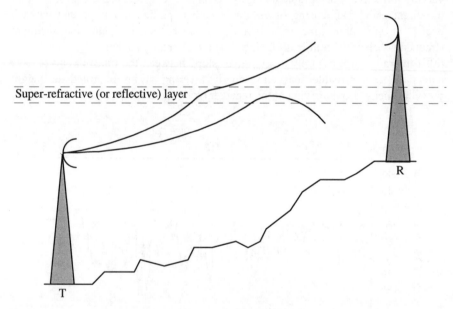

Figure 8.3   High–low situation: rays transmitted below a critical angle are trapped (or reflected) by the layer and do not reach the receiver. Rays above this angle also do not reach the receiver

In this Chapter we are concerned only with single-frequency (narrowband) statistics: the fade level at a given time percentage is the sum of both the flat and the frequency-selective fading components.

8.2.4 Prediction methods for clear-air effects

The prediction methods are taken from ITU Recommendation P.530-6 [16] for the design of terrestrial line-of-sight systems. They can be used to predict the level of propagation loss (or enhancement), relative to the median signal level, expected to be exceeded for a given time percentage. When calculating a system-link budget, these losses must be added to the free-space basic transmission loss to give the total basic transmission loss of the radio link. For reference, the free-space basic transmission loss of eqn. 3.7 can be written in terms of practical units as

$$L_{bf} = 92.5 + 20 \log f + 20 \log d \qquad (8.1)$$

where f is the frequency in GHz, and d is the path length in km.

8.2.4.1   Path-clearance criteria

For a path to be line-of-sight, it is necessary that the link be designed to maintain adequate clearance between the radio path and the terrain being crossed. It is not sufficient to plan for ground clearance simply on the basis of a 4/3-Earth-radius-path profile plot, for two reasons:

(i) even if there is clearance between the radio path and the terrain, up to 6 dB diffraction loss can still be experienced if this clearance is small;

(ii) subrefractive conditions must be allowed for; for example, subrefraction resulting in a k-factor of only 2/3 may occur for 0.1% of the worst month.

The path-clearance criteria are specified in terms of the radius of the first Fresnel ellipsoid F_1 at the most significant path obstruction (see Chapter 4 and eqn. 4.9); i.e. the path obstruction that would cause loss of line-of-sight if the antenna heights were reduced.

The antenna heights should be chosen to satisfy both of the following:

(a) there should be 1.0 F_1 clearance under conditions of median k factor (in the absence of detailed k-factor statistics, use $k = 4/3$);

(b) there should be αF_1 clearance over a path profile drawn with an effective k factor k_{eff} appropriate for 99.9% of the worst month. α and k_{eff} depend on climate and path geometry. In a temperate climate, $\alpha = 0$ (the radio path grazing the obstacle) should be adequate for a single isolated path obstruction but, if the obstruction is extended along a portion of the path, $\alpha = 0.3$ is advised. A value of $k_{eff} = 2/3$ is appropriate on shorter paths (less than 30 km), rising to unity on long paths (greater than 100 km).

If adequate path clearance is not possible, the methods of Chapter 4 can be used to determine the diffraction loss that will be incurred.

8.2.4.2   Multipath fading and enhancement
The complete multipath fading and enhancement prediction methods given in [16] are fairly complicated and will not be reproduced in full here. Instead an outline will be given, with parameters appropriate to the UK. The prediction method is applicable only to narrowband systems, although narrowband-fade statistics are the basis of the 'net' or 'effective'-fade-margin method for digital-system-performance calculations. (Wideband digital systems are discussed in Chapter 11.)

The cumulative distribution of *worst-month* signal-level fading and enhancement is derived in four stages:

(i) The deep fading part of the distribution is calculated (for small time percentages corresponding to fade depths greater than about 15 dB).
(ii) The fading distribution is extrapolated to larger time percentages using an empirical or graphical method.
(iii) The percentage of occurrence of large enhancements (greater than 10 dB) is derived from the deep fading distribution.
(iv) The enhancement distribution is extrapolated to small enhancement levels using an empirical or graphical method.
If necessary,
(v) The *average annual* distribution can be derived from the *worst-month* distribution.

The prediction model for the deep-fading part of the worst-month distribution is the key part of the calculation. In this region, the cumulative distribution is linear on log-log paper: the fade depth exceeded for a given time percentage p increases by 10 dB for each decade decrease in p; for example, if a link experiences fades exceeding 20 dB for 0.1% of the worst month, the fades will exceed 30 dB for 0.01% of the worst month. This 'universal' behaviour is approximately borne out by measurements made on paths over a wide range of frequencies, path geometries and climates. The 10 dB/decade part of the cumulative distribution is characteristic of Rayleigh fading.

A Rayleigh distribution of received signal level is expected when a large number of paths link the transmitter and receiver, each path having a random amplitude and phase. Rayleigh fading typically occurs in scattering phenomenon, but at first sight it does not seem appropriate to a line-of-sight path where there is normally one dominant, relatively stable, ray—the direct ray. In this case, Ricean statistics should apply. However, it can be shown that, as long as more than about 20% of the received signal power is carried by the nondirect ray paths, then the tail of the distribution (for time percentages less than about 0.1%) has a 10 dB/decade slope. This part of the distribution is commonly (though incorrectly) referred to as Rayleigh fading.

The fade depth $A(p_w)$ (in decibels) exceeded for a given time percentage p_w of the worst month can therefore be derived from the fade depth exceeded for 0.01% of the worst month $A_{0.01}$:

$$A(p_w) = -10 \log p_w + A_{0.01} - 20 \qquad (8.2)$$

Although the 10 dB/decade model is used for all paths, the value of $A_{0.01}$ varies with path geometry and climate. ITU Recommendation P.530-6 [16] gives two methods of calculation. The method for detailed link design requires the path-terrain profile to be known in order to calculate an 'average grazing angle' for a ground reflected ray on the path. The simpler method for initial planning purposes is used here:

$$A_{0.01} = 20 + 10 \log K + 36 \log d + 8.9 \log f - 14 \log\left(1 + |\varepsilon_p|\right) \qquad (8.3)$$

f is the frequency in GHz and d is the path length in km.

ε_p is the path inclination (in mrad); the path inclination term expresses the fact that a path that is more inclined to atmospheric layers (assumed horizontal) experiences less refractive effects and therefore less fading, where:

$$\varepsilon_p = \left(h_t - h_r\right)/d \qquad (8.4)$$

h_t and h_r being the heights of the transmitter and receiver antennas (in metres) above some reference height.

K is a geoclimatic factor. The status of this term is not entirely satisfactory at present. It is suggested that K be estimated from fading data obtained from line-of-sight links in the vicinity of the planned radio link, if available; the value of K for the average worst month is then estimated by inverting eqn. 8.3. If data are available for several paths in the region of interest, an average geoclimatic factor should be obtained by averaging the values of $\log K$. Otherwise K can be estimated from formulae given in [16]. It depends on

(*a*) a parameter which indicates the level of occurrence of 'anomalous' propagation; currently, this is p_L: the percentage of time that the average refractivity gradient in the lowest 100 m of the atmosphere is less than -100 N/km for the worst month. In the UK, p_L is always less than 10%.

(*b*) a term which takes account of the orography of the path (whether the path is mountainous, overland, oversea, coastal or mixed).

(*c*) a term which takes account of global factors (latitude/longitude dependent).

Work is in progress to develop better models for the dependence of multipath fading on orographic and climatic factors. For the UK, 10 log K typically lies in the range −40 dB for overwater paths to −55 dB for overland paths in mountainous terrain.

It is emphasised that eqns. 8.2 and 8.3 are only valid in the deep-fading tail of the cumulative distribution. They are considered valid for fade depths greater than about 15 dB or the value exceeded for 0.1% of the worst month, whichever is greater. Formulae and a graphical method for extrapolating the fading distribution to the larger time percentages (lower fade depths) are given in [16].

Average worst-month enhancement levels $E(q_w)$ (greater than 10 dB) exceeded for q_w % of the worst month can be predicted from the deep fade depth $A_{0.01}$:

$$E(q_w) = -3.5\log q_w + 0.2 A_{0.01} - 1.7 \tag{8.5}$$

Formulae and a graphical method for extrapolating to enhancement levels between 0 and 10 dB are given in [16].

If the average annual distribution is required, rather than that for the worst month, it can be obtained from the worst-month distribution by converting annual time percentages p to the equivalent worst-month time percentages p_w:

$$p_w = p \, \Delta g \tag{8.6}$$

where Δg is calculated from the formula

$$10\log \Delta g = 10.3 - 5.0\log\left(1 \pm |\cos 2\zeta|^{0.7}\right) - 2.8\log d + 1.8\log\left(1+|\varepsilon_p|\right) \tag{8.7}$$

and should be limited to a maximum value of 12. d is the path length in km and ε_p is the path inclination in mrad. ζ is the latitude of the path (north or south); the positive sign should be used in the ζ term for $\zeta \le 45°$ and the negative sign for $\zeta > 45°$. The value of $A_{0.01}$ used in eqns 8.2 and 8.5 for average annual predictions must be the fade depth exceeded for 0.01% of the time in the average year; this is obtained from the worst-month value of $A_{0.01}$ by subtracting 10 log Δg:

$$A_{0.01}(\text{average annual}) \;=\; A_{0.01}(\text{worst month}) - 10\log\Delta g \tag{8.8}$$

8.2.4.3   Attenuation due to atmospheric gases
Some attenuation due to absorption by oxygen and water vapour is always present, and should be included in the calculation of total propagation loss at frequencies above about 10 GHz. The attenuation A_a on a path length d (in km) is given by

$$A_a = (\gamma_0 + \gamma_w) d \qquad (8.9)$$

γ_0 and γ_w are the specific attenuations (dB/km) at the surface for dry air and water vapour, respectively. These can be estimated from Figure 6.8 for the frequency of interest, assuming that default values of temperature and water vapour are adequate. If site-specific values of these parameters are available, formulae for γ_0 and γ_w that take account of them are given in [23].

8.2.4.4   Angle-of-arrival variations
Abnormal gradients of the clear-air refractive index along a path can cause considerable variation in the angles of launch and arrival of the transmitted and received waves.   This variation is substantially frequency independent and primarily in the vertical plane of the antennas. The range of angles is greater in humid coastal regions than in dry inland areas. No significant variations have been observed during precipitation conditions.

The effect can be important on long paths in which high-gain (narrow-beamwidth) antennas are employed. If the antenna beamwidths are too narrow, the direct-wave path can be sufficiently off axis that a significant fade can occur (see Section 8.2.3).   It is also important to ensure that antenna alignment on critical paths is not carried out during periods of anomalous propagation.

8.2.4.5   Scintillation
Although scintillation fading due to small scale turbulent irregularities in the atmosphere is always present on line-of-sight links, its effect on the overall fading distribution is not significant at frequencies below about 40 GHz, and can be ignored.

8.2.4.6   Reduction in crosspolarisation discrimination
To maximise spectrum occupancy on line-of-sight links, it is desirable to allow systems using horizontal and vertical polarisations to share the same channel. This obviously can only work if each system can discriminate sufficiently between the two polarisation states to reduce the unwanted (orthogonally polarised) signal to a negligible level. The ratio of the power received from the wanted signal in the copolar channel to that received in the crosspolar channel is known as the *crosspolarisation discrimination* (XPD).

Under unfaded, stable, clear-air conditions, a system's crosspolarisation discrimination XPD_0 is determined by the crosspolar performance of the antennas (typically 30–40 dB). Since sub- and superrefractive conditions can give rise to considerable changes in the angle of launch/arrival in the vertical plane, the antennas must not only be designed to have a good XPD performance on axis, but must maintain a good XPD over the range of off-axis angles normally expected.

During multipath fading, however, high-angle off-axis rays may arise from multipath propagation, and in particular from ground-reflected rays. In addition, vertically and horizontally polarised waves undergo different attenuations and phase shifts on reflection, and this will result in different fading patterns for the two polarisations. In particular, the copolar signal may be deeply faded at a time when the crosspolar signal is unattenuated. This can give rise to much reduced XPD.

There is no general XPD-prediction model for multipath propagation. However, the arguments above, and data from a large number of paths, have led to an empirical relation between the XPD probability distribution, and that of the copolar attenuation CPA:

$$XPD = XPD_0 - CPA + Q \tag{8.10}$$

Q is an improvement factor which shows strong dependence on the slope of the crosspolarised antenna patterns in the vertical plane. Q typically lies in the range 0–10 dB.

8.2.5 Performance-protection methods

The effects of slow, relatively nonfrequency selective fading (flat fading), and faster frequency-selective fading due to multipath can be reduced both by nondiversity and diversity techniques. The effects of frequency-selective fading can also be ameliorated by the use of adaptive equalisation methods (but this is not discussed in this Chapter).

8.2.5.1   Nondiversity methods
Careful siting and design of nondiversity links to take advantage of the terrain along the path can help to minimise the occurrence of degradations due to clear-air effects:

(*a*) Increasing the path elevation will reduce the effects of defocusing, surface and atmospheric multipath; see eqn. 8.3.
(*b*) Links should be planned where possible to reduce the level of surface reflections in order to minimise the occurrence of multipath fading and distortion. Reflections from water and large, flat land surfaces are particularly to be avoided. A slight upward tilt of the antennas can help to reduce the effect of surface reflections, but this entails a trade-off with the loss in antenna gain under normal propagation conditions.
(*c*) In some cases, a reduction in path clearance can reduce the effects of multipath fading. However, a trade-off must be made between the reduction of multipath fading and the increased fading due to subrefraction.

8.2.5.2 Height diversity

In the simplest, and most common, case of two-path multipath propagation, the fading arises from the destructive interference between the two transmission paths; the deepest instantaneous fades occur at the frequency for which the effective path-length difference is an odd multiple of half wavelengths. If two receive antennas are mounted at different heights, the path-length differences will be different for the two antennas, and the probability of them simultaneously experiencing the maximum fade depth at a given frequency is very much less than for the single-antenna situation.

As in all diversity systems, the degree of improvement depends on the extent to which the signals in the diversity branches of the system are uncorrelated. Since atmospheric layers tend to be horizontally stratified, correlation lengths are much smaller in the vertical than in the horizontal, and so height diversity is much more effective than horizontal diversity as a protection against multipath fading. (Note that the opposite is true for diversity protection against rain fading: because rain attenuation does not depend strongly on the station height, horizontal diversity is much more effective than vertical diversity.)

For narrowband systems, advantage can be taken of height diversity either by switching the receiver between the two antennas to select the strongest signal, or by combining the two signals by phase aligning and summing. For wideband systems, a more complex method of combining the two systems to minimise the distortion can be implemented.

For narrowband systems, the degree of improvement afforded by diversity techniques is simply related to the improvement in the statistics of fade depth at a single frequency. The diversity-improvement factor I for fade depth A is defined as

$$I = p(A) / p_d(A) \tag{8.11}$$

where $p_d(A)$ is the percentage time in the combined-diversity signal branch with fade depth larger than A and $p(A)$ is the time percentage for the unprotected branch.

An expression for the height-diversity improvement factor for narrowband systems on overland paths is given in [16]:

$$I = \left\{ 1 - \exp\left(-3.34 \times 10^{-4} S^{0.87} f^{-0.12} d^{0.48} P_0^{-1.04} \right) \right\} 10^{(A-V)/10} \tag{8.12}$$

S is the vertical separation (centre-to-centre) of the receiving antennas in metres (valid for the range 3–23 m). f is the frequency in GHz (valid for the range 2–11 GHz). d is the path length in km (valid for the range 25–240 km). P_0 is the fading-occurrence factor given by $P_0 = p_w 10^{A/10} / 100$ where p_w is the percentage of the time that fade depth A (dB) is exceeded. V is the difference of the two antenna gains (dB).

It should be emphasised that diversity will not usually have any significant effect on the mean depression component of fading, although the provision of a height-diversity antenna can, in certain circumstances, lead to some increase in the mean received signal level due to the different positions of the receive antennas relative to an atmospheric layering structure.

Apart from the reduction of multipath fading, height diversity leads to other improvements:

(i)   The path-clearance criteria given in Section 8.2.4.1 for nondiversity links can be relaxed somewhat. For example, under median conditions, the path clearance for the lower antenna can be reduced significantly below one Fresnel zone [16].
(ii) The XPD degradation will be reduced.

8.2.5.3   *Frequency diversity*

Since multipath propagation is a frequency-selective mechanism, it follows that, if system performance is degraded on a given frequency, then moving the traffic to an alternative frequency may alleviate the problem. Frequency diversity is, however, spectrally inefficient and should be avoided whenever possible in order to conserve spectrum.

8.2.5.4   *Angle diversity*

Vertical angle diversity normally employs two or more antenna feeds spaced in the vertical direction with a common reflector. Since the various multipath components have different angles of arrival, angle diversity can be used to discriminate between them and to prevent the deep nulls that occur when two, nearly equal-amplitude, out-of-phase, components cancel. Angle diversity can be used in situations in which adequate space diversity is not possible or to reduce tower heights.

Whenever space diversity is used, it is recommended that two-antenna angle diversity is also used, by tilting the antennas at different upward angles.

The angle-diversity principle can be extended to include more complex adaptive-antenna techniques. In principle, an antenna array could be phased in such a way as to place a radiation-pattern null in the direction of the principal delayed component of multipath fading. If automated, this could offer a system which could constantly optimise signal level while minimising distortion. Methods such as this should provide significant improvements in reducing the effects of multipath in the future.

8.2.6 Worked example

Figure 8.4 shows the path profile of a 4 GHz line-of-sight radio link. The basic system parameters are:

Frequency f	4	GHz
Path length d	60	km
Height of transmitter antenna AMSL h_t	232	m
Height of receiver antenna AMSL h_r	339	m
Latitude of path ζ	52	°N

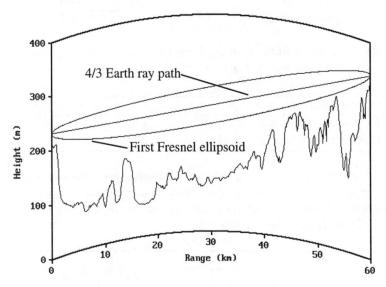

Figure 8.4   *Path profile for line-of-sight example*
Drawn on a 4/3-Earth-radius curvature

(a) Path clearance (refer to Section 8.2.4.1)
Figure 8.4 shows that the radio link has full Fresnel-zone clearance under median conditions ($k = 4/3$). For this path, the effective k factor, $k_{eff} = 0.83$ for 99.9% of the worst month; a path profile based on this k factor would confirm that the path is still line-of-sight. In fact, the path only becomes obstructed at $k = 0.64$ (the obstruction being the hill at 45 km). The path clearance is therefore adequate.

(b) Multipath fading and enhancement (refer to Section 8.2.4.2)
Calculate the following quantities:

Path inclination $|\varepsilon_p|$	1.78	mrad	from eqn. 8.4
Annual–worst month factor Δg	5.2		from eqn. 8.7
Geoclimatic factor $10 \log K$	−51.5	dB	from [16]

Equation (8.3) is now evaluated for $A_{0.01}$:

constant	+20.0	
$+10 \log K$	−51.5	dB
$+36 \log d$	+64.0	dB
$+8.9 \log f$	+5.4	dB
$-14 \log (1+|\varepsilon_p|)$	−6.2	dB
$A_{0.01}$ (worst month)	+31.7	dB

$A_{0.01}$ is the key quantity from which the deep fading and enhancement distributions can be calculated from eqns. 8.2 and 8.5:

Percentage of worst month p_w or q_w	0.1	0.01	0.001
Fade depth $A(p_w)$ exceeded, dB	21.7	31.7	41.7
Enhancement level Eq_w exceeded, dB	{8.1}	11.6	15.1

If average annual distributions are required, the annual time percentages must first be converted to the equivalent worst-month time percentages using eqn. 8.6 and the worst-month value of $A_{0.01}$ converted to the average annual value using eqn. 8.8 and the value of Δg calculated above:

Percentage of the year, p or q	0.1	0.01	0.001
Equivalent p_w or q_w, %	0.52	0.052	0.0052
Average annual $A_{0.01}$, dB		24.5	
Fade depth exceeded, dB	14.5	24.5	34.5
Enhancement level exceeded, dB	{4.2}	{7.7}	11.2

In the above tables, the quantities in curly brackets strictly lie outside the valid range of the large enhancement model. The extrapolation procedure given in [16] should be used to recalculate these values if necessary.

From this it can be concluded that the fade depth will exceed 31 dB for 0.01% of the worst month (about 4 min), while the enhancement level will exceed 11 dB for a similar period. For 0.001% of the average year (about 5 min) the corresponding levels are 35 dB and 11 dB, respectively.

(c) Overall link budget
To obtain the total transmission loss on the link due to clear-air degradations:
(i)   calculate the free-space basic transmission loss (= 140.1 dB from eqn. 8.1);

(ii) add gaseous absorption loss (= 0.4 dB from eqn. 8.9);

(iii) add multipath fading loss (or subtract multipath enhancement loss) depending on the time percentage of interest;

(iv) add losses due to angle-of-arrival and scintillation effects: these are negligible for this path.

If at a given time percentage, the rain attenuation loss (see Chapter 9) is greater than the sum of the clear-air losses (ii)–(iv), rain effects will determine the overall link budget.

8.3   Terrestrial beyond-the-horizon paths

8.3.1 Mechanisms

The only mechanisms causing radio propagation beyond the horizon which occur permanently at frequencies greater than about 30 MHz are those of diffraction at the Earth's surface and scatter from atmospheric irregularities.   Although diffraction can be used to establish communication at short distances beyond the horizon, the path loss increases rapidly at greater ranges (particularly at microwave frequencies). However, beyond a certain distance the rate of increase with distance reduces to a low value. The reason is that the field due to scattering from tropospheric irregularities is more important at these distances than the diffracted signal.

Although scattered-field levels are inherently low, the relatively slow decrease with increasing path length means that links can be established over long distances using the tropospheric-scatter (or troposcatter) mode of transmission; however, the high path-loss values and the severe fading that can be encountered call for special measures to be employed, such as high-power transmitters, large reflector antennas, sensitive receivers and diversity techniques. The frequency range used is around 200 MHz to 5 GHz, limited by the physical dimensions of high-gain reflector antennas at the lower end and by rain attenuation and gaseous absorption at the upper end.

Signals received by means of tropospheric scatter show both slow and rapid variations:

(*a*) *Slow signal variations* are due to overall changes in refractive conditions in the atmosphere associated with weather type, and hence are a function of climate. The slow variations of signal level are well described by distributions of hourly median values which are approximately log-normal. In a temperate climate, the transmission loss varies diurnally and annually. Diurnal variations are most pronounced in summer, with a range of 5–10 dB on 100–200 km overland paths;

the greatest transmission loss occurs in the afternoon, and the least in early morning. Monthly median losses tend to be higher in winter than in summer.

(*b*) *Rapid fading* is due to the motion of small-scale irregularities and has a frequency ranging between a few fades per minute at VHF and a few fades per second at microwave frequencies. Over periods of a few minutes, the rapid variations are approximately Rayleigh distributed, consistent with the level being a vector sum of many components of random amplitude and phase arising from scattering elements in the common volume. In addition, rapid fading from highly reflecting layers of limited horizontal extent in the common volume can be experienced.

8.3.2 Prediction method for tropospheric scatter

8.3.2.1   Overview of the method
The prediction method for the hourly median signal level is taken from ITU Recommendation P.617-1 [19] for the design of trans-horizon radio-relay systems, although certain simplifications are made for ease of presentation. (More details of all aspects of troposcatter-radio-link design are given in [157]). Since scatter paths are not line-of-sight, the contribution of propagation effects to the overall system-link budget is not merely to degrade the free-space basic transmission-loss figure; rather, the scatter provides the means by which the radio link is maintained, as well as causing degradations due to signal fluctuations. In this case it is more appropriate to calculate the transmission loss directly.

The method, as in most scatter models, is based on the concept of a 'scatter volume': the region of the troposphere contributing to the scattering process, defined by the intersection of the transmitter- and receiver-antenna beams. Thus transmission loss (which includes the effects of the transmitter and receiver antennas), rather than basic transmission loss, is used, and it should be remembered that the antenna gains must not be included again in the system-link budget.

The method comprises three parts:

(i) calculation of the average annual hourly median transmission loss not exceeded for 50% of the time $L(50)$;
(ii) derivation of the average annual hourly median transmission loss not exceeded for q% of the time $L(q)$, from the value $L(50)$;
(iii) calculation of the worst-month hourly median transmission loss, if required.

8.3.2.2   L(50): transmission loss not exceeded for 50% of the time
For the design of tropospheric-scatter links with high-gain antennas pointing at each other, and for a restricted range of practical geometries, an empirical formula

has been based on data for the frequency range between 200 MHz and 4 GHz; however the model can be extended to 5 GHz with little error for most applications. (A rather different approach has been used for the evaluation of tropospheric-scatter signals likely to cause interference within intervals much less than an hour—see Section 13.5.4.) In this case, the median transmission loss is given by

$$L(50) = M + 30\log f + 10\log d + 30\log\theta + L_N + L_c - G_t - G_r \qquad (8.13)$$

The basic link parameters are f, the frequency (in Mhz), the great-circle path length d (km) and transmitter and receiver antenna gains G_t and G_r (dB), respectively. Other parameters required are:

(i)   the scatter angle (also known as the angular distance) θ (mrad):

$$\theta = \theta_e + \theta_t + \theta_r \qquad (8.14)$$

θ_t and θ_r are the transmitter and receiver horizon angles, respectively (mrad); that is the elevation of the horizon as seen from the transmitter or receiver. θ_e is the effective angle (mrad) subtended at the centre of the Earth by the great-circle path between transmitter and receiver in median refractivity conditions:

$$\theta_e = 1000 \, d \, / \, ka \qquad (8.15)$$

where $a = 6370$ km is the radius of the Earth and k is the effective Earth-radius factor for median refractivity conditions ($k = 4/3$ should be used unless a more accurate value is known).

(ii)   climate-dependent factors, M and L_N:

   M varies between 19 and 40 dB, depending on the climate. Reference [19] recommends values of the climate-dependent values for nine climatic zones ranging from equatorial to polar. In a maritime temperate climate (such as the UK), $M = 33.2$ dB for overland paths and $M = 26$ dB for oversea paths (where both transmitter and receiver horizons lie on the sea).

   L_N (dB) accounts for the transmission-loss dependence on the height of the common volume. It is obtained from

$$L_N = 20\log\left(5 + \gamma \, H\right) + 4.34\gamma \, h \qquad (8.16)$$

where $H = 10^{-3}\theta d/4$ and $h = 10^{-6}\theta^2 ka/8$. γ is also a climate-dependent parameter, and for a maritime temperate climate has a value of 0.27 km^{-1}.

(iii)   the aperture-to-medium coupling loss L_c (dB), takes account of the fact that, if the antenna beam is narrowed to increase its gain, then the common volume (from which the scattering originates) is reduced and the expected gain in received signal power is not fully realised:

$$L_c = 0.07 \exp\{0.055(G_t + G_r)\} \tag{8.17}$$

8.3.2.3   *L(q): transmission loss not exceeded for q% of the time*
For $q > 50$, $L(q)$ is obtained from $L(50)$ by applying a correction factor $Y(q)$:

$$L(q) = L(50) - Y(q)$$
$$\text{with} \qquad Y(q) = C(q)Y(90) \tag{8.18}$$

$C(q)$ is a scaling factor that relates the q % level of tropospheric scatter to the 90% value. The value of $C(q)$ for practical values of q is as follows:

q	50	90	99	99.9	99.99
$C(q)$	0	1	1.82	2.41	2.90

$Y(90)$ is climate-dependent. For a maritime temperate climate, it is given (in dB) by

$$Y(90) = -2.2 - \left(8.81 - 2.3 \times 10^{-4} f\right)\exp(-0.137h) \qquad \text{for overland paths}$$
$$= -9.5 - 3\exp(-0.137h) \qquad\qquad\qquad \text{for oversea paths} \tag{8.19}$$

where f (in MHz) and h are as before.

For $20 < q < 50$, the average annual transmission-loss distribution can be considered symmetrical (with $50 < q < 80$) and the transmission-loss values estimated from the corresponding values above the median:

$$L(q) = L(50) - \{L(100 - q) - L(50)\} \tag{8.20}$$

This expression should only be used for time percentages down to about 20%. At smaller time percentages, ducting will need to be considered as it may be the dominant propagation mechanism, and the methods described in Section 13.5 become appropriate.

8.3.2.4   *Worst-month predictions*
The average worst-month hourly median transmission loss not exceeded for time percentages greater than 50% can be obtained from $L(q)$ by adding a correction factor which depends on climate, the path length and the time percentage of interest. The correction factors are given graphically in [19]. They are larger for shorter paths and lower time percentages. For a temperate climate, the correction

ranges from 9 dB for a 100 km path at 50% of the time to less than 3 dB for a 1000 km path at any time percentage.

8.3.3 Performance-protection methods

The deep fading occurring with tropospheric-scatter propagation severely reduces the performance of systems using this propagation mode. The effect of fading can be reduced by diversity reception, using two or more signals which fade more or less independently owing to differences in scatter path or frequency. Space, frequency and angle diversity can all be used to reduce the percentage of time for which a given transmission loss is exceeded.

8.3.3.1   Space diversity
The antenna separation required depends on the scale lengths of the refractivity irregularities and on the antenna aperture size. For frequencies greater than 1000 MHz, a suitable value for the antenna separation Δh (m) is

$$\Delta h = 0.36\sqrt{D^2 + C} \qquad (8.21)$$

where D is the antenna diameter in metres; $C = 1600$ for horizontal separation and $C = 225$ for vertical separation.

8.3.3.2   Frequency diversity
For frequencies greater than about 1 GHz, an adequate frequency separation Δf (MHz) is

$$\Delta f = (1.44 f / \theta\, d)\sqrt{D^2 + 225} \qquad (8.22)$$

f is the frequency (MHz), d is the path length (km) and D is the antenna diameter (m). θ is the scatter angle (mrad) calculated in eqn. 8.14. However, as in Section 8.2.5.3, it should be said that frequency diversity is spectrally inefficient and should be avoided whenever possible to conserve spectrum.

8.3.3.3   Angle diversity
Vertical-angle diversity creates vertically spaced common volumes similar to the situation for vertical-space diversity. However, angle diversity is more economical than space diversity (since it only requires a single antenna reflector). The angular spacing $\Delta\theta$ (degrees) required to have approximately the same effect as the vertical spacing Δh (m) in eqn. 8.21 on an approximately symmetrical path is

$$\Delta\theta = 57.3 \arctan(\Delta h / 500d) \qquad (8.23)$$

where d is the path length (km).

8.3.3.4   Polarisation diversity

It is normal to supplement space diversity with polarisation diversity to give quadruple operation.

8.3.4 Worked example

Calculate the transmission loss expected on an overland path, 250 km troposcatter link at 900 MHz, with basic system parameters:

Frequency f, MHz	900	
Path length d, km	250	
Transmitter horizon angle θ_t, deg	−0.5	(taken from a 4/3-Earth-
Receiver horizon angle, θ_r, deg	−0.2	radius path profile)
Antenna sizes, m	15	

The annual median transmission loss $L(50)$ is derived as follows:

G_t		40.4	dB	(15 m parabolic antenna
G_r		40.4	dB	at 900 MHz)
θ_t	−8.7		mrad	
θ_r	−3.5		mrad	
θ_e	+29.4		mrad	from eqn. 8.15
θ		17.2	mrad	from eqn. 8.14
H		1.08	km	from eqn. 8.16
h		0.31	km	from eqn. 8.16
γ		0.27	km^{-1}	from eqn. 8.16
$Y(90)$		−10.4	dB	from eqn. 8.17

Annual median path loss:			from eqn. 8.13
M	+33.2	dB	
$+30 \log f$	+88.6	dB	
$+10 \log d$	+24.0	dB	
$+30 \log \theta$	+37.1	dB	
$+L_N$	+14.8	dB	from eqn. 8.16
$+L_c$	+6.0	dB	from eqn. 8.17
$-G_t$	−40.4	dB	
$-G_r$	−40.4	dB	
$L(50)$	122.9	dB	

From $L(50)$ and eqn. 8.18, the transmission losses not exceeded at various time percentages q are:

q, %	50	90	99	99.9	99.99
$Y(q)$, dB	0	−10.4	−18.9	−25.1	−30.2
$L(q)$, dB	122.9	133.3	141.8	148.0	153.1

The transmission losses should be compared with the free-space transmission loss (i.e. including antenna gains) for this path length and frequency: 58.7 dB from eqn. 8.1. Although the troposcatter transmission loss is much higher, this would be a viable troposcatter link.

8.4   Earth–space paths

8.4.1 Mechanisms

In the design of Earth–space links for communications systems, several effects must be considered:
(i)   attenuation due to rain and other hydrometeors: this is generally the most important mechanism degrading system performance, particularly at the lower time percentages (<0.1%) and for frequencies above about 10 GHz. This topic is covered in Chapter 9 and is not discussed further here;
(ii)   attenuation due to atmospheric gases;
(iii) losses due to random irregularities in the refractive-index structure of the atmosphere: scintillation and wavefront incoherence; and
(iv) losses due to the bulk-refractive-index structure of the atmosphere: angle-of-arrival variations and beam spreading.

Each of these contributions has its own characteristic as a function of frequency, geographic location and elevation angle. As a rule, at elevation angles above 10°, only rain attenuation, gaseous attenuation and possibly scintillation will exceed a few tenths of a decibel, depending on propagation conditions. However, on low-elevation satellite links, because of the longer path in the troposphere and the relatively greater effect of refractive bending, the effects of (iii) and (iv) can be much larger.

8.4.2 Prediction methods for clear-air effects

The prediction methods are taken from ITU Recommendation P.618-3 [20] for the design of Earth–space telecommunications systems. Fade countermeasures for

Earth–space systems are almost exclusively concerned with attenuation due to hydrometeors, and are not discussed in this Chapter.

As for terrestrial line-of-sight paths, the predictions are for the levels of propagation loss expected to be exceeded for a given time percentage, and these losses must be added to the free-space transmission loss (eqn. 8.1) to give the total basic transmission loss of the radio link.

8.4.2.1   Attenuation due to atmospheric gases

Attenuation by atmospheric gases depends mainly on frequency, elevation angle, altitude above sea level and water-vapour density (absolute humidity).   At frequencies below 10 GHz, it may normally be neglected, but its importance increases with frequency above 10 GHz, especially for low elevation angles.

The total Earth–space gaseous attenuation through the atmosphere A_g (dB) for a path elevation angle θ greater than $10°$ is

$$A_g = \frac{\gamma_0 h_0 \exp(-h_s / h_0) + \gamma_w h_w}{\sin\theta} \tag{8.24}$$

In this expression,

(i) h_s is the height (km) of the Earth station above mean sea level;

(ii) γ_0 and γ_w are the specific attenuations (dB/km) at the surface for dry air and water vapour, respectively, obtained from Figure 6.8 for the frequency of interest; formulae for calculating γ_0 and γ_w, taking account of temperature and water-vapour density, are given in [23];

(iii) h_0 and h_w are the 'equivalent heights' (km) for dry air and water vapour, respectively [23].   For frequencies below 57 GHz (i.e. below the oxygen-absorption-line complex around 60 GHz), $h_0 = 6$ km, while h_w can be determined from

$$h_w = h_{w0} \left\{ 1 + \frac{3.0}{(f - 22.2)^2 + 5} \right\} \tag{8.25}$$

$h_{w0} = 1.6$ km in clear weather and 2.1 km in rain.

For path-elevation angles θ less than $10°$ the more detailed expressions given in [20] or [23] should be used.

The attenuation due to gaseous absorption is not constant; this is due principally to the fact that both the surface value of water-vapour density and its vertical profile are quite variable (the oxygen contribution is relatively constant).   The methods of [20] and contour maps of surface water-vapour density [26] can be used to estimate distributions of gaseous attenuation if required.   For low-availability systems, the variations in atmospheric attenuation exceeded for large percentages of the time (when no rain is present) are important. A study at several locations in Europe at 11.4 GHz showed that seasonal variations in the monthly median level of total attenuation did not exceed 0.1 dB, and that the total attenuation exceeded for 20% of the worst month was 0.05–0.15 dB above the monthly median value, depending on location.

8.4.2.2 Scintillation and wavefront-incoherence losses

Scintillations causing rapid fluctuations in signal amplitude can originate in the troposphere and in the ionosphere.

Tropospheric scintillation is caused by small-scale irregularities in the atmospheric refractive index (see Chapter 6). In the absence of precipitation, it is unlikely to cause serious fading at frequencies below about 10 GHz and at elevation angles above 10°. However, at lower elevation angles, and at frequencies above about 10 GHz, tropospheric scintillation can occasionally cause serious degradation in system performance. Amplitude scintillations have a period of a few seconds, or tens of seconds.

Typical maximum values of the RMS variation on a low-elevation (less than 3°) satellite are 1 dB at 10 GHz and 4 dB at 100 GHz, increasing at lower elevation angles. At such low elevations, it may also be necessary to consider the effects of bulk refraction on path geometry. Although not a dominant factor in system design at the higher elevation angles, scintillation levels determine the residual bit-error rate achievable in clear weather conditions.

A procedure for predicting the scintillation fade depth x (in decibels relative to the long-term median value) for the time percentage p is given in [20], valid for frequencies f in the range 4–20 GHz and path elevation angles $\theta \geq 4°$:

$$x = a(p)\, \sigma_{ref}\, f^{7/12} \frac{g}{(\sin\theta)^{1.2}} \tag{8.26}$$

In this expression,

(i) σ_{ref} is a climatological normalisation constant which depends on the monthly or longer-term average value of the water-vapour-dependent part of the surface refractivity. Its value lies in the range 0.004–0.02.

(ii) g is a factor which takes account of antenna aperture averaging (resulting in a reduction of the effective antenna gain). It is a complicated function of the Earth-station-antenna diameter, frequency and elevation angle. For most Earth stations, its value lies in the range 0.1–0.9. The model assumes the presence of a horizontal thin turbulent layer at a height of 1 km.

(iii) for p in the range $0.01 < p \leq 50$, $a(p)$ is given by

p	50	10	1	0.1	0.01
$a(p)$	0	1.3	3.0	4.8	7.2

Ionospheric scintillation can cause significant signal variation in the VHF and UHF bands, and even as high as 10 GHz during conditions of maximum sunspot number for Earth stations near the geomagnetic equator. Peak-to-peak ionospheric-scintillation levels of 10 dB have been observed for periods of several minutes at 4 GHz on a low-latitude path [17]. The effects are very much less in mid-latitude regions. In equatorial regions there is a pronounced diurnal variation with most significant fades around 20.00-02.00 h local time. The dependence of the scintillation amplitude on frequency f is between f^{-1} and f^{-2}. Ionospheric effects are significant at microwave frequencies only for geostationary satellites operating to equatorial regions and for polar orbiting satellites to high-latitude stations.

In addition to scintillation, small-scale irregularities in the refractive-index structure of the atmosphere cause incoherence of the wavefront of a wave incident on a receiving antenna, leading to an antenna-to-medium coupling loss that can be described as a decrease in antenna gain. The effect increases both with increasing frequency and with decreasing elevation angle, and is a function of antenna diameter. For example, measurements with a 31.6 GHz, 7 m-diameter antenna at a 5° elevation angle gave a loss of 0.6 dB. In practice, the effect is likely to be significant only for narrow-beamwidth antennas, high frequencies, and elevation angles below 5°.

8.4.2.3   Angle-of-arrival and beam-spreading losses

The decrease in refractive index with height in the atmosphere produces an increase in the apparent elevation angle of a satellite (see Chapter 6). The effect is only significant for very low elevation paths: the total angular refraction (the increase in apparent elevation) is about 0.65°, 0.35° and 0.25°, for elevation angles of 1°, 3° and 5°, respectively, for a tropical marine climate (which gives the biggest effect).

At 1° elevation, the day-to-day variation in apparent elevation angle is of the order of 0.1° (RMS) and the short-term angle-of-arrival fluctuations, due to changes in the refractivity/height profile, may be of the order of 0.02° (RMS); both effects decrease rapidly with increasing elevation angle. In practice, it is difficult to distinguish between the effect of short-term changes in the height/refractivity profile and the effect of random irregularities superimposed on that distribution.

The mean angle-of-arrival deviations can be allowed for in the design of a low-elevation link. The fading caused by diurnal and short-term effects can be seen to be unimportant except for large antennas (very narrow beamwidths) at very low elevations.

Signal loss on a slant path may also result from additional spreading of the antenna beam caused by the variation of atmospheric refraction with elevation angle. The effect amounts to less than 1 dB at angles above 1°, and is negligible above about 5°.

8.4.2.4   Crosspolarisation effects

In the presence of a magnetic field, the ionosphere exhibits birefringence, splitting the incident wave into ordinary and extraordinary components. Individually, these waves do not change polarisation as they propagate through the ionosphere. However, since they propagate with different phase velocities, the resultant plane of polarisation of the combined wave rotates as it propagates. This effect is known as Faraday rotation. Its magnitude is proportional to the total electron content encountered along the path, and therefore varies with latitude, time of day, and solar activity. It also depends on frequency f as f^{-2}. The estimated maximum Faraday rotation expected on a 30°-elevation path ranges from about 100° at 1 GHz to 1° at 10 GHz.

This rotation of the plane of polarisation would have a serious effect on plane-polarised transmissions (vertical or horizontal) in two ways: first, the copolar signal would experience severe fading due to variations in the Faraday rotation coupled with the crosspolarisation discrimination of the Earth-station antenna; secondly, it precludes the two-fold improvement in spectrum usage that would result from transmitting on both orthogonal polarisations. For this reason, lower-frequency (L-band) satellite systems generally employ circular polarisation which is not affected by Faraday rotation.

8.4.3 Worked example

Calculate the gaseous attenuation at 20 GHz during clear-air conditions between a ground station located at sea level and a geostationary satellite seen at 30° elevation.

Station height AMSL h_s, km	0.0	
Path elevation θ, deg	30.0	
Specific attenuation of dry air γ_o, dB/km	0.01	(at 15 °C and 7.5 g/m^3
Specific attenuation of water vapour γ_w, dB/km	0.10	water-vapour density)
Equivalent height for dry air h_o, km	6.0	
Equivalent height for water vapour h_w, km	2.1	from eqn. 8.25
Total slant path gaseous absorption A_g, dB	0.5	from eqn. 8.24

Chapter 9

Prediction of reliability when degraded by precipitation and cloud

J.W.F. Goddard

9.1 Introduction

The mechanisms whereby precipitating and nonprecipitating water can have detrimental effects on communication systems operating in the microwave and millimetre-wave parts of the spectrum were described in Chapter 7. The purpose of this Chapter is to introduce the prediction tools available to help system designers quantify these effects for a particular link. This includes the prediction of long-term annual attenuation statistics, conversion to worst-month levels, and crosspolar effects. The first point of reference for the engineer is usually the relevant ITU-R prediction model, and the application of these to both terrestrial line-of-sight and Earth–space links is first described, together with the recognised limitations of the current methods. Improvements to the models will come from measurement programmes which provide further insight into the physics of precipitation systems and also data against which their accuracy can be evaluated. These can be obtained from experimental links, which provide accurate data but for only one specific frequency and path geometry. Discussion of other approaches, using radiometers and radars, brings this Chapter to a conclusion.

9.2 Terrestrial line-of-sight radio-relay systems

9.2.1 General

The central ITU-R Recommendation required for the planning of these systems is ITU Recommendation P.530-6 [18], although inputs from several other Recommendations are required, as shown schematically in Figure 9.1. Note that only the precipitation, cloud and water-vapour components are discussed here, although ITU Recommendation P.530-6 also includes an evaluation of the effects of diffraction and multipath fading and enhancement, which have been dealt with in Chapter 8. If, for a given path geometry and frequency, both clear-air and rain attenuation need to be taken into account, then their exceedence statistics can be simply added together.

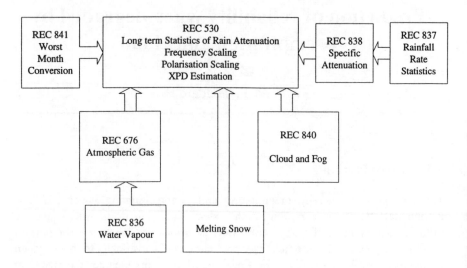

Figure 9.1   Prediction methods for designing terrestrial line-of-sight links

9.2.2 Long-term statistics of rain attenuation

Unless low-availability (>1%) systems are under consideration, attenuation due to rain will be the most significant effect to evaluate. Despite the implementation of many programmes of direct measurements (see Section 9.3), particularly in Europe, the starting point for estimating rain attenuation is still a knowledge of the statistics of rainfall rate. This is because direct measurements can only be obtained for a given frequency and path geometry and, to be a statistically good representation for a particular location, they need to span several years to smooth year-to-year variability. It would be impractical to obtain reliable statistics through an archive of direct measurement for each and every planned link geometry and frequency.

The preferred option is to use locally derived statistics of rainfall rate, specifically the rainfall rate exceeded for 0.01% of the time ($R_{0.01\%}$), with an integration time of 1 min. However, if that is not available, then ITU Recommendation P.837-1 [27] provides global distributions of rainfall rate for various time percentages according to specified zones, as shown in Figure 9.2.

Different parts of the UK fall into three zones: E, F and G. The rainfall rates corresponding to each zone are shown in Figure 9.3.

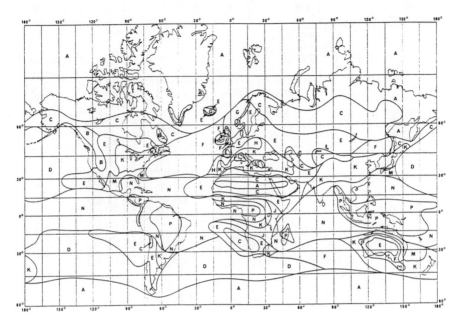

Figure 9.2   Rain climate zones of the world

The second step in the procedure involves computing the specific attenuation γ_R for this rainfall rate, at the frequency and polarisation of the transmission. The units of γ_R are dB/km. ITU Recommendation P.838 [28] provides a method for achieving this using a power-law relationship of the form

$$\gamma_R = kR^\alpha \tag{9.1}$$

The coefficients k and α are frequency dependent, and also have a polarisation dependence due to the oblate shapes of rain drops, discussed in Chapter 7. Coefficient values are given for horizontal and vertical polarisations. They are valid when the elevation angle of the path is zero, but can be converted to other polarisation-tilt angles τ or elevation angles θ using the relations

$$k = \left\{ k_H + k_V + \left(k_H - k_V \right) \cos^2 \theta \cos 2\tau \right\} / 2 \tag{9.2}$$

$$\alpha = \left\{ k_H \alpha_H + k_V \alpha_V + \left(k_H \alpha_H - k_V \alpha_V \right) \cos^2 \theta \cos 2\tau \right\} / 2k$$

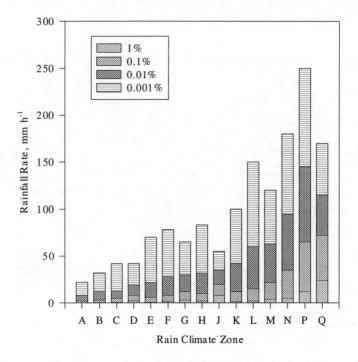

Figure 9.3   Rainfall rates exceeded for 0.01% of time in each climate zone

For circular polarisation, a value of 45° should be used for τ.

Having obtained the specific attenuation corresponding to the reference rainfall rate, we need to estimate the total path attenuation. At this point, it is necessary to recognise that, because of the high degree of spatial inhomogeneity present in rain, it is unlikely that the reference rainfall rate will extend uniformly over the length of the transmission path, unless this is very short. An example of the complexity of the horizontal (and vertical) structure of rain is indicated in the Chilbolton radar vertical sections in Chapter 7. The longer the path, the less likely it is that rain will extend the full length of the path, and so an effective path length d_{eff} is introduced. This is given by

$$d_{eff} = rd \qquad (9.3)$$

where d is the actual path length, and r is a factor which reduces in magnitude as d increases. It is given by

$$r = \frac{1}{1 + d/d_0} \qquad (9.4)$$

The quantity d_0 is a rainfall-rate-dependent factor introduced, in turn, to reflect the fact that the greater the intensity of rainfall in a storm, the smaller the physical dimensions of the storm are. It is given by

$$d_0 = 35 \exp(-0.015 R_{0.01}) \qquad (9.5)$$

whose form is shown in Figure 9.4.

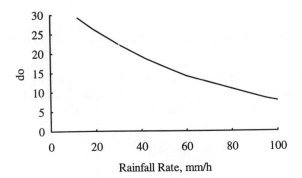

Figure 9.4   Variation of factor d_0 with rainfall rate

The value of d_0 for rain climate F (applicable to the central UK) is 23. Using this value, Figure 9.5 shows the variation of the effective path length d_{eff} as the actual path length increases.

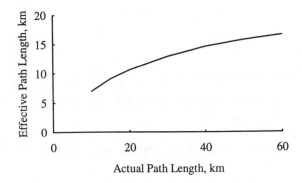

Figure 9.5   Variation of effective path length with actual path length

Then, finally, the path attenuation exceeded for 0.01% of the time is obtained from

$$A_{0.01} = \gamma_R d_{eff} \tag{9.6}$$

If attenuation statistics are required for other time percentages, they can be obtained from

$$A_p = A_{0.01} \times 0.12 p^{-(0.546 + 0.043 \log_{10} p)} \tag{9.7}$$

where p is the required time percentage, in the range $0.001\% \le p \le 1\%$.

9.2.3 Long-term frequency scaling of rain-attenuation statistics

If reliable attenuation statistics are available for a site where a link of a different frequency is planned, rather than build a prediction from rain-rate statistics, it is preferable to scale the existing statistics. This can be achieved with the following empirical formulae, valid in the frequency range $7 - 50$ GHz:

$$A_2 = A_1 (\phi_1/\phi_2)^{1-H(\phi_1,\phi_2,A_1)} \tag{9.8}$$

where

$$\phi(f) = \frac{f^2}{1+10^{-4} f^2} \tag{9.9}$$

and

$$H(\phi_1,\phi_2,A_1) = 1.12 \times 10^{-3} \left(\frac{\phi_2}{\phi_1}\right)^{0.5} (\phi_1 A_1)^{0.55} \tag{9.10}$$

9.2.4 Crosspolar-discrimination reduction in rain

Depolarisation of the transmission can be introduced by raindrops, ice particles or melting particles, as discussed in Chapter 7. In the absence of reliable measurements of *XPD* for a given path, rain depolarisation can be very approximately related to statistics of copolar attenuation (*CPA*) using the equiprobability relation

$$XPD = 15 + 30 \log f - 20 \log(CPA) \quad \text{dB} \tag{9.11}$$

valid in the frequency range $8 \le f \le 35$ GHz. It must be stressed again that this equation is very approximate, with large uncertainties in the coefficients.

9.2.5 Attenuation due to cloud or fog

The effects of cloud or fog (most likely fog on terrestrial paths) only have any relevance at frequencies above 10 GHz, but become very large beyond 100 GHz. Although ITU Recommendation P.840-1 [29] contains some information on fog, it does not provide a complete method for predicting the attenuating effects, mainly through a lack of information on the statistics of liquid-water content and cloud dimensions. Table 9.1 gives some indication of the attenuation to be expected at various frequencies, but the statistics of occurrence will depend on the local climate and the height of the path above sea level.

Table 9.1   Effects of fog as a function of frequency

Frequency	Specific attenuation (assuming LWC of 0.5 g/m³)	Path attenuation (assuming a 5 km path in fog)
GHz	dB/km	dB
10	0.1	0.5
20	0.2	1
50	1.0	5
100	2.5	12.5
200	5	25

9.2.6 Melting snow

As discussed in Chapter 7, modelling of melting ice particles suggests that melting snow can cause significantly higher specific attenuation than the equivalent hydrometers once melted. There is, however, no recommended procedure to estimate the effects, which could be particularly severe at the near-zero elevations of terrestrial paths, if they are located in high-latitude regions, or high above sea level.

9.2.7 Conversion of annual statistics to worst-month statistics

For planning purposes, the average annual worst-month time percentage of excess p_w is often preferred to annual statistics. ITU Recommendation P.841 [30] contains general information which permits the conversion between the two sets of statistics, which are related through a factor, Q

$$p_w = Qp \qquad (9.12)$$

This factor is a two-parameter function of the annual time percentage, and for time percentages appropriate for rain effects, it is given by

$$Q(p) = Q_1 p^{-\beta} \qquad (9.13)$$

Equation 9.13 is valid in the range $(Q_1/12)^{1/\beta} < p < 3\%$. In the absence of more precise information, global values of 2.85 for Q_1 and 0.13 for β may be used. Table 9.2 gives resulting values of Q for various time percentages.

Table 9.2   *Global and local values of Q for various annual time percentages*

Annual time percentage p	Q (global)	Q (north west Europe)
1	2.85	3.0
0.1	3.84	4.0
0.01	5.2	5.5
0.001	7.0	7.4

ITU Recommendation P.841 also contains locally derived values of Q_1 and β for rain effects specifically, for example, on terrestrial paths, which are considered to give a more accurate means of conversion. The resulting Q values for north west Europe are also shown in Table 9.2.

Whichever method for deriving Q is adopted, eqn. 9.12 is then used to obtain the p_w value which corresponds to $p_{0.01}$, and then this value can be substituted in eqn. 9.7 to scale the annual statistic of attenuation $A_{0.01}$, to obtain the corresponding worst-month attenuation.

9.3   Earth–space paths

Many of the procedural elements for evaluating precipitation effects on Earth–space paths are similar to those for terrestrial paths, discussed in Section 9.2 with respect to Figure 9.1. Only the significant differences will be discussed here. In particular, the central ITU-R document is Recommendation P.618-3 [20], rather than Recommendation P.530-5.

9.3.1 Long-term statistics of rain attenuation

Because the paths are no longer horizontal, the major difference from the procedure described in Section 9.2 is that account must be taken of the height at which the change in phase of hydrometeors, from ice to water, takes place, close to the 0 °C isotherm. As discussed in Chapter 7, attenuation due to ice or snow is negligible, at least below 100 GHz. Figure 9.6 illustrates the parameters involved in the prediction process.

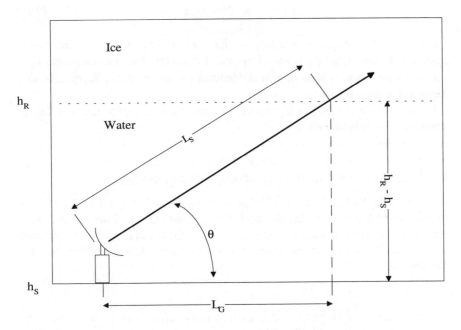

Figure 9.6   Parameters used in Earth–space path-prediction model

The first step in the procedure is to calculate the effective rain height h_R which will represent an annual mean value for the location of the link, as follows:

$$h_R = \begin{cases} 3.0 + 0.028\phi & 0 \le \phi < 36° \\ 4.0 - 0.075(\phi - 36) & \phi \ge 36° \end{cases} \tag{9.14}$$

where ϕ is the station latitude. For a latitude of 52°, this gives a height of 2.8 km, compared with a maximum height of 4 km at 36°.

Having determined h_R, the slant path length L_S must be computed for the specific satellite elevation, using

$$L_S = \begin{cases} \dfrac{\left(h_R - h_S\right)}{\sin\theta} & \theta \ge 5° \\[4mm] \dfrac{2\left(h_R - h_S\right)}{\left(\sin^2\theta + \dfrac{2\left(h_R - h_S\right)}{R_e}\right)^{1/2}} & \theta < 5° \end{cases} \tag{9.15}$$

The horizontal projection L_G of the slant-path length can then be computed from

$$L_G = L_S \cos\theta \quad \text{km} \tag{9.16}$$

The remainder of the procedure follows that for horizontal paths, as described in Section 9.2, except that L_G is used in place of d in eqn. 9.4. The parameter d_0 in eqn. 9.4 (also defined in eqn. 9.5) is identical to the parameter L_0 used in ITU Recommendation P.618.

Frequency scaling from a known set of attenuation statistics can be achieved using eqn. 9.4 in Section 9.2.2.

9.3.2 Attenuation due to other atmospheric constituents

As in the terrestrial-path case, ITU Recommendation P.840 is referred to for the calculation of attenuation due to cloud or fog. The effects of dry air and water vapour are estimated with the help of ITU Recommendation P.676, although 'engineering formulae' are provided in ITU Recommendation P.618 to simplify the procedure.

9.3.3 Long-term statistics of hydrometeor-induced crosspolarisation

The recommended method for obtaining XPD statistics is based on a knowledge of the copolar-attenuation statistics A_p for the required percentage of time. It is valid for elevation angles $\theta \le 60°$ over the frequency range $8\,\text{GHz} \le f \le 35\,\text{GHz}$. The XPD due to rain XPD_{rain} is calculated from

$$XPD_{rain} = C_f - C_A + C_\tau + C_\theta + C_\sigma \tag{9.17}$$

where

$$C_f = 30\log f \tag{9.18}$$

$$C_A = V\log(A_p) \quad \text{with} \begin{cases} V = 20 & \text{for} \quad 8 \le f \le 15\,\text{GHz} \\ V = 23 & \text{for} \quad 15 < f \le 35\,\text{GHz} \end{cases} \tag{9.19}$$

and

$$C_\tau = -10\log\left[1 - 0.484\{1 + \cos(4\tau)\}\right] \tag{9.20}$$

where τ is the polarisation-tilt angle with respect to the horizontal. A value of $\tau = 45°$ is to be used for circular polarisation. The elevation-angle-dependent term

$$C_\theta = -40\log(\cos\theta) \tag{9.21}$$

is valid for elevations angles of $\theta \leq 60°$, and there is also a term that takes account of the spread of canting angles of individual rain drops:

$$C_\sigma = 0.0052\sigma^2 \qquad (9.22)$$

Finally, an ice-crystal-dependent term is calculated:

$$C_{ice} = XPD_{rain} \times (0.3 + 0.1 \log p) / 2 \qquad (9.23)$$

so that the total XPD can be computed from

$$XPD_p = XPD_{rain} - C_{ice} \qquad \text{dB} \qquad (9.24)$$

9.4   Site diversity

Although not widely utilised at present, the concept of site diversity is nevertheless of potential importance for Earth–space paths. It makes use of the fact that intense rain cells usually have quite limited horizontal dimensions, so that it is unlikely that two stations a few kilometres apart will both suffer severe attenuation simultaneously (see Figure 9.7). Then, by rerouting information from the station which is most severely attenuated to the other station, an improvement in performance can be achieved. Two formulations of the benefits to be gained by diversity operation are available.

Diversity improvement factor is the ratio of the single-site time percentage and the diversity time percentage at the same attenuation level. It is defined as

$$I = \frac{p_1}{p_2} = \frac{1}{(1+\beta^2)}\left(1 + \frac{100\beta^2}{p_1}\right) \approx 1 + \frac{100\beta^2}{p_1} \qquad (9.25)$$

where

$$\beta^2 = 10^{-4} d^{1.33} \qquad (9.26)$$

Typical values of p_2 as a function of p_1 and site separation are shown in Figure 9.8.

Diversity gain is the difference between the single-site and diversity attenuation values for the same time percentage. ITU Recommendation P.618 provides an empirical expression for diversity gain G:

$$G = G_d G_f G_\theta G_\Psi \qquad (9.27)$$

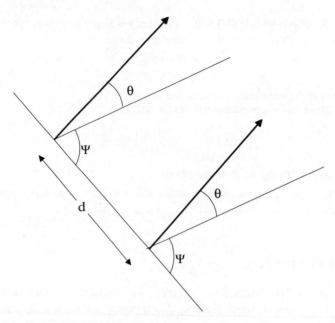

Figure 9.7   Geometry of site diversity

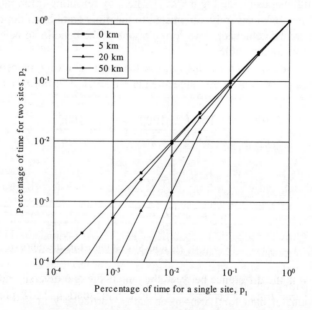

Figure 9.8   Relationship between percentages of time, with and without diversity, for the same attenuation

The spatial dependence G_d is given by

$$G_d = a\left(1 - e^{-bd}\right) \tag{9.28}$$

where d (km) is the separation between the two sites, and

$$a = 0.78A - 1.94\left(1 - e^{-0.11A}\right)$$
$$b = 0.59\left(1 - e^{-0.1A}\right) \tag{9.29}$$

Here, A (dB) is the single-site path attenuation due to rain. The dependence G_f on frequency f is given by

$$G_f = e^{-0.025f} \tag{9.30}$$

The dependence G_θ on elevation angle θ is given by

$$G_\theta = 1 + 0.006\theta \tag{9.31}$$

and the term G_ψ allows for propagation paths at any angle ψ to the baseline between sites:

$$G_\psi = 1 + 0.002\psi \tag{9.32}$$

9.5   Measurement of attenuation and other effects

9.5.1 Direct measurements

Improvements to the models described above will come from measurement programmes which provide further insight into the physics of precipitation systems and also data against which their accuracy can be evaluated. Changes to the ITU-R procedures described above will only be acceptable if they can demonstrate improved prediction capability when tested against the ITU-R data bank. The prime source of data comes from experimental links, which can provide accurate data but with the drawback that it is for only one specific frequency and path geometry.

A large amount of data for Earth–space paths, particularly in the 11–14 GHz part of the spectrum, was assembled during the European COST 205 project [44]. More recently, a major programme of measurements has just been completed, which made use of propagation beacons on the ESA *Olympus* satellite. Three frequencies were transmitted from this very sophisticated satellite: 12.5 GHz, 30 GHz and a polarisation-switched 20 GHz signal. The three frequencies allowed frequency-scaling ratios to be obtained, in particular 12.5/20 and 12.5/30, so that the very widespread database of existing Ku-band data could be extrapolated to the higher frequencies. In addition, the polarisation-switched 20 GHz beacon allowed the effects of differential attenuation and differential phase to be investigated. The satellite signals could be received over western Europe and the east coast of the USA, with elevation angles between 10° and 45°. A good selection of preliminary results can be found in [142].

Figure 9.9 shows data obtained from the 12.5 GHz beacon, recorded at the RAL site at Sparsholt, Hampshire, UK, for 30 November 1992. The ordinate scale is received power in units of 1/10th of a decibel. Several fades can be seen, with a particularly deep fade, exceeding 8 dB, just before 16.00 GMT. The cyclical variation in the 'clear sky' level is due to motion of the satellite with respect to the geostationary position, and requires careful analysis to remove it before the genuine propagation effects of precipitation can be deduced.

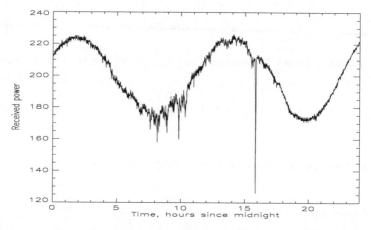

Figure 9.9   Signal strength at 12.5 GHz, recorded on 30th November 1992

Useful information can also be obtained from rain gauges located adjacent to the beacon receivers. Figure 9.10 shows data from a drop-counting rain gauge, obtained on the same day as the data in Figure 9.9. A peak rainfall rate of nearly 50 mm/h coincides with the deepest fade.

An example of statistics from a BT experiment at Martlesham Heath, also making use of the *Olympus* beacons, is shown in Figure 9.11. These data were collected over a 19 month period from November 1989 to May 1991, and show good agreement with predictions for rain zone E, based on ITU Recommendation P.618 (see Section 9.3).

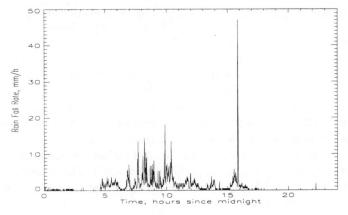

Figure 9.10   Rainfall rate recorded on 30th November 1992

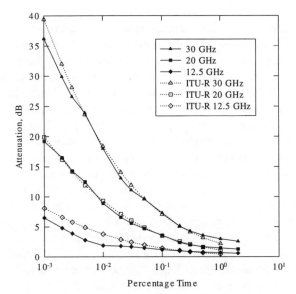

Figure 9.11   Cumulative attenuation statistics at 12.5, 19.8 and 29.7 GHz for Martlesham Heath, UK

9.5.2 Radiometers

Measurements of atmospheric emission noise temperature using passive radiometers may be used to infer total atmospheric attenuation, although they cannot measure it directly. The relationship between the atmospheric-emission noise-temperature measurement T_B and the total atmospheric attenuation α is given by

$$\alpha = 10\log\left\{\frac{T_A}{T_A - T_B}\right\} \tag{9.33}$$

where T_A is the effective temperature of the atmosphere. In practice, because thermodynamic equilibrium does not hold everywhere, and the medium is not a pure absorber, eqn. 9.33 does not hold exactly, and careful calibration procedures are required. These usually involve calibration against a cold (liquid-nitrogen-cooled) and heated load. However, radiometers provide a cheap and convenient method of obtaining propagation information.

9.5.3 Radars

Radar measurements of the backscattered power from hydrometeors can be used to infer attenuation and other propagation properties in the microwave part of the spectrum. As was the case for radiometer techniques, there are inherent uncertainties associated with this inference: drop-size-distribution variations cause significant ambiguity in the relation between reflectivity and attenuation, while in convective storms the phase of the hydrometeors may be uncertain. This latter point is important because, although ice and dry hail particles can scatter quite strongly, they cause little attenuation, as discussed in Chapter 7.

The use of switched-polarisation radars has been very successful in reducing errors in attenuation estimation due to both of these factors. The technique involves rapid polarisation switching between horizontally and vertically polarised transmission and reception, relying on the fact that raindrop shapes become increasingly oblate with increasing size. As a consequence, the difference in received power on the two polarisations can be used as a measure of the mean drop size. In addition, the phase of the hydrometeors also significantly affects the difference in received power (see Chapter 7), so that ice and hail can often be distinguished from water, even in convective storms. Once the precipitation has been characterised in this way, the radar measurements, usually made in the 3–10 GHz region, can be 'extrapolated' to assess attenuation and other propagation phenomena at much higher frequencies, up to at least 30 GHz.

Radars play two roles in the development of propagation models. First, detailed radar studies can increase our understanding of the microphysics of rain. Secondly, if data are collected over a significant period of time, a single databank can be used to produce statistics of, for example, attenuation, for communication systems operating at a variety of frequencies, and with a variety of geometries. Figure 9.12 shows a comparison between radar estimates, obtained using the Chilbolton 3 GHz radar, and beacon measurements at the three frequencies of the

Olympus satellite recorded simultaneously, but at a site 50 km distant from the radar. The radar estimates of path attenuation do not include the effects of water vapour absorption, which is particularly significant around 20 GHz.

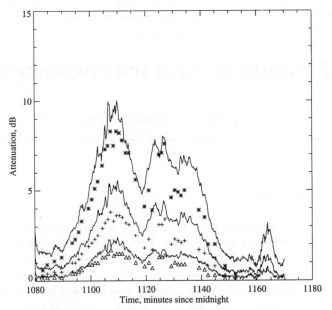

Figure 9.12   *Attenuation at 12.5, 20 and 30 GHz*
Beacon measurements at three frequencies
Radar

Δ 12.5 GHz
+ 20 GHz
* 30 GHz

Chapter 10

Principles of wideband propagation

P.A. Matthews and A.R. Nix

10.1 Introduction

To achieve successful radio communication it is necessary to understand the problems associated with a radio channel. Rather than receiving just a single copy of the transmitted signal, it is common to receive multiple versions, each with their own amplitude, phase, time delay and arrival angle. The summation of these multipath components can result in extreme envelope variations with fades as deep as 40 dB relative to the mean. In addition, for digital transmission the variation in time delay between incoming signals can seriously affect the performance of the system. This Chapter will investigate these difficulties and discuss their impact on typical transmissions.

Wideband propagation has different meanings to different people; in this Chapter the wideband characteristic depends on the relationship between the bandwidth of the transmitted signal and the frequency response of the radio channel. As mentioned above, it is possible for the signal to arrive at a receiver by more than one path; examples are found in HF, troposcatter, microwave-link and mobile-radio systems. At HF a signal may be reflected at different heights in the ionosphere resulting in a signal at the receiver which is the phasor summation of two or more components. Similarly, as shown in Figure 10.1, in a mobile-radio channel the signal may be shadowed by hills and large buildings or scattered, diffracted and reflected by buildings and objects near to the transmitter and receiver.

At the receiving antenna the signal is made up from the vector summation of the multipath components. As shown in eqn. 10.1, the phase of each signal will depend on the physical path length d between transmitter and receiver, the initial phase of transmission θ_0 and the radio wavelength λ:

$$\theta = \theta_0 + \frac{2\pi d}{\lambda} \qquad (10.1)$$

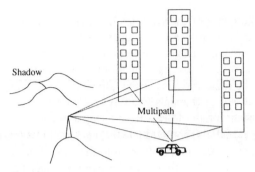

Figure 10.1   Multipath propagation

Equation 10.1 shows that the received phase of each signal varies as a function of frequency and path length. If the mobile is now displaced by a distance x, the kth multipath will be phase rotated by $2\pi x \cos(\alpha - \beta_k)/\lambda$ radians, where α represents the direction of motion and β_k the angle of arrival of the wave. Hence, assuming there is a spread in multipath arrival angles β_k, when the transmitter and/or receiver moves, the phase rotation of each multipath will result in variations in the received signal. If all the paths arrive with approximately equal amplitudes, then the resulting fading is known as Rayleigh and is shown below in Figure 10.2a. However, if a dominant component exists at the receiver, the fading is characterised by a Rician distribution with the K-factor being used to determine the ratio of the power in the dominant component $(A_c^2/2)$ to the power in the random multipath components (σ^2), i.e. $K = 10\log[A_c^2/(2\sigma^2)]$ dB.

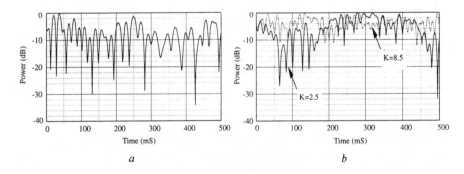

Figure 10.2   Rayleigh and Rician temporal variations
a Rayleigh
b Rician

Interestingly, for a narrowband channel the fading envelope is invariant of frequency, i.e. the magnitude of the channel is frequency flat. From eqn. 10.1 it can be seen that the phase of each multipath varies with λ; however, when their physical path lengths are equal, each multipath experiences a common phase shift. Hence, as the frequency is changed the resulting magnitude remains constant with the phase merely rotating. If the path length of each multipath is assumed equal, since the speed of the radio wave is constant, the time delay for each path will also be equal. For narrowband channels, the physical path length and time delay are assumed to be constant for all of the paths arriving at the receiver.

As frequency is varied, the phase rotation of each multipath is given by $2\pi\Delta f\tau$, where Δf represents the frequency offset from the carrier and τ the time delay of the multipath. For a spread in time delay, τ can no longer be assumed constant and the phase of each multipath will rotate uniquely with frequency. The resulting phasor summation produces *frequency-selective fading* with some frequencies experiencing signal peaks while others undergo signal fades. A typical multipath-propagation channel can therefore be represented mathematically by a complex frequency-transfer function which varies with frequency and with the relative positions of the transmitter, receiver and scattering obstacles.

The simplest multipath environment consists of just two paths with time delays of t_1 and t_2, respectively. For a carrier frequency f_c, the received signal can then be written as

$$s(t) = A_1 \cos\left(2\pi f_c t_1\right) + A_2 \cos\left(2\pi f_c t_2\right) \qquad (10.2)$$

The amplitude of the received signal is given by

$$|s(t)|^2 = A_1^2 + A_2^2 + 2A_1 A_2 \cos(2\pi f_c \tau) \qquad (10.3)$$

in which $\tau = t_2 - t_1$. The amplitude of the signal varies in a way which depends on the product $f_c\tau$. At HF the carrier frequency might be in the region of 10 MHz and the relative time delay might be 1 ms; however, at UHF the carrier frequency might be 1 GHz and the time delay 1 μs. The amplitude of the received signals as a function of frequency for signal amplitudes $A_1 = 1$ and $A_2 = 1.5$ are shown in Figure 10.3.

Clearly in the HF case ($\tau = 1$ ms) the dips in the amplitude response cover a narrower band of frequencies than in the UHF case ($\tau = 1$ μs). At HF the dips are less than 1 kHz wide but at UHF they are in the region of 50 MHz. The consequence is that, for the HF environment considered above, signal bandwidths greater than 200 Hz must be considered as wideband since there is a high probability of encountering frequency-selective fading. However, for the UHF scenario a signal bandwidth up to 160–170 kHz can be considered narrowband and does not suffer from frequency-selective fading unless one of the frequency notches falls directly into the operating band of the system. The actual bandwidth that can be considered narrowband is governed by the time delay τ between the rays.

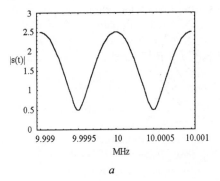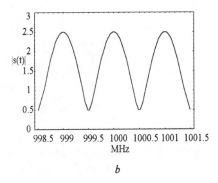

a *b*

Figure 10.3 Two-path channel-frequency/amplitude response
    a Two-path response at HF
    b Two-path response at UHF

10.2 Multipath propagation

When a radio signal propagates from a transmitting antenna to a receiver it can often travel by more than one path. For example, on a microwave link the signal may be reflected from the ground as well as travel on a direct line to the transmitter. Alternatively, it may travel by more than one path due to refraction effects in the atmosphere. The signals on a troposcatter link are scattered form many points in the atmosphere. At HF, signals propagating by ionospheric reflection may travel by two or more paths. In mobile-radio systems the signals are received after reflection, transmission, diffraction and scattering in the local environment and the total signal at the receiver is the phasor summation of the many components. In satellite communication to ships, reflection from the sea may cause problems.

In all these cases the electrical path length of the separate components will be different. If the path lengths change (see Section 10.1) then the overall phasor addition will vary. Since the received signal is made up from the phasor summation of individual multipaths, it follows that the amplitude and phase of this signal will change as a function of position and/or time. This can be illustrated diagrammatically as shown in Figure 10.4. This diagram shows a two-dimensional distribution of scatterers lying around the path from transmitter to receiver. The path between transmitter and receiver with the smallest time delay is the direct path. Paths by way of particular scatterers will have longer time delays depending on the positions of the scatterers. The locus of points with a particular time delay lie on an ellipse with foci at the transmitter and receiver.

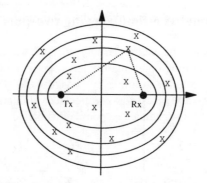

Figure 10.4   Ellipses of constant time delay around the line-of-sight from transmitter to receiver

The area can be divided by confocal ellipses into regions each of which contain scatterers giving a certain range of time delays (Figure 10.4). The signal at the receiver for a particular range of time delays will depend on the phasor sum of several components. If a short pulse is transmitted, the signal at the receiver will be an extended pulse, the spread in time depending on the positions and strengths of the scatterers. If the transmitter, receiver and all the scatterers are stationary in position, the time-delay spread of the signal will not vary with time. However, if there is any movement the time-delay spread function will vary with time as the various path lengths change. The effect on the time-delay-spread function is that it appears to change its form as time passes. At a particular time delay, the signal amplitude bubbles up and down as the phasor sum of the signal components with this time delay go in and out of phase with each other. Also, as there is movement of the transmitter, receiver and scatterers, there will be Doppler shifts on every signal component at the receiver.

The Doppler shift will depend on the position of the scatterers relative to the transmitter and receiver. Some scatterers may give an increase in frequency, some a decrease, so that at any particular time delay there will be both positive and negative Doppler shifts. The result is that the original transmitted pulse is spread in both time and frequency. We can expect that there will be inter-relationships between the time-delay spread or impulse response of the radio channel and the frequency response and also between the response at different time delays and at different points in the frequency spectrum.

If we are to describe the multipath channel we have to be able to describe this complicated pattern of the combination of signals. We want to be able to describe the channel mathematically so that we can simulate the channel when we are designing and assessing the performance of systems.

10.3   The radio channel as a time-varying two-port network

The analysis of the response of a time-varying channel has been presented by several authors. A most thorough presentation has been given by Bello [56], and re-presented by other authors, for example by Proakis [149] and Parsons [145]. The analysis for the time-varying case is complicated. For many problems the assumption that the short-term channel response is static, or quasistationary, is often made.

10.3.1 Response of a time-invariant two-port network

We can make the definition of wideband more precise by considering what happens to a signal transmitted through a two-terminal-pair transmission channel, Figure 10.5. The two terminal pair network has a frequency transfer function $H(\omega)$ and an impulse response $h(t)$. The input and output signals are related in the frequency and time domain by the multiplication and convolution

$$Y(\omega) = X(\omega)H(\omega) \qquad y(t) = x(t) \otimes h(t) \tag{10.4}$$

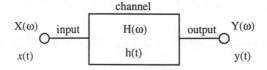

Figure 10.5   Two-terminal-pair network

The frequency-transfer function and the impulse response are related by the Fourier-transform pair

$$H(\omega) = \int_{-\infty}^{\infty} h(t)\exp(-j\omega t)\,dt \qquad h(t) = \frac{1}{2\pi}\int_{-\infty}^{\infty} H(\omega)\exp(j\omega t)\,d\omega \tag{10.5}$$

For our purpose the terminal pairs can be taken to be the input port to a transmitting antenna and the output port of a receiving antenna. The response of the network may vary with time and, in general, it will be time-varying when the transmission channel is a radio link.

The response of such a multipath channel can be expressed analytically by setting up expressions for the correlations between the various factors and between the correlations and the spectral response. The radio channel is viewed as a time-varying two-port network. Following Proakis [149], suppose that the transmitted signal is represented by

$$s(t) = \text{Re}\{u(t)\exp(j2\pi f_c t)\} \tag{10.6}$$

Along each of the multiple propagation paths there will be a different delay and a different attenuation which, in general, is time varying. The received signal can be written

$$y(t) = \sum_n \alpha_n(t)s\{t - \tau_n(t)\} \tag{10.7}$$

in which $\alpha_n(t)$ is the attenuation and $\tau_n(t)$ the propagation delay of the signal arriving through the nth path. Substituting for $s(t)$ gives

$$y(t) = \text{Re}\sum_n \alpha_n(t)\exp\{-j2\pi f_c \tau_n(t)\}\, u\, \{t - \tau_n(t)\}\, \exp(j2\pi f_c t) \tag{10.8}$$

The equivalent lowpass signal $r(t)$ is given by

$$r(t) = \sum_n \alpha_n(t)\exp\{-j2\pi f_c \tau_n(t)\}\, u\, \{t - \tau_n(t)\} \tag{10.9}$$

The time-varying impulse response of this equivalent lowpass channel is therefore given by

$$h(\tau,t) = \sum_n \alpha_n(t)\exp\{-j2\pi f_c \tau_n(t)\}\, \delta\, \{t - \tau_n(t)\} \tag{10.10}$$

For a sinusoidal carrier at frequency f_c the corresponding lowpass signal is $u(t) = 1$. Assuming this input, the resulting received signal is

$$r(t) = \sum_n \alpha_n(t)\exp\{-j2\pi f_c \tau_n(t)\} = \sum_n \alpha_n(t)\exp\{-j\theta_n(t)\} \tag{10.11}$$

where $\theta_n(t)$ is given by $2\pi f_c \tau_n(t)$ and represents the phase difference due to a time delay τ_n and a carrier frequency f_c.

The received signal is the phasor sum of the multipath components. Each component differs in amplitude and phase. Usually the amplitude changes relatively slowly with space or time compared with the changes in phase. This is because the phase change depends on spatial variations at the carrier-frequency wavelength.

When the number of multipath components is large, the phases of the different components at the receiver will be random and follow a uniform distribution between 0 and 2π. If the receiver, transmitter or scatterers move, the phasor summation will change. The probability distribution of the phasor sum is that of a large number of randomly distributed components. By the central limit theorem the received signal $r(t)$ is then expected to be a zero-mean complex-valued Gaussian random process. The phase distribution is uniform over the range 0–2π. The envelope distribution is a Rayleigh distribution.

If there is a constant component in the received signal, for example a line-of-sight component, as well as randomly distributed components then the envelope will have a Rician distribution.

10.4   Mathematical description of the wideband channel

The statistical properties of a multipath channel can be characterised by a number of correlation functions and power density spectra.

10.4.1   Uncorrelated scattering

If we assume that the equivalent lowpass impulse response of the channel, $h(\tau;t)$, is wide-sense stationary, then the autocorrelation function of this response is

$$\phi_c(\tau_1, \tau_2, \Delta t) = 1/2\, E\{ h^*(\tau_1;t)h(\tau_2;t + \Delta t)\} \qquad (10.12)$$

This function determines the correlation in the channel impulse response at delays τ_1 and τ_2 with the latter value being derived at a time shift of Δt seconds. When the attenuation and phase shift on the path with delay τ_1 are uncorrelated with those on the path with delay τ_2 the scattering process for the two paths is said to be uncorrelated.

10.4.2   Multipath delay profile and RMS delay spread

In a multipath channel, signals arrive with various powers and time delays. For a channel with uncorrelated scattering, eqn. 10.12 is reduced to

$$1/2\, E\{ h^*(\tau_1;t)h(\tau_2;t + \Delta t)\} = \phi_c(\tau_1; \Delta t)\delta(\tau_1 - \tau_2) \qquad (10.13)$$

where the right-hand side is zero unless $\tau_1 - \tau_2 = 0$. When $\Delta t = 0$ the autocorrelation function $\phi_c(\tau, 0) = \phi_c(\tau)$. This parameter is often referred to as the average-power-delay profile and expresses received signal power relative to time delay τ.

This function is a measure of the time-dispersive measure of the channel and can be obtained by transmitting a wideband signal through the channel. The wideband signal may be a narrow pulse in the time domain, a swept-frequency signal or another type of signal which has a narrow, ideally an impulse, autocorrelation function.

Figure 10.6 shows two typical power-delay profiles. The first represents a large outdoor cell with time delays up to 10 μs (normalised to the first multipath component). The second profile was generated using a ray-tracing computer-modelling technique (see Section 10.6.4) and illustrates the expected profile for line-of-sight reception in a 30 m by 30 m room. It is common to process these profiles mathematically to generate a measure for the signal spreading or variance around the mean arrival time. The most commonly used parameter is the root-mean-square (RMS) delay spread T_{RMS} (see Section 10.4.3).

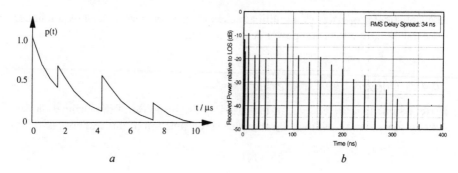

Figure 10.6   *Typical delay profiles for outdoor and indoor radio channels*
   a Outdoor
   b Indoor (simulated)

10.4.3 Definition of delay spread

A definition of the delay spread is needed because it is clear from Figure 10.6*a* that the power-delay profile may have several peaks and that the profile takes a long time to finally fall to a low value. It is usual to use the second central moment of the profile as a measure of the delay spread. However, when the delay spread is to be evaluated from practical measured profiles, noise in the measuring system must be taken into account. A definition of delay spread has been given by Lorenz [125]. A typical profile is shown in Figure 10.7.

In Figure 10.7 the following definitions are assumed:

μ_{nf} = noise floor
μ_p = peak of delay profile
a_n = lowest level evaluated, 3 dB above the noise floor

t_m = span of delay profile $(t_1 - t_0)$
t_0 = time of arrival at which the signal first rises 3 dB above the noise floor
t_1 = time of arrival at which the signal last falls 3 dB above the noise floor
t_p = time of arrival at which the signal is a maximum.

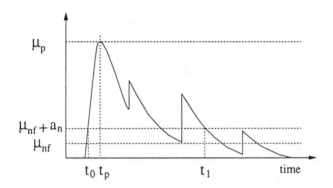

Figure 10.7   Calculating RMS delay spread from the power-delay profile

For a sampled version of the power-delay profile, the average delay τ_a and the RMS delay spread can be calculated as below:

$$\tau_a = \frac{\displaystyle\sum_{k=1}^{L} \tau_k P_k}{\displaystyle\sum_{k=1}^{L} P_k} \qquad T_{RMS} = \sqrt{\frac{\displaystyle\sum_{k=1}^{L} (\tau_k - \tau_a)^2 P_k}{\displaystyle\sum_{k=1}^{L} P_k}} \qquad (10.14)$$

For a continuous delay profile the RMS delay spread can be calculated using eqn. 10.15.

$$\tau_a = \frac{\displaystyle\int_{\tau=t_0}^{\tau=t_1} \tau p(\tau)d\tau}{\displaystyle\int_{\tau=t_0}^{\tau=t_1} p(\tau)d\tau} \qquad T_{RMS} = \sqrt{\frac{\displaystyle\int_{\tau=t_0}^{\tau=t_1} (\tau - \tau_a)^2 \, p(\tau)d\tau}{\displaystyle\int_{\tau=t_0}^{\tau=t_1} p(\tau)d\tau}} \qquad (10.15)$$

For a simple equal-amplitude two-spike delay profile, eqn. 10.14 can be applied to obtain the following result

$p_1=1$ $p_2=1$

$\tau_a = \dfrac{\tau}{2}$ $\sigma_{RMS} = \sqrt{\dfrac{\left(-\tau/2\right)^2 + \left(\tau - \tau/2\right)^2}{2}} = \dfrac{\tau}{2}$

time

0   τ

It follows for the above profile that the resulting RMS-delay spread will be $\tau/2$ s with τ representing the relative time delay between the signals.

While the RMS-delay spread is a useful measure for wideband channels, since the calculation avoids the use of phase information the resulting value can only be used as an indicator of average system performance. It is also possible for channels with the same RMS-delay spread to have different power-delay profiles. In practice, the shape of the delay profile and the presence of a line-of-sight component are also strong factors in determining wideband system performance.

10.4.4 Spaced-frequency spaced-time correlation function

The wideband channel can be characterised by its frequency response as well as its time response. The time-varying equivalent lowpass frequency-transfer function can be obtained from the Fourier transform of the equivalent lowpass impulse response:

$$H(f,t) = \int_{-\infty}^{\infty} h(\tau,t)\exp\left(- j2\pi f_c t\right)d\tau \tag{10.16}$$

When the channel impulse response is a complex-valued zero-mean Gaussian random process the frequency-transfer function is a similar process. For a wide-sense-stationary channel the autocorrelation function of the frequency response at frequencies f_1 and f_2, the latter shifted by Δt seconds is given by

$$\phi_c(f_1,f_2,\Delta t) = 1/2\,E\!\left(H^*(f_1;t)H(f_2;t+\Delta t)\right) = \phi_c(\Delta f;\Delta t) \tag{10.17}$$

The above equation is the frequency-domain equivalent of eqn. 10.12. This autocorrelation function depends only on the difference in frequency Δf and the difference in time Δt and hence it may be referred to as the *spaced-frequency spaced-time correlation function* of the channel. It can be obtained as the Fourier transform of the multipath power-delay profile.

10.4.5 Coherence or correlation bandwidth

If $\Delta t=0$ then the frequency and time-domain correlations may be simplified as follows: $\phi_c(\Delta f;0) = \phi_c(\Delta f)$ and $\phi_c(\tau;0) = \phi_c(\tau)$. The Fourier transform then becomes

$$\phi_c(\Delta f) = \int_{-\infty}^{\infty} \phi_c(\tau)\exp\left(-j2\pi\Delta f\tau\right)d\tau \qquad (10.18)$$

The above autocorrelation function is a measure of the frequency coherence in the channel, i.e. it determines the similarity of the channel's output over a frequency range of Δf Hz. If significant amplitude and/or nonlinear phase variation is observed in the transmit band, the received signal can become seriously distorted (see Section 10.5). In general, as the frequency separation is increased, the degree of correlation will reduce. For a given value of correlation, the maximum bandwidth that can be achieved is known as the *coherence* or *correlation bandwidth* Δf_c. The coherence bandwidth is often quoted for correlations of 0.5 and 0.9, the latter being more representative for modern digital-communication systems.

It is possible to estimate the coherence bandwidth from the RMS delay spread of the power-delay profile using a simple two-ray construction as shown in Figure 10.8. If we assume two unit-amplitude rays R_1 and R_2 delayed by Δt seconds and separated by Δf Hz, the phase difference between these rays can be calculated from eqn. 10.11:

$$\theta_1 = 2\pi f_1 t_1 \quad \theta_2 = 2\pi f_2 t_2 \quad \text{hence} \quad \Delta\theta = \theta_2 - \theta_1 = 2\pi\Delta f\Delta t \qquad (10.19)$$

Figure 10.8 shows graphically the phasor summation of these two rays at a frequency separation of Δf Hz and a relative time delay of τ seconds. When $\Delta f=0$ and $\tau=0$ the phase difference between the two rays is zero and the resultant amplitude will have a value of two. If we define the coherence bandwidth as the frequency separation at which the resultant envelope falls to a value of 0.5 times its initial amplitude, from Figure 10.8 it can be seen that $\cos\{(\Delta\theta)/2\} = 0.5$, i.e. $\Delta\theta = 2\pi/3$ radians (at this angle the resulting magnitude is one as opposed to its initial value of two).

From eqn. 10.19 it now follows that

$$\Delta\theta = 2\pi\Delta f\Delta t \quad \text{hence} \quad 2\pi/3 = 2\pi\Delta f_c\Delta t \quad \text{therefore} \quad \Delta f_c = 1/(3\Delta t) \qquad (10.20)$$

Hence, for a simple two-ray construction the coherence bandwidth can be estimated using $1/(3\Delta t)$, where Δt represents the time delay between the two rays. As was shown in Section 10.4.3, for such a model the RMS delay spread T_{RMS} is given by $\Delta t/2 = \tau/2$. Using this expression, the coherence bandwidth can now be approximated using $1/(6T_{RMS})$.

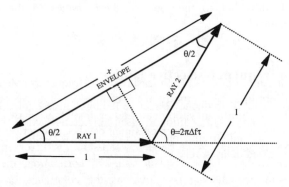

Figure 10.8   Estimating coherence bandwidth using simple two-ray phase summation

Table 10.1   Typical values of RMS-delay spread and coherence bandwidth (UHF)

Environment	Approximate RMS delay spread	Approximate coherence bandwidth
	ns	kHz
Indoor	100	1670
Microcell (<500 m)	500	334
Small cell (<2 km)	1000	83
Large cell (>2 km)	4000	21

If a modulated carrier signal is transmitted through the channel, the signal will suffer little distortion if the signal bandwidth is less than the coherence bandwidth. If the signal bandwidth is greater than the coherence bandwidth, the signal will be distorted by the frequency-selective response of the channel. For large-cell systems, data rates are therefore restricted to around 20 kbit/s unless some form of equalisation is applied. However, for microcellular and indoor systems, data rates as high as 334 kbit/s and 1.67 Mbit/s are possible before wideband distortion causes serious problems.

The fading encountered in a channel is described as narrowband or wideband depending on both the signal bandwidth and the particular channel being investigated. For example, for a 200 kHz transmission in a large-cell environment (such as GSM) the fading would clearly be of the wideband variety. This arises since the spread in time delay is far greater than the effective bit period of the system. However, for a 1.5 MHz indoor transmission (such as DECT), the relatively low values of expected delay spread allow the channel to be characterised as narrowband in all but the most hostile environments. It is therefore obvious that the terms narrowband and wideband do not apply to transmission bandwidths but to the ratio of these values to the coherence bandwidth of the channel.

10.5 Impact of frequency-selective fading

In this Section the impact of frequency-selective fading is investigated for typical digital-transmission schemes. In addition to discussing areas such as intersymbol interference and wideband irreducible bit-error rate, it will introduce the importance of normalised delay spread.

10.5.1 Intersymbol interference

It was shown in Section 10.4 that there is a limit to the maximum bandwidth that can be supported in a time dispersive channel. Figure 10.9 shows the resulting frequency-domain variations for the power-delay profile given in Figure 10.6*b*. From this plot the frequency-selective nature of the channel can be clearly observed with deep notches being seen with widths of approximately 5–7 MHz. The actual bandwidth which can be supported in this channel depends on the frequency chosen for transmission. If the transmission is centred on a null in the frequency domain, the resulting distortion will seriously limit the bandwidth of the signal. Alternatively, if the signal is placed at the midpoint between frequency nulls, a far higher bandwidth (and therefore data rate) can be achieved. Unfortunately, as the user moves through the standing-wave pattern the position of these nulls tend to vary in sympathy with the phase of the received rays. This means that at some point nulls will occur in the centre of the band and therefore introduce the possibility of error bursts in the received data.

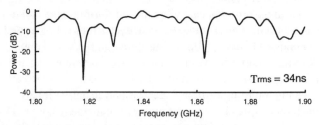

Figure 10.9   Wideband frequency-selective fading

For a digital transmission the introduction of frequency-selective fading can result in intersymbol interference (ISI), i.e. the spreading of a symbol into its adjacent symbol slots. The generation of ISI is shown graphically in Figure 10.10 for a simple three-path-channel model (for simplicity a baseband model is assumed).

The received signal prior to detection is formed by the phasor sum of many time-delayed versions of the transmitted data stream. For a narrowband channel, the spread in time delay is very much smaller than the bit period of transmission. However, when the delay spread becomes an appreciable fraction of the bit period it becomes possible for error bursts to be generated during signal fades. These error bursts are known as *irreducible* since it is not possible to reduce their value by increasing the transmitter power. It is common for the RMS delay spread to be normalised relative to the bit rate of transmission, and a system is generally regarded as wideband when this value d becomes significantly larger than 0.01. As a rule of thumb, the average irreducible bit-error rate is around 1 in 1000 for normalised delay spreads in the region of 0.1.

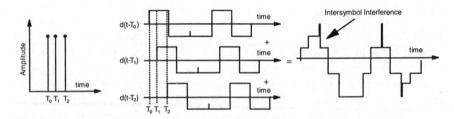

Figure 10.10   Delay spread and the introduction of ISI

Figure 10.11 shows the typical impact of frequency-selective fading on a digital-modulation scheme. In a narrowband channel, the error rate reduces as the average received energy per bit increases. However, for a wideband channel an irreducible error floor can arise whose value is directly dependent on the value of the normalised delay spread. For low values of delay spread (d=0.074 in Figure 10.11) the average irreducible error floor has a value of around 2e-4. For many applications this error floor may seem negligible, although it should be remembered that these errors occur in bursts and during these bursts the average error is extremely high. For example, in systems such as CT2 (second generation cordless telephone) and DECT (Digital European Cordless Telephone) these error bursts can result in short irreducible outages that correspond to the position of narrowband fades. As the value of normalised-delay-spread increases, irreducible error bursts become more frequent and are less correlated with narrowband fades. For normalised delay spreads of 0.6 or higher, irreducible errors tend to occur at all fading levels, and for reliable reception to be maintained some form of adaptive equalisation becomes necessary.

It is now apparent that a system is described as narrowband or wideband depending on the channel response and the data rate or bandwidth being used for transmission. For a 273 kbit/s transmission in an outdoor environment, i.e. T_{RMS} = 4 μs, the fading would be described as wideband since the normalised delay spread is much greater than 0.1 (d = 4 μs × 273 kbit/s = 1.09). However, for a

1.1 Mbit/s indoor transmission, for T_{RMS} up to 100 ns, the relatively low value of time-delay spread allows the fading to be considered narrowband ($d = 100\,\text{ns} \times 1.1\,\text{Mbit/s} = 0.11$).

In practice, various techniques exist for improving the delay-spread tolerance of a system. One simple technique is to use directional antennas effectively to reduce the time-delay spread. However, when no line-of-sight path exists, the positioning of such antennas may prove difficult to determine. An alternative technique is to employ antenna diversity to increase the delay-spread tolerance of the system while maximising the received signal strength. For systems such as CT2 and DECT, diversity can be added at the basestation rather than the handset since both systems use a common transmit and receive frequency. This technique is widely used in DECT basestations to improve the delay-spread performance in larger cells.

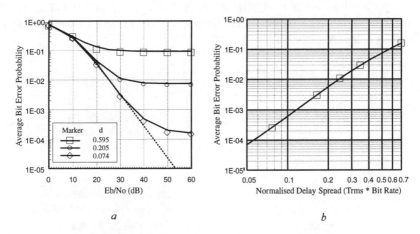

a   *b*

Figure 10.11   *Bit-error-rate performance in a wideband channel*
    a Typical BER performance (E_b/N_0)
    b Typical BER performance (irreducible)

10.5.2 Wideband-envelope variations

While for narrowband transmissions the received signal statistics clearly follow either a Rayleigh or a Rician distribution, for wideband signals this condition no longer holds. If the signal bandwidth is narrow with respect to the frequency notches introduced by delay spread then the resulting amplitude and phase distortion will take values corresponding to the block of frequencies it currently occupies. For a wideband signal, part of the transmission may be suffering a fade while the remaining frequencies experience a signal peak. The resulting amplitude and phase distortion in the time domain then becomes an average of the values experienced in the frequency domain.

Figure 10.12 shows the resulting wideband averaging for various signal bandwidths in a channel with an RMS delay spread of 34 ns. It follows that as the bandwidth expands, the severity of the amplitude and phase variations reduce owing to increased averaging (frequency diversity). For a 500 kHz transmission ($BT_{RMS} = 0.017$), the fading appears relatively narrow and produced statistics which approximate to a Rician distribution ($K = 3$ dB). The actual distribution falls away from Rician for deep carrier fades where a lower probability of occurrence was observed. This can be attributed to the fact that, whereas a single tone can fall directly into a narrow frequency notch, even with a 500 kHz-wide spectrum, significant averaging will occur. As the bandwidth increases, the channel appears to suffer less and less from signal variation, indeed, assuming a given probability, for the 20 MHz scenario ($BT_{RMS} = 0.68$) the fading can be reduced by as much as 10 dB. These results become important when making the choice of reception technique.

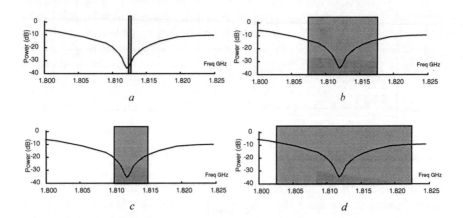

Figure 10.12   Wideband averaging in typical fading channel
   a $1/(BT_{RMS}) = 60$
   b $1/(BT_{RMS}) = 3$
   c $1/(BT_{RMS}) = 6$
   d $1/(BT_{RMS}) = 1.5$

For coherent demodulation to be effective, the channel used for transmission must be frequency flat. If this is not the case, the fading which effects the carrier tone is no longer representative of the entire bandwidth. This can result in a situation where it becomes impossible to remove the amplitude and phase effects of the fading channel, and alternative techniques, such as adaptive equalisation, may be required to receive the transmission correctly.

10.6   Channel models

Channel models may be based on the physical nature of the channel, on measurements and on mathematical convenience. Because most radio channels are continuously varying with time and position, it is difficult to set up models unless simplifications and idealisations are made. In this Section, models for some different types of channel will be considered to illustrate some typical conditions.

10.6.1 Two-path model

The simplest case of multipath propagation is the case in which there are only two paths. This case was assumed in Figure 10.3 to illustrate the frequency/amplitude response for typical time delays found at HF and UHF. It should also be remembered that the phase response of the channel can also very significantly with time; in the case of an equal-amplitude two-ray model, the phase response can flip by π radians during a signal fade (this phase flip will produce irreducible errors when digital-modulation schemes are used). Because of this rapid phase change a two-path channel can severely distort both narrowband and wideband signals. This simple model provides some feeling for the magnitudes of the quantities involved, although should be used with caution in a system simulation.

10.6.2 Mobile-radio model

A mathematical description of propagation in mobile-radio systems has been presented in [70,89]. The wideband characterisation of the mobile-radio channel has been described in [173] and [145]. The latter paper describes the tapped-delay-line model for the channel. It also illustrates the relations between the various functions describing the channel. These are the time/time-delay, the frequency/time, the frequency/Doppler and the delay/Doppler functions. These functions are related through Fourier and inverse Fourier transforms. Each of these functions describes a random process. The nature of each random process can be described by taking the autocorrelation function of the process. Because of the symmetrical relationship of the transforms, the channel can be characterised by measurements in either the time domain or the frequency domain making complex measurements of the time/time-delay function or the frequency/time function. Both methods of measurement have been used but measurements in the time/time-delay domain have been most common.

Measurements have shown that the channel can be modelled by exponentially decaying functions for the channel's power-delay profile. From measurements, models for the channel can be set up [101]. Typically, the channel response has the form shown in Figure 10.6a. Each exponential represents the scattered signal from a particular area. Physically, this corresponds to scattering from particular buildings or features in the terrain. Each exponential is the result of the phasor combination of a number of signal components. To model the time variations, a sequence of profiles must be generated varying one to the next with the appropriate correlation.

10.6.3 GSM models

As part of the research carried out in preparation of the specification of the GSM mobile-radio system, numerous measurements were made throughout Europe to characterise the mobile-radio channel. From these measurements, a number of empirical models for the time-delay spread to be expected in different situations were set up. These models are of use in the simulation testing of the GSM system and in hardware simulators for testing equipment. These models were set up by the collective work carried out under COST 207 [45].

Propagation models were formulated for rural, nonhilly areas, urban nonhilly areas, hilly urban areas, and hilly terrain. As the bandwidth of a GSM channel is about 250 kHz for nonhilly areas, the channel may be regarded as being narrowband; however, in hilly areas the channel must be considered wideband. One consequence of this is that adaptive equalisers must be used in a GSM receiver if the system is to operate under all scenarios.

For simulation purposes, a digital representation of the continuous power-delay profiles is required; see Figure 10.6*b*. Four standard GSM models have been defined for testing and simulation purposes and these are shown in eqns. 10.21–10.24.

$$\text{rural} \qquad p(\tau) = \begin{cases} \exp(-9.2\tau) & 0 \ \mu s < \tau < 0.7 \ \mu s \\ 0 & \text{elsewhere} \end{cases} \qquad (10.21)$$

$$\text{urban} \qquad p(\tau) = \begin{cases} \exp(-\tau) & 0 \ \mu s < \tau < 7 \ \mu s \\ 0 & \text{elsewhere} \end{cases} \qquad (10.22)$$

$$\text{hilly urban} \quad p(\tau) = \begin{cases} \exp(-\tau) & 0 \ \mu s < \tau < 5 \ \mu s \\ 0.5\exp(5-\tau) & 5 \ \mu s < \tau < 10 \ \mu s \\ 0 & \text{elsewhere} \end{cases} \qquad (10.23)$$

$$\text{hilly terrain} \quad p(\tau) = \begin{cases} \exp(-3.5\tau) & 0 \ \mu s < \tau < 2 \ \mu s \\ 0.1\exp(15-\tau) & 15 \ \mu s < \tau < 20 \ \mu s \\ 0 & \text{elsewhere} \end{cases} \qquad (10.24)$$

The frequency response of the channel can be found by taking the Fourier transform of the complex amplitude-delay profile. The power-delay profile gives the amplitudes of the components but not the relative phases. In practice, the phase of each component varies rapidly with position. To illustrate the effect of changes in phase on the frequency spectrum, two cases have been assumed. In the first case all the components in the power-delay profile are in phase; in the second the phase of the major component have been rotated by π radians.

Figure 10.13 shows that the form of the frequency response depends on the relative phases of the multipath components. As a mobile moves through the local environment, all the phases change relative to each other and so the frequency response of the channel will continuously vary with frequency notches passing through the received signal (it is this phase dependency that results in the bursty nature of the wideband channel).

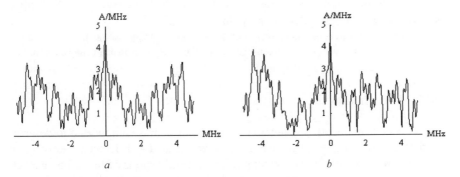

Figure 10.13   Frequency response with varying phase
   a Frequency domain (all rays in phase)
   b Frequency domain (π phase shifts)

10.6.4 Ray-traced channel models

Recently there has been a trend towards the use of ray tracing to computer generate the required channel models for use in simulation [121]. Ray tracing is a technique which analytically determines the multiple paths by which signals travel from transmitter to receiver. Figure 10.14 shows the type of output that can be produced for an indoor environment.

The computer calculates the amplitude, time delay, phase and arrival angle for each of the determined multipaths and, as confirmed by eqn. 10.10, this information can be used to construct a suitable representation of the radio channel. Indeed, this technique has the advantage of producing departure- and arrival-angle information at the transmitter and receiver. This information is particularly useful if antenna patterns or adaptive antennas are to be studied in the simulations. It is also possible to process the channel model to determine the mean received signal level, the RMS delay spread and the statistics of the fading envelope. If necessary, results can be calculated over a predetermined area to allow the statistics for received signal level and delay spread to be determined.

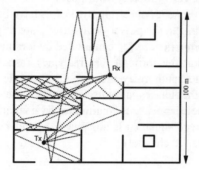

Figure 10.14   Typical 'ray-traced' output
For each ray: amplitude; time delay; arrival angle and arrival phase
For each location: mean signal level; RMS delay spread and fading distribution

10.7   Conclusions

The effects of multipath propagation on a radio system is complicated to describe mathematically. Different models have to be set up for different types of radio system. Most models do not attempt to describe all aspects of the radio channel; they concentrate on the parameters which can be measured and which are appropriate to the types of signal to be transmitted through the channel.

In this Chapter we have seen that there are two main mechanisms that lead to the generation of an irreducible error floor. First, for the narrowband scenario, the amplitude and phase variations associated with the fading channel produce an error rate which is directly proportional to the user's speed and the frequency of transmission. As the data rate is increased, the problems due to Doppler shift reduce, although these become replaced by intersymbol interference resulting from time dispersion in the channel. The data rate at which a system can be considered wideband has been shown to depend on the particular environment being considered. For example, GSM is regarded as a wideband system and operates with a data rate around 270 kbit/s. However, the DECT system is narrowband and transmits with a gross data rate of approximately 1.1 Mbit/s. The major difference between these two systems is the typical RMS-delay spread expected in their operating environments (DECT and GSM operate with delay spreads as high as 100 ns and 4 μs, respectively). If we then multiply these delay spreads by their system data rate, we obtain values of 0.11 and 1.08. Generally, for a normalised delay spread less than 0.1, a system can be considered narrowband and the majority of the irreducible errors will arise as a result of the time variations in the channel. Once this normalised value exceeds 0.1, the channel will start to exhibit strong wideband characteristics and some form of equalisation may be required to maintain a low overall error rate.

10.8 Acknowledgments

The authors are grateful for the many comments and suggestions received during the preparation of this chapter, in particular Dr Andrew Nix would like to thank Mr Yuk Chow for his help and advice relating to the mathematical aspects of this work.

Chapter 11

Wideband on terrestrial trunk routes and Earth–space paths

G.D. Richman

11.1 Introduction

This Chapter describes the important aspects of wideband radio systems on terrestrial trunk routes and Earth–space paths. It begins by reviewing some of the important propagation effects which influence the performance of wideband digital-radio systems, reports on modelling techniques and describes performance-prediction methods.

The reader is introduced to the evaluation of radio-link budgets and is provided with an understanding of methods which can be used to measure bit-error ratios and signatures of radio systems. Finally, a process to estimate the performance of a radio link is described.

11.2 Propagation characteristics

11.2.1 General

Although this Chapter is not intended to deal with fundamental propagation aspects, it is appropriate to highlight some of the propagation problems encountered in digital systems, and to define two of the terms in use.

It is assumed that the basic link engineering has been carried out according to the principles outlined in Chapter 8. This will ensure that the link will not incur excessive nonselective fading owing to inadequate ground clearance producing excessive diffraction loss. This criterion applies equally to narrowband and wideband systems.

Chapter 10 addresses the basic principles of fading in a wideband channel, giving an introduction to multipath fading and its characterisation. A simple channel model, based on two paths between the transmit and receive antennas, is described. Chapter 11 develops this model and discusses the effect of multipath fading on wideband digital radio systems.

11.2.2 Single-frequency fading

11.2.2.1 General

Various equations are given in the ITU-R Recommendations to estimate single-frequency fading on terrestrial radio paths, taking into account geoclimatic factors. This Chapter extends this to wideband systems, and considers two-path fading.

The phase of the resultant vector in a two-path-fading situation, will vary considerably across the frequency span of a digital channel. Since group delay is proportional to rate of change of phase, there could be outage caused under selective fading conditions owing to group-delay distortion. The sense of the group delay will be dependent on the nature of the fading, and the terms minimum-phase and nonminimum-phase fading are used in this context. The explanation of these two terms is as follows.

11.2.2.2 Minimum-phase fading

Consider the situation in which we have two-path propagation with the amplitude of the signal arriving via the direct path being unity, and that via the indirect path being, say, 0.9. Figure 11.1 shows the relevant vector diagram.

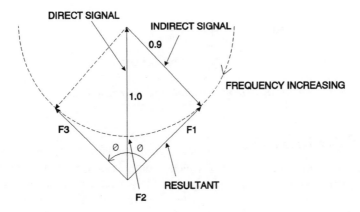

Figure 11.1   Vector diagram for a minimum-phase fade

Since the phase angle between the resultant and the reference (direct) signal is dependent on the relative delay between the direct and indirect paths expressed in wavelengths, the locus of the resultant can be scaled in terms of frequency. If we examine this locus and move round it in a clockwise direction, the maximum phase shift is at frequency F_1, following which it rapidly decreases to zero at F_2, before increasing to the maximum in the opposite direction at F_3, as shown by the arrow. Thus the phase excursion is between limits of $\pm R$, and the condition is referred to as being a minimum-phase fade.

11.2.2.3 Nonminimum-phase fading

If, as shown in Figure 11.2, the relative amplitudes of the direct and indirect signals is reversed, then between F_1 and F_3 the phase rapidly increases as far as F_2 and then decreases. The group delay produced is in the opposite sense to that of the minimum-phase situation and this condition is known as a nonminimum-phase fade.

Although in the above explanation the signals were referred to as the direct and indirect, the type of fade is determined by the amplitude of the signal to which the receiver is locked, relative to that of the second component. Thus, if the receiver loses lock during a deep fade, it is possible for it to re-establish lock on either signal so that the nature of the fade can be reversed.

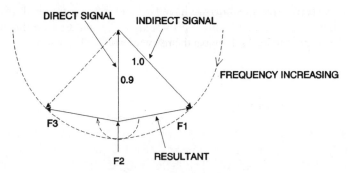

Figure 11.2   Vector diagram for a nonminimum-phase fade

11.2.3 Wideband fading

Figure 11.3 shows the situation in which a multipath notch is located within the channel bandwidth of a wideband system. The depth of the notch is shown as 25 dB.

The AGC line will, however, indicate a fade somewhat less than this, since it equates to the reduction in received power averaged across the channel bandwidth.

This immediately points to the fact that the cumulative distributions of AGC indicated fading in a wideband channel will be different from that determined from the equation to estimate single-frequency fading. The term *system fade distribution* may be used, indicating that such distributions depend on the channel bandwidth.

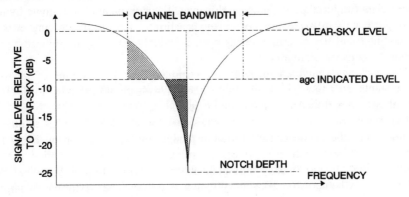

Figure 11.3   Relationship between multipath-notch depth and wideband fade depth

11.3 System comparisons

This Section compares narrowband and wideband systems and compares their more important parameters.

11.3.1 Narrowband systems

In a narrowband analogue radio system, the multipath delay that produces deep notches in the frequency domain is such that the variation in signal level across the channel width is minimal. The fading can be considered identical to a spot-frequency situation. Since analogue systems are mainly sensitive to received signal level, the multipath-fade-prediction model mentioned previously is sufficient for most of the planner's needs.

11.3.2 Wideband systems

With the advent of high-capacity digital radio, however, we can find ourselves involved with systems having bandwidths of many tens of megahertz, and we must consider the effects this may have on fading data. The wideband nature of digital systems means that the variation of signal level, and the associated phase, across the channel bandwidth becomes more important than the indicated signal level.

Wideband digital-communication links are relatively insensitive to mean-depression fading, and the flat fade component will not, in isolation, have any serious effect on the system performance until the received signal level approaches the receiver threshold. However, flat fades cannot be ignored since their presence can make the system more sensitive to selective fading effects, generating transmission errors owing to the additional distortion introduced by the frequency-selective component.

This fact has generated interest in determining the distributions of the two components of fading, especially in connection with the urgent need to develop digital-system performance-prediction models.

Experience on digital systems indicates that the absolute depth of multipath fading is not the prime cause of system outage, multipath distortion being the dominant problem with crosspolar discrimination degradation being a further consideration for cofrequency crosspolar systems. Thus, the methods used for determining whether an analogue system will achieve the performance targets cannot always be used for accurately assessing digital-system performance.

11.3.3 Satellite paths

Unlike terrestrial line-of-sight paths, satellite paths are characterised, in general, by limited system-fade margin and precipitation-induced flat fading. The use of higher-order modulation schemes is limited by the system's increased sensitivity to noise and interference. However, for link-engineering purposes, a similar approach to that taken for rain-dominated terrestrial links may be adopted.

11.4 System characterisation

An indication of the ability of a particular digital-radio system to withstand the distortion effects of a multipath notch close to, or within, the channel bandwidth is given by the system signatures. A method is required to allow radio-system performance to be estimated, given link and equipment parameters.

11.4.1 Performance objectives

Telecommunication networks are designed to meet internationally agreed performance and availability objectives. When radio systems are introduced into a network, planning guidelines are followed to ensure that these systems will meet the design objectives even under adverse propagation conditions.

We shall begin by examining the definition of these objectives before describing how to calculate radio link budgets.

Errored second: any second which contains one or more transmission errors

Severely errored second: any second in which the average error ratio is equal to or greater than 10^{-3}, but excludes unavailable time

Degraded minute: any minute in which the average error ratio is equal to or greater than 10^{-6}, but excludes all severely errored seconds

Unavailability: a period of unavailability is entered when ten consecutive severely errored seconds occur, and is exited when ten consecutive nonseverely errored seconds occur.

The percentage of time for which a given error-performance objective must not be exceeded is laid down in ITU-T and ITU-R Recommendations for a 64 kbit/s circuit. The ITU-T is producing a new Recommendation G.826 for broadband circuits.

11.4.2 Measurement of bit-error ratios

The method used for the measurement of bit errors is shown in Figure 11.4. A pseudorandom test sequence is transmitted over the radio channel and compared on a bit-by-bit basis with a locally generated sequence. Signals, which indicate the presence of errors, can be used to calculate bit-error ratio.

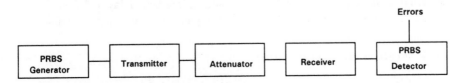

Figure 11.4   Measurement of system BER

To characterise the receiver, it is customary to measure the characteristic of bit-error ratio against input-signal power level. This can be performed by varying the attenuator in the signal path. The minimum power level necessary to produce a bit-error ratio of 10^{-3} is an important parameter and is used extensively in link-budget calculations.

11.4.3 Measurement of signatures

Another important aspect is the tolerance of the radio system to distortions in the channel, e.g. amplitude nonlinearity and variations in the frequency and phase response.   The latter, which usually arises from anomalous propagation conditions, e.g. multipath, can be simulated using a multipath fade generator, and system signatures measured as shown in Figure 11.5.

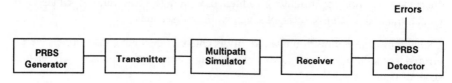

Figure 11.5   Measurement of signature characteristic

These are derived by driving the equipment under consideration from a multipath simulator, and measuring the notch depth at which the system-error ratio exceeds a given threshold at a number of frequencies across the channel (for both minimum- and nonminimum-phase conditions).

The system signature is a measurement, reproduced in pictorial form, of the tolerance of the radio system to multipath fades. The locus of these notch depths is then the system signature for the given set of conditions. An example of the system signature is shown in Figure 11.6 for various modulation schemes measured at an error ratio of 10^{-3}. Such a signature is sensitive to both the multipath delay and the available fade margin.

Figure 11.7 shows a range of signatures for a typical 64 QAM digital-radio system, measured at an error ratio of 10^{-3} and an interpath delay of 2.3 ns. Such a signature is also sensitive to both the multipath delay and the available fade margin.

11.4.4 Use of signatures in comparing system performance

A simple comparison of the robustness of different digital equipment to multipath activity can be made by comparing their signatures. The smaller the area under a signature, the more robust the system. This is fine for comparing equipment, but, if we want to predict the performance of equipment set in a particular location, we need to develop a model based on the signature set together with measured fading statistics for the area concerned.

Examination of a full set of signatures reveals several interesting facts:

(*a*) The width of the signature is not sensitive to multipath delay when the full system-fade margin (defined as the flat fade that can be tolerated before the system-error ratio exceeds 10^{-3}) is available, although this is not the case once the fade margin is reduced.

(*b*) The signature width is sensitive to the available fade margin, especially for short multipath delays.

(*c*) The height of the signature has some delay dependence, especially under the reduced-fade-margin conditions.

We can see therefore that the prediction of system performance based on signatures is difficult unless a detailed database for propagation in the area under consideration is also available.

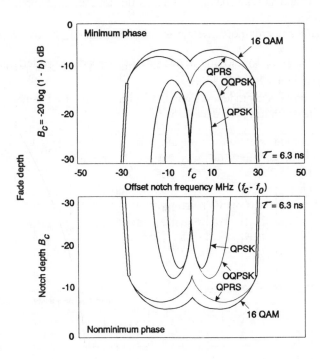

Figure 11.6   Example of radio-system signatures

11.5   Performance modelling using measured propagation statistics

11.5.1 Basic assumptions

In addition to the propagation assumptions made previously, there are others which must be made regarding the behaviour of the system under adverse conditions. These are:

(*a*) All interchannel interference can be treated as noise.
(*b*) Adjacent channels (crosspolar or copolar) do not fade in sympathy with the wanted channel.
(*c*) If a cofrequency crosspolar system is being considered, the two cofrequency channels fade together.
(*d*) XPI degrades on a one-to-one basis with CPA, once a predetermined threshold of fading has been exceeded.
(*e*) System signatures degrade in height and width with reduced signal level, after a threshold fade depth has been exceeded.

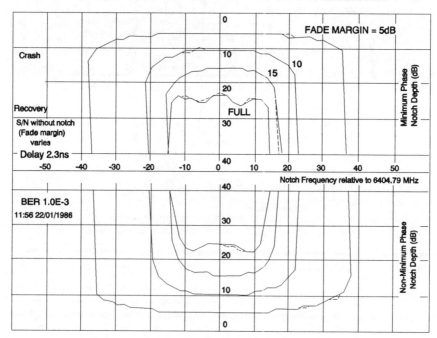

Figure 11.7   Example of radio-system signatures

Assumption (*b*) was adopted for the sake of simplicity in the initial modelling procedure. As it stands, this is somewhat more severe than will be met in practice, and now that the mode has been fully proved this area has almost certainly been modified. Recent studies indicate that fading on adjacent channels on the same routes correlated with flat fading, and was uncorrelated with frequency selective-fading.

Assumption (*d*) mirrors the current representation within the ITU. In practice, it is believed that, except for anomalous XPI events, the degradation is in sympathy with the selective component only, but until separate databases of flat and selective fading become widely available this cannot be proved.

11.5.2 Multipath-notch-probability approach

A method, that has been developed by BT Laboratories, models the effect of known distributions of flat and selective fading, which are considered to be uncorrelated, on the dynamically adjusted signatures for the equipment concerned. Thus, the method takes account of the fact that the signature will vary with the multipath delay and the degree of flat fading present, which are factors not truly considered in other, empirical models.

It is worth understanding just how a model such as that developed by BT Laboratories operates, since it gives a good insight into the ways in which a system degrades when operating in a multipath environment. First, let us look at some of the fundamental assumptions that are utilised within the basic model, and then note how these may be changed as it becomes more sophisticated.

(*a*) Flat and selective fading are statistically independent.
(*b*) Flat fading has a log-normal distribution defined by a mean value and a standard deviation.
(*c*) Narrowband selective fading has a Rayleigh probability distribution.
(*d*) An activity factor scales the overall joint probability.
(*e*) Multipath delays have an exponential probability-distribution factor.
(*f*) Minimum- and nonminimum-phase fades are equiprobable.
(*g*) Space diversity does not affect the flat-fade component but reduces the selective fading so that the cumulative distribution of fading has a slope of approximately 5 dB/decade.

Most of the above assumptions are derived from observation on an experimental link which was set up for this purpose, but two of them may require further explanation.

In (*f*) the assumption was made for simplicity of the basic model. It is only of interest for systems which do not have a symmetrical set of signatures — usually as a result of some form of adaptive equalisation. In the real situation it is reasonable to assume that, for very deep fading, this condition will be true, since for a deep fade the direct and delayed components of the signal will be almost the same level and any slight enhancement in their relative amplitudes may bring about a change in the type of fade being experienced. The smaller the fade, the more likely it is to be minimum phase. Figure 11.8 shows the form of modelling of this function which is likely to be used in later versions of the model.

Assumption (g) can be supported on the grounds that most flat fading is as a result of antenna defocusing or similar phenomena and that the provision of a diversity antenna is unlikely to change the situation. There is the possibility that during ducting situations, in which single transmit and receive antennas are screened from one another by an intense layer, the diversity receive antenna may not suffer this problem and the flat fade will be reduced. However, this is considered to be likely in only a few cases and cannot be catered for in a general model.

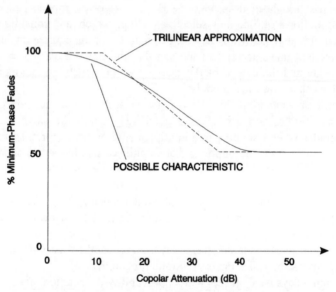

Figure 11.8   Modelling of % minimum-phase fades against copolar fade

Figure 11.9 is a very basic representation of the outage model. The system signature for an error ratio of 10^{-3} has been idealised as a rectangular block. The system will go into outage if the multipath notch intrudes into the signature. Thus the probability of outage is the product of the probability P_n of the notch lying within the frequency limits of the signature and the probability P_c of the notch being deeper than the signature.

Moving on, the model considers the situation for all values of flat fading less than the system fade margin, all values of selective fading less than a given upper limit and all delay values within a given range. The model says: let us take a value of flat fade and then calculates the idealised dimensions of the signature taking account of factors such as adjacent and cochannel interferences which are treated as noise-like and hence cause a reduction of the system fade margin. The next step is shown in Figure 11.10 in which for a given multipath delay a certain narrowband selective fade depth in the centre of the signature is considered. This fade can be the result of a multipath notch at the centre frequency equal in depth to the selective fade or one of a family of notches offset in frequency (either side) but yielding an identical fade at the centre frequency. The model then determines the range of such notches which fall within the frequency range of the signature and also are deeper than the signature. The sum of the probabilities is then multiplied by the probability of the flat fade, the selective fade and the multipath delay before being stored in an accumulator. The whole process is then repeated for the full range of flat and selective fades and the delays. The final figure in the accumulator is then scaled by an occurrence factor for the link being considered to give an outage probability. The process can, of course, be repeated for different error ratios by using the appropriate signature sets. One of the strengths of the model is that it makes use of the statistics of narrowband fading for which there is a vast amount of data available.

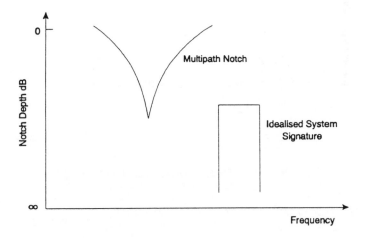

Figure 11.9   Basic modelling approach

11.5.3 In-band amplitude-distortion-based approach

There are several different approaches to outage prediction, starting with simple methods which translate single-frequency-fade predictions to link outage. A more comprehensive approach is based on the statistics of the difference between received signal level at two points symmetrically spaced within the system bandwidth. These statistics give a more accurate link to the corresponding error ratio prevailing. Considerable work was carried out in this area, although this approach does not seem to have been widely adopted owing to the lack of suitable statistics.

11.6 Example calculation

This Section identifies the performance objectives which are commonly used for link planning purposes. In view of the complexity of the task, this Section concentrates on the stages usually followed in link design.

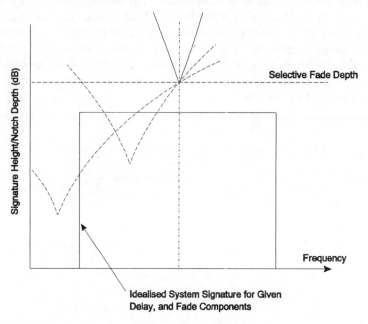

Figure 11.10   Idealised-system-signature approach

11.6.1 Link parameters

This Section describes how to evaluate the link budget of a radio system. Although we will consider a terrestrial radio link, the method is equally applicable to satellite or mobile radio links provided that the appropriate propagation models are used.

We shall first consider what data are required to calculate the link budget. Figure 11.11 shows a block diagram of a typical line-of-sight radio system. From

Figure 11.11, it can be seen that the information required to produce a link budget is listed in Table 11.1.

11.6.1.1 Transmitter output power

The transmitter output power is obtained from the specification for the radio system. Typical values for a high-capacity digital radio system range from 0.5 to 5 W. It usually refers to the power of the modulated signal, as this parameter can easily be measured.

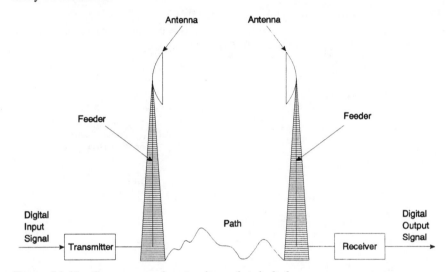

Figure 11.11   Components forming line-of-sight link

Table 11.1   Parameters required in link budget calculation

Transmitter output power
Feeder losses to transmit antenna
Transmit antenna boresight gain vs isotropic
Free-space path loss
Receive antenna boresight gain vs isotropic
Feeder losses from receive antenna
Receiver 10^{-3} BER threshold

11.6.1.2 Antenna-feeder losses

Various components make up the connection between the transmitter output flange and the antenna input. These components, which are application dependant, are shown in Figure 11.12. The circulators permit channel combining with minimum insertion loss. Various types of feeder, e.g. coaxial cable, elliptical or circular waveguide (with decreasing loss/metre) are available to convey the signals to the antenna.

11.6.1.3   Antenna-boresight gain

It is customary in line-of-sight radio systems to use directional antennas to combat the free-space path loss. The gain of the antenna G_i, relative to that of an isotropic radiator, can be specified as shown in Figure 11.13.

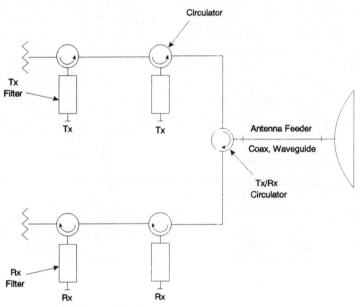

Figure 11.12   Channel-branching components

This ratio can be calculated from the mechanical dimensions of the antenna (e.g. aperture, illumination efficiency), and verified by measurement, in either an anechoic chamber or a test range.

$$G_i = 20 \log D - 20 \log \lambda + 10 \log \eta + 9.943$$

where D = antenna diameter, m
λ = wavelength, m
η = aperture efficiency

The free-space path loss for a 'normally' planned link can be evaluated from eqn. 2.9.

Proprietary PC-based software packages are available and can be used to demonstrate the features of performance prediction. The specific details of performance prediction are too complex to be described in detail here, but through the use of the following example, it should be possible to gain an insight into the process.

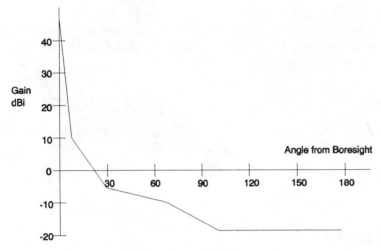

Figure 11.13   Typical antenna-radiation-pattern envelope

Table 11.2   Parameter values used to calculate performance

Transmitter power, dBm	30
Carrier frequency, MHz	6000
Receiver 10–3 BER threshold, dBW	−105
Path length, km	56
Terrain factor	40
Climate factor	1
Mean temperature, deg C	12.8
Transmit-antenna gain, dBi	45
Receive-antenna gain, dBi	45
Transmit-feeder loss, dB	5
Receive-feeder loss, dB	5
Field margin, dB	0
Dispersive-fade margin, dB	35
Adjacent-channel interference, dB	50
External interference, dB	50
Countermeasure option	
Space-diversity spacing, m	9.1
Frequency-diversity spacing, MHz	80
Result without diversity, s	2556.5
Result with space diversity, s	79.6
Result with frequency diversity, s	12.8

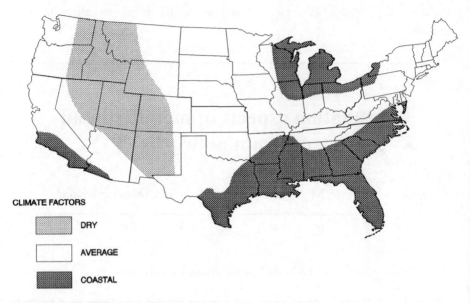

CLIMATE FACTORS

☐ DRY

☐ AVERAGE

☐ COASTAL

Figure 11.14   Climatic factors for the USA

Figure 11.15   Mean annual temperatures for the USA

Chapter 12

Propagation aspects of mobile spread-spectrum networks

M.A. Beach and S.A. Allpress

12.1 Introduction to spread-spectrum techniques

The use of wideband air-interface techniques is regarded by many as a suitable means for obtaining high capacity and flexibility of service provision for future-generation wireless systems. This is certainly true in Europe where there are numerous research initiatives currently underway addressing the selection of the primary air-interface technique for the Universal Mobile Telecommunication System (UMTS). Relevant examples include the RACE II [68] ATDMA and CODIT projects, as well as the UK DTI/SERC LINK CDMA [167] programme.

Networks employing *spread-spectrum* [79] access methods are extremely good examples of wideband radio access. Here a narrowband message signal is spread to a much wider bandwidth than is strictly necessary for RF transmission; then at the receiver the wideband representation is despread to yield an estimate of the original narrowband message signal. Ratios of 10 kHz message bandwidth to several MHz RF-channel bandwidth are commonplace in such systems.

There are principally two techniques employed to spread the spectrum. In *frequency hopping* (FH) spread-spectrum systems the narrowband message signal is modulated with a carrier frequency which is rapidly shifted in discrete increments in a pattern dictated by a spreading function. This spreading function is also available at the receiver, and enables it to 'retune' to the correct channel at the end of each hop. Although this is the simplest spread-spectrum technique to visualise, only a few have attempted to exploit this method commercially for personal-communication networks. The Geonet [124] system is one such embodiment, and an analysis of FH propagation in urban cellular environments is given by Purle [151]. However, much of the current interest in spread-spectrum systems is now focused towards the implementation and deployment of *direct-sequence* systems.

Direct-sequence (DS) spread-spectrum systems take an already modulated narrowband message signal and apply secondary modulation in the form of pseudorandom noise (PN), thereby spreading the spectrum.   This secondary modulation process is usually implemented in the form of phase-shift keying (PSK), and the PN sequence is known as the *spreading waveform* or *sequence*, or *code*. At the receiver, the incoming spread-spectrum waveform is multiplied by an identical synchronised spreading sequence, thus resulting in the recovery of the information or message signal. By associating a unique spreading code with each user in the network, multiple users can be simultaneously overlaid in both the time and frequency domains, and the concept of *code-division multiple access* (CDMA) realised. Given the considerable interest and investment by companies such as Qualcomm [91], InterDigital [163] and others in this technology, the generic term CDMA has become synonymous with cellular systems specially employing the DS spread-spectrum technique.

The user capacity of all cellular-radio access techniques is *self-interference limited*; however, CDMA differs from frequency and time multiple-access methods in as much as the primary source of this interference arises from users simultaneously occupying the same frequency and time allocations. The resultant capacity of such systems can be significantly increased by employing signal-processing techniques at the receiver which are able to exploit the wideband propagation characteristics of the DS-CDMA channel.   It is this fundamental aspect of high-capacity CDMA cellular networks that is described in this Chapter.   Techniques such as *synchronous* CDMA [143] can also be employed to virtually eliminate self interference; however, such techniques are restricted in application to small isolated cell systems.

12.2   Properties of DS-CDMA waveforms

To fully understand the propagation and sensitivity analysis of wireless networks employing CDMA access techniques, it is first necessary to consider the correlation properties of the spreading codes frequently employed in such systems. In the context of this discussion it is only necessary to consider the *autocorrelation* properties [94] of binary pseudorandom or *m-sequences*, as this gives a measure with which the code can be distinguished from a time-shifted version of itself. M-sequences display an almost ideal two-valued autocorrelation function $R_{ac}(\tau)$, as given in eqn. 12.1, where N is the period of the sequence measured in chips*.

---

* Each binary component of the pseudorandom sequence is termed a chip to distinguish it from the message data bits, or symbols.

$$R_{ac}(\tau) = \sum_{i=1}^{N} a_i (a_{i+\tau}) = \begin{cases} 1, & \text{for } \tau = kN \\ -1/N, & \text{for } \tau \neq kN \end{cases} \qquad (12.1)$$

Further, the normalised form of this function is illustrated in Figure 12.1. Here, it can clearly be seen that the autocorrelation function contains large triangular correlation peaks corresponding to the synchronisation of the waveforms, thus aiding both code acquisition and synchronisation functions which are necessary in the receiver in order to despread the DS-CDMA waveform.

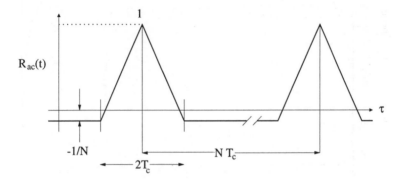

Figure 12.1   Normalised autocorrelation function of an m-sequence

In addition, it will be shown in the following Sections that this technique can be used to resolve multipath energy present at the receiver in a manner similar to that employed in some wideband channel sounders [45].

The spectral occupancy $S(f)$ of the DS-CDMA waveform can be derived from the autocorrelation function given above using the Wiener–Khintchine relationship. This is given by the following expression, where T_c is the chip duration of the spreading code, and it can be seen to consist of individual spectral lines within a $\sin^2 x/x$ envelope variation as illustrated in Figure 12.2.

$$S(f) = -\frac{1}{N}\delta(f) + \frac{N+1}{N^2}\text{sinc}^2(\pi f T_c) \sum_{m=-\infty}^{\infty} \delta\left(f + \frac{m}{NT_c}\right) \qquad (12.2)$$

The waveform illustrated in Figure 12.2 would indicate that DS-CDMA systems require an infinite RF-bandwidth allocation; however, in order to permit interworking with other systems, the transmitted spectrum is filtered. Filtering does result in a reduction of the correlation function, as illustrated in Figure 12.3, and distortion of the triangular in-phase correlation peak, as given in Figure 12.4.

Practical realisations of CDMA systems employ filtering constraints similar to that shown in Figure 12.4*b*. For example, the Qualcomm IS95 CDMA cellular system [86] employs a 1.2288 Mchip/s spreading-code rate operating within a baseband channel mask of 590 kHz passband and an ultimate stopband attenuation of 40 dB at 740 kHz.

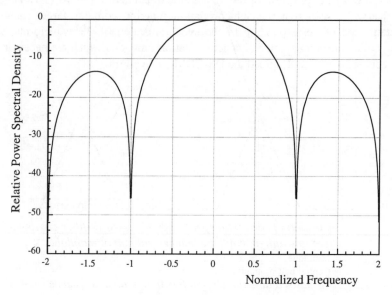

Figure 12.2   *Envelope of the power-spectrum density of an m-sequence*

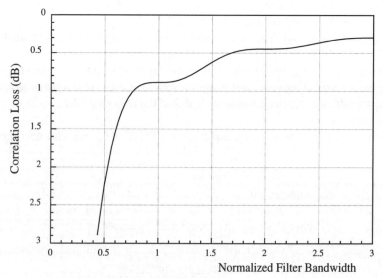

Figure 12.3   *Correlation loss due to filtering*

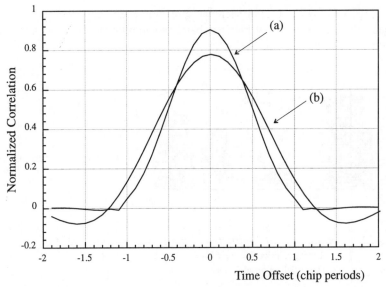

Figure 12.4   *Distortion of correlation peak with a brick-wall-filter passband*
equal to width of mainlobe and central half of mainlobe
a Mainlobe
b Central half of mainlobe

12.3 Impact of the mobile channel

The long-term-fading statistics, path and shadow loss, of both narrowband and wideband signal propagation in the mobile-radio channel have been shown to exhibit very similar characteristics according to theory and also by measurement [122]. However, the short-term statistics or the multipath-fading characteristics of these channels differ significantly, since CDMA systems tend to operate using channel bandwidths far in excess of the coherence bandwidth of the mobile channel.

The impact of the mobile-radio channel upon the CDMA waveform is most readily understood by considering the form of the correlation function after transmission through an environment similar to that shown in Figure 12.5(i). The correlator output [see Figure 12.5(ii)] shows three peaks corresponding to the time synchronisation between the local m-sequence in the mobile receiver and the multipath components 'a', 'b' and 'c'. The relative delay between these peaks corresponds to the different path delays in the channel, and thus, if the total multipath delay T_m is known, then the maximum number of multipath components L_m which can be resolved by the receiver is given by [166]:

$$L_m \leq \frac{T_m}{T_c} + 1 \qquad\qquad (12.3)$$

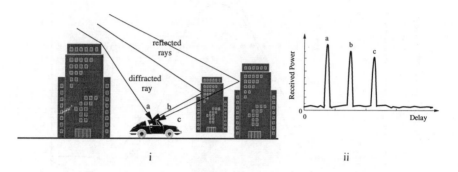

Figure 12.5   *Resolution of multipath components using DS-CDMA*
  (i) Urban cellular environment
  (ii) Correlator output

Thus, it can be seen that the degree of multipath resolution is directly related to the chipping rate of the CDMA waveform. With reference to Figure 12.1, resolution of individual multipath rays would be possible if an infinite chipping rate were employed. As already stated, it is necessary to bandlimit such systems, and thus the multipath activity observed is clustered together in discrete time intervals, or *bins*, separated by T_c. Hence, as the mobile moves within the environment the signal energy within each bin of the correlator output will fade.

The severity of fading will depend upon the amount of multipath activity present in each bin, and this is related to both the physical nature of the scattering volume and the time resolution of the system. Since each bin of the correlator contains energy contributions from different multipath components, these will tend to fade independently and also have different instantaneous Doppler components. Thus, as shown in Section 12.5, the bin-level-fading statistics in an urban cellular environment vary from 'Rayleigh-like' characteristics through to 'Rician' distributions with relatively high K-factors as the chipping rate is increased.

The inherent ability of the CDMA technique to produce multiple replicas of the same information bearing signal leads to the important concept of *path-diversity* reception. Here multiple despreading circuits can be used to decode the wanted signal contributions contained within the active bins of the correlator, and then combining these using the familiar techniques of *switched*, *equal-gain* or *maximal-ratio combining* to obtain a better estimate of the message signal. This is also known as *rake* reception based on the analogy with a garden rake being used to rake the impulse response of the channel, with each prong, branch or finger corresponding to a time bin in the correlator. This architecture is illustrated in

Figure 12.6, where $r(t)$ is the received signal, $d(t)$ is the message estimate and $A_0\exp(j\phi_L)$ corresponds to amplitude and phase offsets of the spreading waveform $c(t)$.

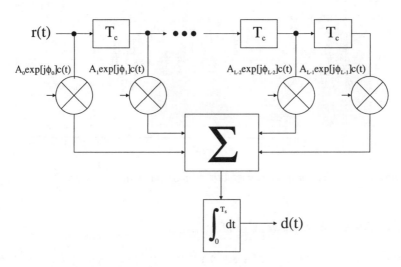

Figure 12.6   Conceptual diagram of a rake receiver

In addition to the time domain or temporal distortion caused by the multipath channel, it also follows from Figure 12.5(ii) that the channel must also inflict severe frequency-domain distortion. The received spectra for a 10 Mchip/s system operating in a simulated urban environment is illustrated in Figure 12.7 using channel parameters as given in Table 12.1. Despite the numerous frequency-domain nulls in the sinc envelope of the CDMA signal, data demodulation is still possible because of the inherent frequency diversity associated with spread-spectrum systems.

Table 12.1   Fading statistics of the synthesised urban channel

Ray	Excess delay	Mean amplitude	Fading	Fade rate
	ns	dB		Hz
1	0	0	Rician, $K=7.0\,dB$	100
2	60	−1.0	Rayleigh	31
3	125	−2.0	Rayleigh	96
4	250	−4.2	Rayleigh	35
5	500	−8.3	Rayleigh	94
6	1000	−16.7	Rayleigh	80

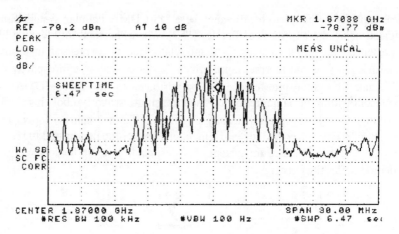

Figure 12.7   Typical received spectrum in a multipath environment[†]

12.4 DS-CDMA-system performance

The capacity analysis of CDMA cellular networks is a complex process and, since capacity can be traded for quality, these systems exhibit a property known as *soft capacity* [91]. Test-bed or field-trial validation [72,78,171] can be used to provide a rigorous analysis of a subset of teleservices, but lacks flexibility if a system-wide perspective is required. Bit-level simulations require many hours of CPU time owing to the high signalling rates involved, but can yield useful results if techniques such as *importance sampling* are employed, as well as directly exploiting propagation data available from field experiments. Mathematical modelling [54] offers by far the most computationally efficient approach for determining the relative performance and sensitivity analysis of DS-CDMA.

To develop a simulation or mathematical model of a cellular CDMA system, it is initially advantageous to make a number of simplifying assumptions. Here it has been assumed that ideal power control exists within the network in order to circumvent the near–far problem. This is particularly acute in DS systems, since all users share the same frequency band for either the uplinks or downlinks of the system, and thus the power-control scheme [90] must ensure that the cell-site receivers are not captured[‡] by a single user who may be in close proximity to the basestation. Furthermore, in DS-CDMA it has been found that the power-control loop must compensate for both the slow and fast fading present in the channel.

[†] Analyser parameters have been selected in order to emphasis the frequency-selective fading characteristics of the mobile channel.

[‡] Recent work on 'near–far' resistant techniques [191] potentially offers an alternative solution.

The capacity or bandwidth efficiency of a DS-CDMA cellular system can be evaluated using the approach of Pursley [152], where the total interference level N_I present at the input of a user's demodulator can be expressed in terms of mutual interference and thermal noise effects N_0 for a total system capacity of K CMDA users. Mutual interference can be attributed to the $(K-1)$ CDMA users also present within the same frequency band transmitting at a symbol energy of E_s, and further the power spectral density of this interference can be shown to be Gaussian. The actual power level of this interference is scaled by the *processing gain* G_p of the spread-spectrum system after despreading and prior to the input of the demodulator, thus giving

$$N_I = \frac{(K-1)E_s}{G_p} + N_0 \tag{12.4}$$

For multiple-access systems it can be shown§ that $G_p = 3T_s/T_c$, where T_s is the coded symbol duration of the information signal.

The bandwidth efficiency η of the system can be expressed in terms of symbol rate of the vocoder R_s, number of symbol states employed in the RF modulation scheme M, and the bandwidth of the spread-spectrum transmission W_{ss}:

$$\eta = \frac{R_s K (\log_2 M)}{W_{ss}} \tag{12.5}$$

For a large number of users, $K-1 \cong K$, $W_{ss} = 1/T_c$ assuming that the transmitted spectrum has been filtered to half of the mainlobe (see Figure 12.4b), and $R_s = 1/T_s$. Also, provided that a sufficient interleaving depth has been employed to ensure a discrete memoryless channel, the benefits of convolutional error-protection coding can be evaluated as given below. Here the term $(E_s/N_I)_{REQ}$ is the ratio of coded symbol energy to interference energy required to support a given error rate to the vocoder when using a rate-r_c convolutional code, with E_s/N_0 representing the thermal-noise level in the receiver. In the following results the thermal-noise power has been assumed to remain at a constant level of 10 dB, and QPSK modulation ($M=4$) with a half-rate, constraint-length-7 convolutional code has been employed for the air-interface throughout:

$$\eta \approx 3r_c (\log_2 M) \left\{ \left(\frac{N_I}{E_s} \right)_{REQ} - \frac{N_0}{E_s} \right\} \tag{12.6}$$

---

§ Assuming asynchronous random spreading codes

Equation 12.6 gives the bandwidth efficiency of a single isolated CDMA cell. In a cellular CDMA system, users in the adjacent cells also contribute to the mutual interference level since they operate within the same CDMA frequency band, and thus reduce the capacity when compared with the single cell system. This is termed the *frequency re-use efficiency* L_{fr} of the system, and if the interference signals are modelled as zero-mean Gaussian noise sources uniformly distributed within an hexagonal cellular floor plan, then it can be shown that $L_{fr}=0.67$ [91] assuming a fourth-order path-loss exponent.

To reduce the impact of mutual interference on the overall capacity of DS-CDMA systems, several techniques can be employed. First, it is directly possible to exploit the fact that users are only engaged in active conversation for approximately 35% of the time during a call, and thus users can be gated out of the system during the numerous periods of silence. This is known as the *voice-activity factor* G_{va}, and is equal to the reciprocal of the voice activity of the users. In addition, directional or sectorised base-station antennas can be employed to reduce the amount of interference received by a cell site receiver. If an idealised three-sector base-station antenna system is assumed with a *tophat* response, then the antenna gain G_{sa} is 3. Combining these effects gives

$$\eta \approx 3r_c(\log_2 M)G_{va}G_{sa}L_{fr}\left\{\left(\frac{N_I}{E_s}\right)_{REQ} - \frac{N_0}{E_s}\right\} \qquad (12.7)$$

The expression for the bandwidth efficiency of the system is dominated by the term $(N_I/E_s)_{REQ}$, and diversity-reception techniques can be employed to reduce the symbol energy necessary to maintain a given bit error rate (BER), and thus enhance the capacity of the network. In particular, CDMA receivers make use of path-diversity reception techniques as illustrated in Figure 12.5, and using the concept of the coded channel described by Simon [166], the bandwidth efficiency of a CDMA system can be calculated including the effects of diversity-reception methods operating in both log-normal Rayleigh and log-normal Rician mobile-radio channels [48].

Figure 12.8 gives the BER performance against E_s/N_I for varying orders of internal diversity while operating in a log-normal Rayleigh channel with a standard deviation of 8 dB. In these results the diversity branches have been combined using maximal ratio combining (MRC), and it can be seen that, for a fixed BER, increasing the number of diversity branches results in a considerable reduction of the required E_s/N_I.

Figure 12.9 illustrates the BER performance against E_s/N_I when equal-gain combining (EGC) is employed for a system operating under the same channel conditions as given in Figure 12.8. Again, it can be seen that there are considerable benefits to be obtained from employing diversity-reception techniques, and when comparing MRC against EGC there is an approximate power saving of 3 dB in favour of MRC for fifth-order path diversity at a BER of 10^{-3}.

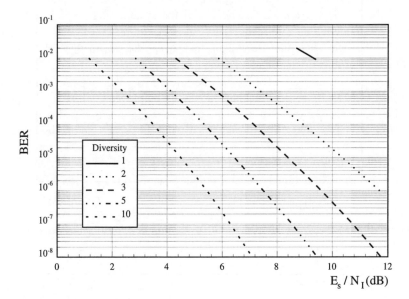

Figure 12.8   BER against E_s/N_I for MRC operation in a log-normal Rayleigh channel

Figure 12.10 gives the bandwidth efficiency against diversity order for both MRC and EGC architectures operating in the log-normal Rayleigh channel for a BER of 10^{-3}, assuming no base-station sectorisation ($G_{sa}=1$). Again, the benefits of employing MRC combining rather than EGC can be clearly seen, and also that there are considerable benefits to be attained in terms of capacity enhancement when employing up to four diversity branches, whereas increasing the diversity order beyond this level tends to result in diminishing returns with respect to complexity and cost for any further increase in capacity.

Results giving the BER performance of a CDMA rake receiver with MRC processing for varying numbers of diversity branches operating in a log-normal Rician channel with a K-factor of 3 dB and a standard deviation of 2.77 dB are given in Figure 12.11. When compared with the results given for the log-normal Rayleigh channel, there is a considerable reduction in the required E_s/N_I to maintain a given BER because of the deterministic nature of the Rician channel. Further, it can also be seen that the BER performance is less sensitive to the diversity order of the MRC architecture.

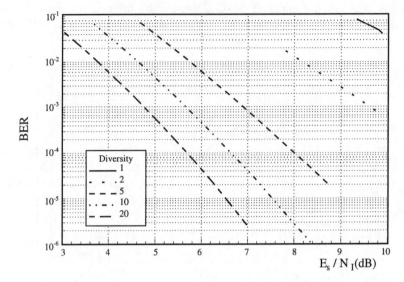

Figure 12.9   BER against E_s/N_I for EGC operation in a log-normal Rayleigh channel

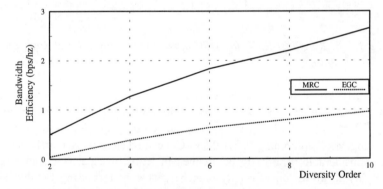

Figure 12.10   Bandwidth efficiency against diversity order for MRC and EGC operating in a log-normal Rayleigh channel

Figure 12.12 illustrates the relative bandwidth efficiencies of MRC-rake-reception techniques whilst operating in both log-normal Rayleigh and Rician channels for a BER of 10^{-3}, again assuming no basestation sectorisation. The operational characteristics obtained for these channels are extremely different, with the system performance being very sensitive to the diversity order for the Rayleigh case, and with considerably reduced sensitivity shown for the Rician channel.

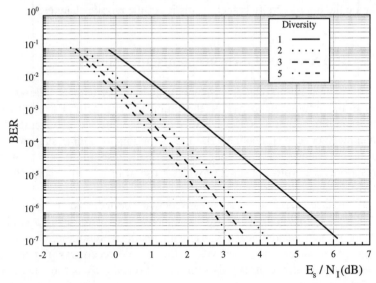

Figure 12.11 BER against E_s/N_I for MRC operation in a log-normal Rician channel

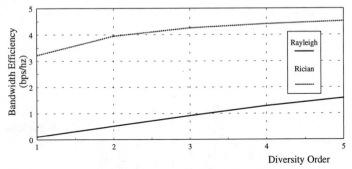

Figure 12.12 Bandwidth efficiency for a MRC rake architecture operating in both log-normal Rayleigh and Rician channels

The sensitivity of DS-CDMA to the assumptions made regarding the propagation characteristics of the mobile fading channel has prompted several investigations [49,50] into the characteristics of this medium when viewed through a wideband rake CDMA receiver. Based upon the discussion so far, the key issues to be resolved by engineers considering such systems are:

(i)   can sufficient independent diversity branches be resolved from the channel?
(ii)   what are the bin (diversity branch) statistics of the mobile channel?

12.5   DS-CDMA channel measurements and models

The use of wideband-correlation channel-sounding techniques allows the extraction of the impulse response of the multipath channel. By employing continuous-sampling techniques, the statistics directly related to the operation of CDMA rake receivers can be obtained. Before examining the fading statistics of a wideband urban channel, a discussion on multipath resolution based on the impulse response of a typical channel provides a useful insight into the problems of the CDMA-system designer.

A typical impulse response of an urban cellular service area in the city of Bristol is illustrated in Figure 12.13. This response was taken using a sliding-correlator [73] sounder operating at 1.8 GHz with a path resolution of 50 ns. It can be seen that most of the energy arrives at the mobile receiver within a 1 µs excess-delay window, although there is some low-level activity out to 2 µs. Also shown in this Figure are the discrete-time bins of a 1 µs-resolution rake receiver. If such an architecture were employed in this situation, then only 1 path or branch of an internal diversity-processing architecture could be supported if a 10 dB-diversity power window is assumed. To support the degree of diversity processing as suggested in the previous Section, it is necessary to increase the bin resolution to 200 ns, as illustrated in Figure 12.14. Here it can be seen that up to four diversity branches can be supported.

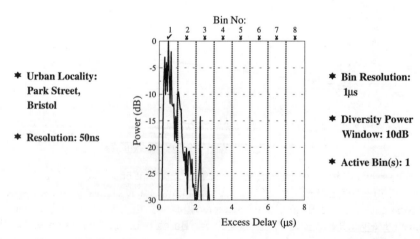

Figure 12.13   Impulse response of a typical urban mobile channel with 1 µs rake bin resolution

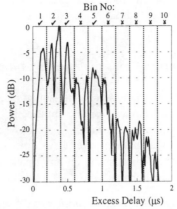

Figure 12.14 Impulse response of a typical urban mobile channel with 200 ns
rake bin resolution
Urban locality: Park Street, Bristol
Resolution: 50 ns
Bin resolution: 200 ns
Diversity power window: 10 dB
Active bin(s): 1,2,3 and 5

To investigate the multipath resolution and the bin statistics of a DS-CDMA system, modifications were made to a sliding-correlator wideband channel sounder operating within the 1.8 GHz PCN band, thus emulating a multifingered rake receiver. In the following sets of results (Figures 12.15 – 12.19, a 255-chip m-sequence was clocked at 1.25, 2.5, 5, 10 and 20 Mchip/s while the receiving system moved within an urban locality.

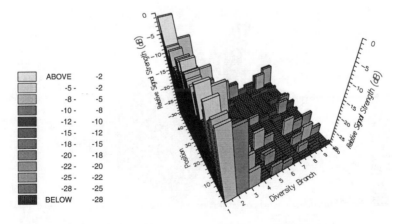

Figure 12.15 Urban DS-CDMA rake emulation at 1.25 Mchip/s

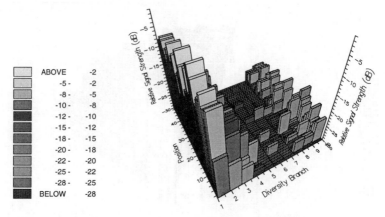

Figure 12.16   Urban DS-CDMA rake emulation at 2.5 Mchip/s

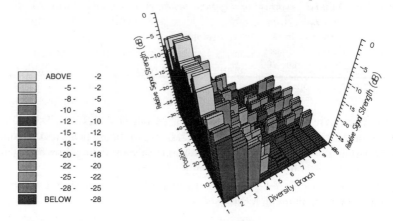

Figure 12.17   Urban DS-CDMA rake emulation at 5 Mchip/s

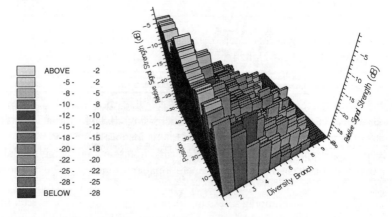

Figure 12.18   Urban DS-CDMA rake emulation at 10 Mchip/s

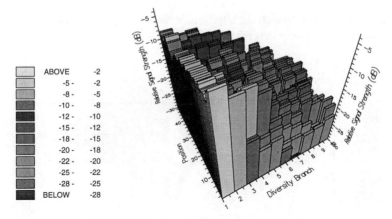

ABOVE	-2
-5 -	-2
-8 -	-5
-10 -	-8
-12 -	-10
-15 -	-12
-18 -	-15
-20 -	-18
-22 -	-20
-25 -	-22
-28 -	-25
BELOW	-28

Figure 12.19   Urban DS-CDMA rake emulation at 20 Mchip/s

From these results it can be clearly seen that, as the chipping rate is increased, the multipath activity spreads to more diversity branches or rake bins, thus making more independently fading samples available for diversity combining in the receiver. Also, increasing the chipping rate or spread-spectrum bandwidth has a most notable affect on the signal-strength variability of each branch, but in particular that of the first bin, as illustrated in Table 12.2. Increasing the chipping rate reduces the signal variability, and this has a significant impact upon the E_s/N_I required to support a given error rate. In addition, higher chipping rates reduce the sensitivity of the system to power-control errors, since the dynamics of the fast-fading component have been significantly reduced.

Table 12.2   Rician-K factor of dominant bin

Chipping rate, Mchip/s	1.25	2.5	5	10	20
K-factor, dB	−6.04	0.78	3.22	3.99	6.68

The practical results presented here can be shown to have a good agreement with the theoretical work of Holtzman [107], as given in Figures 12.20 – 12.22. In Figure 12.20 the mean signal power of the measured data is shown as a function of the diversity-bin number for the varying chipping rates considered during the field-trial experiments. It can be seen that the mean signal power contained within each branch reduces with increasing diversity order, but this occurs with a diminishing rate as the chipping rate is increased.

Increasing the chipping rate of the system results in less energy being available on a per-diversity bin basis for the same operational environment. This is shown in Figure 12.21 in terms of the available power as additional branches are combined for the various chipping rates considered. From here it can be seen that low-chipping-rate systems (~1 Mchip/s) extract most of the available multipath power with a single-branch architecture, whereas higher-chip-rate systems can make use of higher-order diversity.

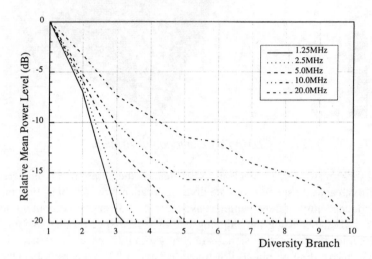

Figure 12.20   Mean signal power against diversity branch

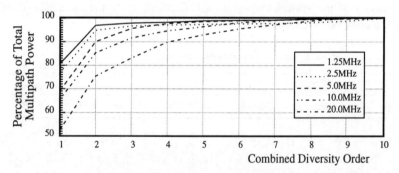

Figure 12.21   Available power against combined diversity order

The signal-strength variability can also be considered in terms of the coefficient of variation [134] and the normalised standard deviation. The former is defined as the ratio of the standard deviation relative to the mean, whereas the latter is given by first normalising the received powers measured in dBs, and then calculating the standard deviation relative to the mean. These results are illustrated in Figure

12.22, where it can be seen that as the chip duration is increased** the coefficient of variation of the first diversity branch tends to 1, and also the value of the normalised standard deviation approaches 5.57 dB. These values correspond to the Rayleigh case.

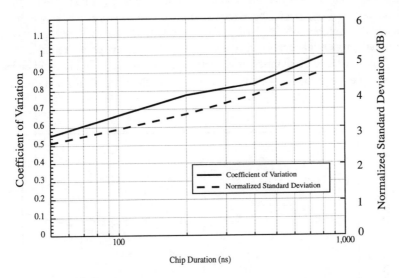

Figure 12.22   Coefficient of variation and normalised standard deviation against chip duration

12.6   Support of handover in cellular networks

Cellular CDMA networks should not be considered as single isolated cell structures. This has already been discussed in terms of mutual interference arising from users in adjacent cells operating on the same CDMA frequency band. In addition, mobile users will frequently move from cell to cell, and thus it is necessary to *handover* service provision from one cell site to the next. Currently, all operational cellular networks employ a technique called *hard* handover, where the air-interface is physically switched between cell sites. Hysteresis is normally employed to prevent the switching pingponging between two cell sites at the radio-coverage boundaries; however, service quality is often poor and calls are frequently dropped during handover.

If the path-diversity process inherent in the CDMA technique is again considered as illustrated in Figure 12.5, the multipath components *a* and *c* could

---

** Increasing the chip duration corresponds to a decrease in the chipping rate or bandwidth of the CDMA transmission.

be considered to originate from one cell site, and *b* from another.   Now by designing the network such that multiple base stations can transmit and receive information associated with the same mobile call within the handover regions of the system, the concept of *soft* handover is realised [78].   There are numerous benefits associated with the soft-handover process when compared with that of hard handover; first, it can be engineered as a *make-before-break* switch, thus reducing the number of dropped calls.   In addition, the diversity gain of the rake-signal processing can be directly exploited in terms of a reduction of the E_s/N_I required at the boundaries of the service area, thus resulting in a net capacity increase when compared with hard handover CDMA, as illustrated in Figure 12.23.

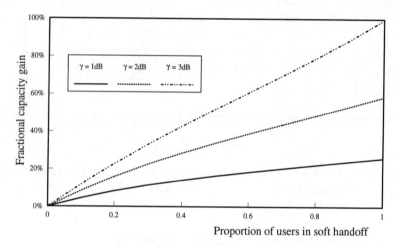

Figure 12.23   Capacity gain against diversity gain γ due to soft handover [69]

12.7 Conclusions

From the results presented in Section 12.4 it can be seen that the capacity of a CDMA network can be traded against the complexity of the diversity signal-processing architecture employed within the receiver, and that the gain available from path-diversity processing is extremely sensitive to the short-term propagation characteristics of the channel. Furthermore, the chipping rate of these systems must be sufficient to ensure that multipath energy can be extracted from the channel (see Figures 12.12 and 12.14) for use in the diversity process, with the RF bandwidth being directly related to this value.

    It has also been shown that the channel statistics of the mobile-radio channel vary when observed through a DS-CDMA receiver as the chipping rate is varied, and this has a significant impact on the overall system performance.   For low chipping rates (≤1.25 Mchip/s) the measured channel displayed 'Rayleigh-like' statistics with a Rician *K*-factor of –6 dB. As the chipping rate was increased to 20 Mchip/s, strong Rician statistics were observed. The terms narrowband and wideband CDMA are now becoming associated with these systems, with chipping rates of approximately 1 Mchip/s, and rates in excess of 5 Mchip/s, defining the respective categories.

A common figure of merit used to compare channel impulse responses and air-interface performance is the *RMS delay spread*. This measure has not been discussed here in the context of CDMA systems, since it provides no useful information regarding system performance. This is because no phase information of multipath signals arriving within a particular time window is given, and also there is too much ambiguity in the form of the impulse-response data. This is illustrated in Figure 12.24, where the RMS-delay spread of both impulse-response measurements is 200 ns. The response shown in Figure 12.24*a* contains two rays of equal amplitude arriving within a 400 ns time window, whereas Figure 12.24*b* again contains two rays, but with the second path arriving 700 ns after the first at a power level of –10 dB.

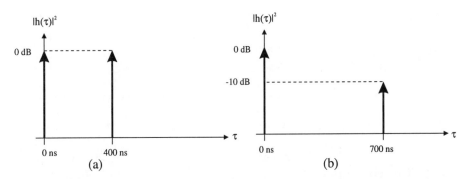

Figure 12.24   Illustration of ambiguity of RMS-delay spread

Although both impulse responses have the same RMS-delay spread, when processed by a rake receiver they produce very different results. For example, if a chipping rate of 2 Mchip/s is employed, Figure 12.24*a* would produce severe bin fading in the form of the two-ray model scenario, whereas no bin fading would be observed by the receiver for the environment of Figure 12.24*b*.

12.8 Acknowledgments

The authors thank colleagues at the Centre for Communication Research (University of Bristol), in particular Prof. Joe McGeehan, Simon Swales, Andy Nix and Joseph Cheung, for invaluable help and support of this work. They also greatly value the early contributions from Amnon Hammer on CDMA capacity analysis.

Chapter 13

Prediction of signal levels likely to cause interference for frequencies above 1 GHz

M.P.M. Hall

13.1 Introduction

As the demand for radio-frequency usage continues to grow, so does the necessity for more accurate prediction of signal levels likely to cause cochannel interference. This is needed by those responsible for developing national services, as well as those concerned with the international planning of sharing between services. Most of the Chapters in this book have been concerned with *reliability* of services and the prediction of low signal levels which will be exceeded for a large percentage of time, say 99% or even 99.99%, according to the specific radio service. In this Chapter we will be concerned with high signal levels which may cause cochannel *interference* by 'anomalous propagation'. These levels are due to certain meteorological conditions, the effects of which will be disruptive for a small percentage of time, e.g. 0.01%. 0.1%, 1 or even 10%, according to the specific radio service. Both of these situations (loss of (signal) reliability and interference) need ultimately to be considered as a composite duration of loss, but just as losses due to different effects have to be accounted for separately (e.g. attenuation due to rain, or losses due to multipath) and then the times added (if they do not occur concurrently), so the durations for permissible interference levels due to different effects (e.g. ducting or hydrometeor scatter) have to be predicted separately and then the times added (again if they are not concurrent events).

Much work on prediction methods has been conducted over the years, but the primary limitation to improvement of these prediction methods has been the lack of good measured data against which to test results. This situation was, until recently, so acute for rain scatter (a special case of hydrometeor scatter) that there was widespread doubt that it has any significant effect. The present Chapter comments on major improvements now implemented; it is generally concerned with fixed terrestrial stations (but not necessarily fixed services) and with frequencies above about 0.7 GHz. Methods extending from lower frequencies and oriented to point-to-area services are considered in Chapter 18.

Within Europe there has been a special concern for improving prediction techniques, because of the rapid technological and market growth in services, and because a co-ordination area around an Earth station may cover several small countries so that the practical, administrative and political implications are major. The *co-ordination* area is that within which national administrations must agree arrangements for sharing frequencies by the Earth station and any terrestrial stations, in order to avoid mutual interference. This area is determined so as to be that for which interfering signals outside it come below a permissible level.

This European concern to predict interference and co-ordination areas accurately led to the setting up of the COST Project 210 (where COST is concerned with European Co-operation in Science and Technology) [46], a project which produced a new data bank of measured data. It also produced a new procedure for the prediction of interference levels likely to occur during clear-air conditions (i.e. in the absence of hydrometeors) and a major extension of the ITU method for the prediction of levels due to scatter from various hydrometeor forms (i.e. rain, hail, snow, ice crystal clouds etc.). The primary objective of these studies was to recommend improved procedures for the prediction of interference levels, and for establishing co-ordination distances, so as to minimise the safe distance between radio systems.

These developments have now been incorporated in the ITU Radiocommunications Sector (ITU-R) Recommendations that will be commented on in this chapter. It is possible here only to outline the procedures available and refer the reader to fuller texts.

13.2   Investigations relating to interference and co-ordination

The need for sharing of radio frequencies within and between services is still increasing rapidly. About 10% of the 3600 satellites in orbit are geostationary, and the majority of these are employed in the fixed-satellite service. In the ITU's Region 1, almost half the spectrum from 3 to 30 GHz has an allocation to the fixed-satellite service, and the majority of this spectrum is shared on an equal primary basis with others, in particular the terrestrial fixed service.

Potentially, there is a wide range of technical investigations needing to be carried out, where a radio transmission intended for one specific receiver location, or service area, may produce interference at other receiving stations. Some of these situations are illustrated in Figure 13.1. 'Wanted' signal paths are shown between terrestrial stations and between satellite and Earth stations. 'Unwanted' (interference) paths are seen to be caused by various propagation mechanisms dividing generally into clear-air and hydrometeor-scatter conditions. The former comprise diffraction, tropospheric scatter, scatter from terrain (i.e. from open land, trees, buildings, lighting towers etc.), surface super-refraction, ducting, and reflection and refraction from elevated tropospheric layers. Paths with low elevation angles at both terminals are most affected by the clear-air-propagation mechanisms. Any precipitation occurring on such a path is unlikely to fill a large proportion of the common volume shared by the transmitter and receiver beams. However, precipitation (or ice-crystal clouds) occurring in the narrow elevated beam of an Earth station may well fill most or all of the common volume shared with the terrestrial-station antenna and give high levels of scatter signals. In producing prediction procedures, it is necessary to examine separately the conditions of hydrometeor-scatter elements (rain, ice crystals and snow, and melting), but the users of such predictions need not be aware of this).

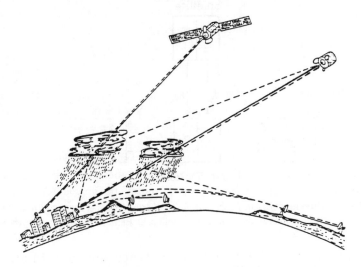

Figure 13.1   Paths of possible interference
——————— wanted paths
– – – – – unwanted paths

The main situations for which predictions are required are potential interference between:

(i)   a terrestrial link terminal and the Earth station of an Earth–space link (the spacing between the two typically being up to several hundred kilometres), also one terrestrial line-of-sight terminal and another (again over a long path);

(ii)   a terrestrial link terminal and the satellite of an Earth–space link; and

(iii)   two Earth stations using reverse-band working, i.e. one transmitting and the other receiving on the same frequency, (possibly both using relatively small antennas with quite small spacing between them).

As indicated above, the first of these cases is currently the most crucial, and recent work has concentrated on this. However, the range of separations between terminals has been extended downwards to also accommodate the shorter distances necessary for new systems. This first case is covered by ITU-R Recommendation P.452-7 [12] which is primarily concerned with interaction between the terrestrial link and the Earth station, but can also be applied to two terrestrial terminals. The second case above (a terrestrial terminal and a satellite) is covered by ITU-R Recommendation P.619-1 [21]. These two cases are indicated in Figure 13.2. The third case (an up-link operating at the same frequency as a nearby down-link) is covered by ITU-R Report 1010 [40].

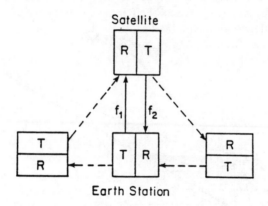

Figure 13.2   *Interference paths between fixed satellite and terrestrial services*
Transmitters and receivers on a satellite, an Earth station and two terrestrial stations

—————   wanted signal
— — — —   interfering signal

Among the features sought in the prediction procedure for the first (most crucial) of the three cases indicated above are (*a*) to have only small differences between observed and predicted interference levels, with a view to minimising distances between radio systems (and consequent increase in re-use of the radio spectrum), and (*b*) adaptability to computer implementation. The totality has to have (c) flexibility (to cover a wide range of frequencies, path lengths, climates, time percentages and geometries), (*d*) integrity (to be operational without any need for intervention of external engineering judgement) and (*e*) completeness (to ensure that there are no 'dead ends' in which the procedure will fail to provide a solution in cases previously not considered). With the application to new services, the time percentages for which signal-level exceedences are to be computed have widened to cover from 20% of a year (ITU-R Recommendation SF. 558) [34] to 0.03% of any month (ITU-R Recommendation SF. 615) [35]. For many purposes, the priority frequency range of concern has been about 0.7 – 30 GHz, as this range has a large number of frequency-sharing and co-ordination issues. The procedures have to be in a form for practical use, founded on improved models arising from a better understanding of the physical processes involved. The term cochannel is taken to mean interference from transmitters at frequencies within the receiver passband, and not interference due to sources of either natural or man-made noise, which may be equally important and is usually considered as 'electromagnetic compatibility' or 'EMC'.

As well as the primary objective of providing prediction procedures, consideration has been given to means available to reduce their effects, noticeably by site shielding. This may be sought deliberately through the use of buildings and screens normally situated in the local environment of the receiver, and as such is conceptually different from that associated with high horizons occurring on transhorizon paths (see Section 13.7).

International planning is achieved at ITU-R Radio Conferences on a world-wide, regional or service basis through the Radio Regulations [1], (see Chapter 1). The ITU Radiocommunications Bureau maintains a list of transmissions from which nations must agree whether projected use of the frequencies which countries assign to their stations will or will not cause harmful interference to existing links. Radiowave-propagation aspects of prediction of signal levels likely to cause interference are considered in the ITU-R Recommendations mentioned above. However, there is first the need to use a procedure for evaluating *co-ordination distance*, which is the distance from an Earth station beyond which any interference to or from a terrestrial station may be neglected in planning for a new Earth station on a shared-frequency basis, or a new terrestrial station which could interfere with an Earth station. The propagation aspects on which the co-ordination area (the locus of a point at the end of the co-ordination distance) around the Earth station must be evaluated, are covered in ITU-R Recommendation P.620-2 [22]. Before another terminal can be located within this co-ordination area (i.e. within the co-ordination distance), a more detailed calculation of predicted interference levels has to be performed to the satisfaction of the two national administrations concerned, usually using the methods set out in Recommendation 452. Recommendations 620 and 452 cover the propagation aspects of evaluating co-ordination area and levels of signal likely to cause interference, respectively. The material in them is incorporated with system aspects in ITU-R Recommendations IS. 847-1 [4] and SF. 1006 [36], respectively. The former is the basis for future change to Appendix 28 of the Radio Regulations [1], which is the legal basis for international agreement.

Figure 13.3 shows two examples of co-ordination area computed using the procedures of Recommendation 620: one for the UK's Radiocommunications Agency site at Baldock, UK, with a satellite over the Indian Ocean and another for BT's Earth station at Goonhilly, UK, with a satellite over the Atlantic Ocean. It will be seen that the contour for clear-air conditions is very sensitive to horizon elevation angle and whether the path is largely over sea or over land. By contrast, the contour for hydrometeor-scatter conditions is a circle (centred on the common volume of scatter). The co-ordination area is defined by the largest of these two, or by a distance of 100 km if this turns out to be greater. The fact that the contours include parts of several countries outside the UK (five for Figure 13.3*a*, seven for Figure 13.3*b*) emphasises the need for accuracy in prediction. Figure 13.3*b* also includes auxiliary contours corresponding to successive 5 dB increases in transmission loss corresponding to certain avoidance angles in terrestrial-station antenna pointing. These are illustrative of those used in Recommendation 847 and Appendix 28 (and will be in the next version of Recommendation 620); they can be used to eliminate from further consideration all low-threat stations which are known to be angled away from the Earth station. The figure also shows a maximum coordination distance of 1027 km for clear-air conditions. (This is not the 1200 km or 900 km referred to in Section 13.4.2, but a value interpolated in Appendix 28 for interference levels to be expected for 0.005% of time).

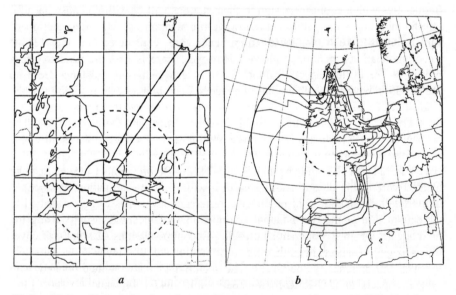

a	*b*

Figure 13.3   Two examples of co-ordination contours for an 11 GHz Earth station

      ————————   clear-air condition
      – – – – – –   hydrometeor-scatter condition

This chapter covers developments up to the versions of ITU-R publications available in 1995, but some further developments have already been agreed by

ITU-R Study Group 3 which may be expected to be published in 1996. The developments for Recommendation 452 include (i) improvement of predictions on long inland interference paths, (ii) unification of previous European and global prediction methods, (iii) improved prediction of worst-month results; and (iv) incorporation of a height/gain function for urban situations which allows for additional protection of terminals within such cluttered areas. In addition, the physical basis for the hydrometeor-scatter model is indicated. The development of the clear-air model for Recommendation 620 closely parallels that of Recommendation 452 to harmonise results obtained from the two Recommendations and to remove some over-cautious distance dependence. In addition, auxiliary contours for hydrometeor scatter are added (as mentioned in the previous paragraph). For 1997 it is hoped to improve the radiometeorological basis for taking account of effects that vary with climate, and for 1998 it is intended to take better account of terrain/building scatter. Developments will continue as results become available.

This chapter will now be concerned with evaluating co-ordination distances and the probability of occurrence of certain interference levels, but it is first appropriate to consider propagation mechanisms common to both.

13.3 Propagation mechanisms

Following the principles outlined above, it is necessary for evaluation of co-ordination distances and prediction of interference-signal levels to examine each likely propagation mechanism in turn, and then to sum the percentage times for the signal level giving concern (if the mechanisms do not occur concurrently). Normally, one mechanism will strongly dominate over others (but, should two have near equal values at some point of a distribution, then in principle the time percentages would be approximately doubled near this signal level). Such dominance may depend on climate, radio frequency, time percentage of interest, distance and path topography.

For *terrestrial interference paths* at frequencies above about 0.7 GHz, it is necessary to consider the following propagation mechanisms, of which the first four may be regarded as long-term (or continuous) interference mechanisms (Figure 13.4), and the remaining four as short-term (or anomalous) mechanisms (Figure 13.5), which usually generate much higher interfering signal levels.

(i) *Line-of-sight*: The most straightforward interference propagation mechanism is line-of-sight under well mixed atmospheric conditions. An additional complexity can, however come into play when subpath diffraction causes an increase in signal level (see item (v) below).

(ii) *Diffraction*: The accuracy to which this mechanism can be modelled often determines the density of microwave systems that can be achieved in a given area.

The diffraction-prediction capability must have sufficient utility to cover smooth-Earth, discrete-obstacle and irregular (unstructured) terrain situations. It is most significant at short ranges, say up to 50 km, and will show statistical time variability as the large-scale refractive-index-against-height profile (or effective-Earth-radius factor) changes.

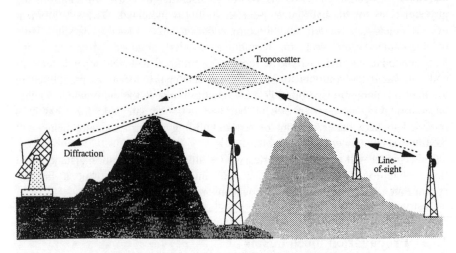

Figure 13.4   Long-term interference propagation mechanisms

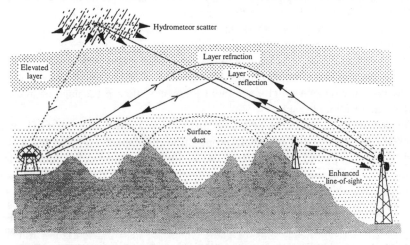

Figure 13.5   Short-term (enhanced) propagation mechanisms

(iii) *Tropospheric scatter* from refractive-index irregularities in clear-air conditions: This mechanism defines the 'background' interference level for longer paths (e.g. 100–150 km). The prediction requirement for this mode in the interference context is somewhat different from that needed for troposcatter-link design, as it is time percentages below 50% that are of interest, often on paths with a highly asymmetrical geometry and with the common volumes being generated by antenna sidelobes. This mechanism may be the dominant one and the cause of interference levels exceeded between about 1 and 50% of time on over-land and over-sea paths, or for shorter time percentages on predominantly over-land paths when site shielding by obstacles near either of the terminals reduces the effect of ducting.

(iv) *Scatter from terrain and buildings* (not illustrated in Figure 13.4): This may become important as crossing of paths and high-density networks of paths become more commonplace. This mechanism is not currently covered by a prediction procedure, but there are plans to take account of it in the future in Recommendation 452. Scatter from aircraft may also be a problem, particularly if they fly frequently close to one of the path terminals, but few data exist.

(v) *Enhanced line-of-sight*: Interference-propagation mechanism of unobstructed transmission on line-of-sight paths may sometimes have levels raised by multipath diffraction effects or focusing effects.

(vi) *Surface super-refraction and ducting*: This is the most important short-term interference mechanism over water and in flat coastal land areas. It may be present for time percentages up to about 1% in temperate climates or 10% in some equatorial or subequatorial climates.

(vii) *Elevated-layer reflection and refraction*: The treatment of reflection and/or refraction from layers at heights up to a few hundred metres is of major importance as these mechanisms enable signals to bypass the diffraction loss of the terrain very effectively under favourable path-geometry situations.

(viii) *Hydrometeor scatter*: Hydrometeor scatter is particularly important as a possible source of interference because it may act virtually omnidirectionally. As with mechanisms (iii) and (vii), it may carry signals over deliberately placed or fortuitous diffraction screening obstacles.

For *Earth–space paths*, the direct, line-of-sight interference path is generally the most important. However, an indirect path via hydrometeor scatter can occasionally be significant and, for paths which are just 'transhorizon' from the Earth-based terminal, ducting may very occasionally produce near-free-space conditions.

13.4   Evaluation of co-ordination distance

13.4.1 General

Evaluation of co-ordination distance uses a simplified procedure (compared with prediction of interference levels) which, so far as path geometry is concerned, requires information only on the location of the Earth station. Distances from this Earth station are determined for each azimuth for which interference levels would exceed a certain criterion; this is done for two modes of propagation (clear air and hydrometeor scatter). The larger of the two at each azimuth defines the co-ordination distance, and the locus at the end point for each azimuth defines the co-ordination area. The criterion for acceptable loss is given in Recommendation IS. 847.

13.4.2 Clear-air conditions

For evaluation of co-ordination distances around an Earth station for clear-air conditions, Recommendation 620 considers the world categorised into three climatic zones: land (subdivided into coastal and other land), 'cold' seas (mainly those at latitudes above 30°) and 'warm' seas (mainly at latitudes below 30°). Coastal land lies below 100 m height and within 50 km of nearest sea (which includes inland water covering a circle at least 100 km in diameter). For each zone, the clear-air co-ordination distance d_1 (km) is found using values of the minimum permissible basic transmission loss $L_b(p)$ appropriate to the time percentage of interest p; thus:

$$d_1 = \left\{ L_b(p) - 120 - 20\log f - A(p) - A_h \right\} / \gamma \qquad (13.1)$$

where f = radio frequency (GHz)
  γ = specific attenuation (dB/km)
  A_h = horizon angle correction factor (dB) for the Earth station being
      considered
and $A(p)$ (dB) is a simple function of p.

Equation 13.1 is largely empirical, with the various factors in the expression for γ adjusted to fit available measured data for $0.001 < p < 1\%$, taking account of gaseous absorption and attenuation due to other effects, the magnitude of which depends on climatic zone, frequency and time percentage. Also, the value of A_h is derived from a simplified model which requires only a knowledge of the radio frequency and the elevation of the local horizon. It is also assumed that distant stations being considered have their (standard-gain) antenna beams directed towards the Earth station (i.e. the 'worst case'). The value of L_b (p) is obtained from Recommendation 847. To simplify the procedure further and ensure conservative or 'safe' values of co-ordination distance, the *minimum* value of this distance (below which co-ordination is always required) is assumed to be 100 km and land paths are assumed to be over relatively flat terrain. A *maximum* distance up to which co-ordination may be required is assumed to be 1200 km (in Recommendation 620; Recommendation 847 uses this value for a warm sea and 900 km for a cold sea). For land paths, practical values of co-ordination distance are much less than this. For mixed land–sea paths, calculations of loss for each successive path section are made, using the appropriate values of γ for each section, until the required total loss is achieved.

13.4.3 Scattering from hydrometeors

For evaluation of co-ordination distances for hydrometeor-scatter conditions, we are concerned with interference which may result from large scattering angles and beam intersections outside the great-circle plane, since the scattering is (to a first approximation) omnidirectional (for vertical polarisation). However, to ensure a conservative or 'safe' margin, the worst case of a backscatter geometry is assumed in the co-ordination calculations. Here both terminals are on the same side of a rain cell with beams intersecting in it and so attenuation owing to rain is at a minimum.

The basic model for hydrometeor scatter used by ITU-R enables signal level (transmission loss) to be calculated for rainfall rates corresponding to appropriate time percentages for any given distance, using a world-wide set of rain climatic zones. However, for ease of practical use, the results are inverted to produce probability distributions of transmission loss, with distance as a parameter. For simplicity, transmission-loss thresholds have been tabulated for the various rain climates for frequencies between 1 and 40 GHz, and a full evaluation is required only if the threshold proves less than the required transmission loss criterion of Recommendation 847 paralleling that used for clear-air conditions in Section 13.4.2 above.

The full procedure involves several assumptions. Since the interference signal results from a complex combination of scatter in the common volume and attenuation both inside and outside that volume (including gaseous-absorption effects), the validity of the basic model is vital. Indeed, models and empirical representations of rainfall which are adequate for predicting total attenuation along a horizontal terrestrial path of, say, 50 km (as used in Chapter 9) are not generally directly applicable to hydrometeor-scatter interference. For a given small time-percentage, the total-attenuation statistical rain model will tend to underestimate the net signal in the interference mode. Moreover, the variation of maximum rain-cell height with climate has to be allowed for. This leads to *maximum* distances for the hydrometeor scatter mode of about 350 km in equatorial latitudes and about 280 km in high latitudes (>60°). The *minimum* distance is again 100 km.

As a rough guide, hydrometeor scatter is of particular significance in the frequency range, say, 6–30 GHz. At lower frequencies the scatter is reduced, while at higher frequencies the effect of attenuation increases. But note that, for frequencies above about 20 GHz, the *backscatter* model (as used for co-ordination purposes) predicts significantly greater signals than a *forward-scatter* model.

For hydrometeor-scatter conditions, the position is somewhat more complex than that shown in eqn. 13.1 for clear-air conditions. First we have to deal with transmission loss L rather than basic transmission loss L_b (between ideal loss-free isotropic antennas) because the antenna gains are somewhat crucial to the common volume illuminated. (Strictly this should also be the case for clear-air troposcatter, but this mechanism is considered secondary to ducting, for which the beamwidth is not of primary consequence.) The loss $L(p)$ (dB) not exceeded for p% of time, to be compared with the acceptable loss, is given in Recommendation 620 by

$$L(p) = 168 + 20\log d - 20\log f - 13.2\log R(p) - G_T + A_b - C + \Gamma + E + A_g \qquad (13.2)$$

where f = radio frequency (GHz)
  d = path distance (km)
  R = surface-rainfall rate (mm h^{-1})
  G_T = gain of the terrestrial station antenna, assumed to be 42 dB
A_b is a term for the difference between Mie (higher-frequency) and Rayleigh (lower-frequency) scatter (zero for f < 10 GHz); C is a term for rain attenuation inside the scatter common volume (zero for f < 4 GHz); Γ is a term for rain attenuation outside the common volume; E is a term for loss in scatter coupling for heights above the melting layer; and A_g is the attenuation due to atmospheric gases.

Means of calculating these terms are given in Recommendation 620. Studies within COST Project 210 showed it to be crucial as to whether the scatter occurs below or above the melting layer; hence the term E to allow for the 6.5 dB/km drop in radar reflectivity above the melting layer. This term involves d, and the consequence of this, together with the term 20 log$d$, is that eqn. 13.2 cannot be inverted into the form of eqn. 13.1. As a result, a series of relationships has to be stored linking $L(p)$ with p through the rainfall rate for the climate of concern, which appears in the term A_b, C (twice), Γ and, of course, log$R(p)$. The distance d is a labelling parameter for each of the $L(p)$ relationships, and these then have to be inverted to obtain d_2 applying to the critical value of $L(p)$ obtained from

Recommendation 847. The co-ordination distance is the lowest of the value of d_1 (from eqn. 13.1) and d_2.

13.5 Prediction of interference levels

13.5.1 General

The ITU-R procedure for prediction of signal levels likely to cause interference in *clear-air* conditions (normally within the co-ordination area evaluated by the simplified procedure mentioned in Section 13.4) derives from the activity of the European COST Project 210 which produced many new measured data and collated these together with some earlier data which had needed very considerable attention (searching out of missing descriptions of the radio paths) before they could be used. These data were then assembled in a data bank and 54 paths (approximately 110 path years) were used to test various prediction methods. This led to the basis for a new ITU-R method which was adopted in 1992. Again, evaluation has to be made of each propagation mechanism in turn, and the worst case taken as dominant.

Recent improvements to the procedures for prediction for *clear-air* conditions ensure that its description is unambiguous, that it takes account of the important interference propagation mechanisms, and that it provides prediction values of signal level exceeded for 0.001 – 50% of time. In contrast to ten identified problems with the earlier procedure of Recommendation 452, it now: (i) has an integrated procedure designed for computer implementation, (ii) includes a reliable worst-month model for land, sea and mixed paths, (iii) has no significant discontinuities with time percentages or distance, (iv) embraces all significant propagation mechanisms verified by radio-meteorological research, (v) has the 'Δh' concept replaced by an improved treatment of terrain roughness, (vi) does not use standard climates for the troposcatter model, (vii) contains its own diffraction-calculation facilities, including a new 'k-factor' model for time percentages less than the median, (viii) has a solution to the earlier failure mode in the 'site-shielding' model, (ix) takes account of antenna height through an 'angular distance parameter' and height-gain function for sea duct coupling, and a full diffraction solution allowing for variation in antenna heights, and (x) has a new troposcatter model with an improved cumulative distribution.

Measurements were also made within COST Project 210 for *hydrometeor-scatter* conditions, and subsequent improvements to the procedure for prediction ensure that it now: (i) takes account of the variability of height of the melting layer, (ii) takes account of short and very short paths (down to a few kilometres) by using more detailed geometrical formulae which also allow for antenna elevations that can be well above the local horizons, (iii) uses Gaussian-shaped antenna patterns rather than simple cones, and (iv) introduces a well defined procedure to convert the transmission loss into the corresponding cumulative distribution. In addition, a number of the earlier ambiguities have been removed.

13.5.2 Line-of-sight

When calculations indicate that there is adequate clearance from ground (i.e. there is no obstacle in the first Fresnel zone, see Section 8.2.3.1), the basic transmission loss L_b (dB) not exceeded for p % of the time is given by

$$L_b = 92.5 + 20\log f + 20\log d - E_s(p) + A_g \qquad (13.3)$$

where f, d and A_g are as in eqn. 13.2 and $E_s(p)$ is the sum of the instantaneous focusing effects and multipath interference between rays arriving over different paths in the atmosphere. In some circumstances, $E_s(p)$ may reach 10 dB, giving rise to interference from a line-of-sight path which would not normally do so.

13.5.3 Diffraction

Once a path is known (or thought) to extend beyond the horizon, it will be appropriate to evaluate the diffraction loss for a classical single knife-edge diffraction, a rounded edge, double edges or multiple edges (by the cascaded-cylinder stretched-string method) according to ITU-R Recommendation P.526-4 [14]. Some information on this is given in Chapter 4, and is also given in other texts (e.g. [58] and [99]).

When the surface no longer has any distinguishable edges, the signal levels at a distance will generally be considerably lower, but they may still be significant for time percentages exceeding, say, 20% of the time. An approximation of spherical diffraction may be used, for which the diffraction loss will depend on the effective radius of the Earth, or k factor, i.e. on the refractive-index lapse rate. Again this is considered in Recommendation 526.

13.5.4 Tropospheric scatter

The procedures of Recommendation 452 are based on those tested and used by COST Project 210. For higher-level interfering signals caused by tropospheric scatter, the following equation has been found to be more effective than the more general case of eqn. 8.13 (which is for median values only). The basic transmission loss L_{bs} (dB) for troposcatter time percentage, $0.001 \le p \le 50\%$ (i.e. not to include low signal levels) is given by

$$L_{bs} = 190 + k(f) + 20\log d + 0.57\theta - 0.15N_0 + L_c + A_g - 10\{-\log(p/50)\}^{0.7} \quad (13.4)$$

where $k(f) = 30 \log f$ for $f \geq 2\,\text{GHz}$
           $= 3 + 20 \log f$ for $f < 2\,\text{GHz}$ (13.5)

f, d and A_g are as in eqn. 13.2, θ is the scatter angle (angular distance) (mrad), N_0 is the sea-level surface refractivity at the path centre, and L_c is the aperture-to-medium coupling loss.

13.5.5 Ducting and layer reflection

This is the primary mechanism for small time percentages on longer paths. The ITU-R procedures of Recommendation 452 are taken essentially from the work of COST Project 210. Prediction is based on a cumulative distribution L_{br} of basic transmission loss L_{ba} with a reference point β which undergoes independent translations along the time percentage and basic transmission-loss axes, depending on the path-input parameters. Here, β is referred to as the reference time percentage and L_{br} as the reference basic transmission loss. Figure 13.6 shows the principle of operation of the model and the factors influencing the alignment of the reference point against the two axes. In practice, the shape of the distribution is also slightly modified by the path parameters. Derivation of a prediction for this mode is straightforward, and involves calculations for the parameters β, L_{br} and the cumulative distribution L_{ba}.

The term β % is dependent on the path geometry, the terrain roughness and a term $\beta_0\%$ which is the time percentage for which super-refractive lapse rates exceeding 100 N units/km can be expected. These are provided on world maps for annual and worst-month conditions, but a more precise value may be calculated for north-west Europe, dependent only on the longest continuous land section (inland and coastal) and longest continuous inland section of the great-circle path. For Europe, the worst-month case is predicted from the case for the whole annual cumulative distribution. As mentioned in Section 13.2, a global model now incorporates this approach.

Calculation of L_{br} (dB) is achieved using

$$L_{br} = 102.5 + 20\log f + A_c + \gamma_d\theta' + A_g \quad (13.6)$$

where A_c = antenna-to-propagation mode-coupling correction, taking account of site-shielding diffraction losses at the terminals

γ_d = angular specific attenuation (dB/mrad), which depends on the refractive-index lapse rate and radio frequency

θ' = modified angular distance between the terminal antenna beams (angular distance plus modified terminal horizon elevations)

A_g = gaseous attenuation (dependent on the proportion of the path over sea).

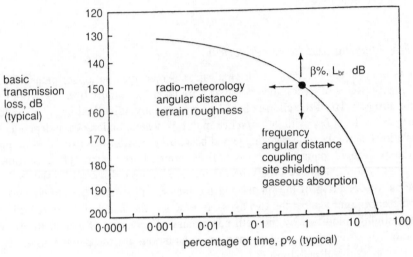

Figure 13.6   Principle of operation of the ducting and layer-reflection model

13.5.6 Hydrometeor scatter

For the propagation mechanisms considered above, the signal is propagated in the forward direction in the great-circle plane containing the transmitter and the receiver. By contrast, scattering from hyrometeors can be considered to be omnidirectional (for vertical polarisation), i.e. interference may be caused for large scattering angles (as well as small) with antenna beams intersecting outside the great-circle path. A basic relationship for hydrometeor scatter is eqn. 7.15. The procedure of Recommendation 452 covers the main problem of importance: i.e. interference between an Earth station and a terrestrial radio-relay station (see Figure 13.7). The developments of COST Project 210 were incorporated by ITU-R.

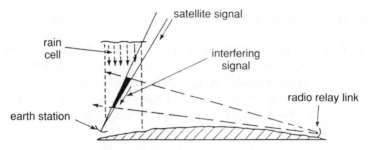

Figure 13.7   Interference by hydrometeor scatter between a terrestrial station and an Earth station

From Recommendation 452, the transmission loss L (dB) not exceeded for $p\%$ of time may be expressed by

$$L = 197 - 10\log\eta_E + 20\log d_T - 20\log f - 10\log Z_R + S + A_g - C \qquad (13.7)$$

where f and A_g are as in eqn. 13.2,

d_T = distance from the terrestrial (wide-beam) station to the scattering volume

η_E = antenna-efficiency factor (<1) of the Earth station

Z_R is the rain-scatter reflectivity factor per unit volume (mm^6m^{-3}) assumed independent of height below the mean $0\ ^\circ C$ isotherm height (or melting level) during rainy conditions (above this height, the reflectivity is assumed to decrease by 6.5 dB/km of additional height, a decrease allowed for in the factor C, below). It is given by

$$Z_R = 400 R(p)^{1.4} \qquad mm^6\ m^{-3} \qquad (13.8)$$

where $R(p)$ = point rainfall rate exceeded for $p\%$ of time

S = allowance for the deviation from Rayleigh scattering at frequencies above 10 GHz (becoming zero below 10 GHz)

C is the effective scatter transfer function which includes attenuation effects inside and outside the raincell. This requires integration over the whole of the scattering volume common to the two antenna beams. The worst case of polarisation is assumed, i.e. the same for transmission and reception.

To produce a cumulative distribution of transmission loss due to hydrometeor scatter, first the transmission loss is calculated for each combination of rainfall rate and rain height as if there was a deterministic dependence on these parameters, but assuming their statistical independence. Secondly, the probabilities of all rainfall-rate/rain-height contributions leading to the same transmission-loss values are summed to yield the total probability of that transmission loss.

Worst-month scaling factors Q (see Chapter 3) for hydrometeor scatter have been found to be about 3 at 1% of time, 4.5 at 0.01 % time and about 7 at 0.001 % time in Europe. These are based on a limited number of data at frequencies of 11 GHz and above, and are found to be very close to the European Q factors for rainfall rate. Therefore, in the absence of further measured Q factors for hydrometeor scatter, values for rainfall rate are used.

13.6   Interference on Earth–space paths

As well as interference between a terrestrial station and an Earth station, covered in Section 13.5, interference may exist between either of these and a space station, the interference being in either direction. The three principal propagation conditions which give rise to these interference paths are clear-air propagation, hydrometeor scatter and differential attenuation on adjacent Earth–space paths. Predictions are covered in Recommendation P.619-1 [21].

For clear-air conditions, the basic transmission loss L_b (dB) can be expressed as

$$L_b = 92.5 + 20\log f + 20\log d + A_g + A_D - G_s \qquad (13.9)$$

where f, d and A_g are as in eqn. 13.1. A_D is the attenuation due to beam spreading, which is dependent on elevation angle and refractive index lapse rate. It is less than 4 dB (at zero elevation angle) and is negligible at elevation angles greater than $4°$. G_s is the 'gain' (dB) due to scintillation, which is a function of frequency, antenna diameter of the Earth station, elevation angle and local climate. It may reach 5–6 dB for elevation angles of 1–$2°$ at about 7 GHz for time percentages less than 1%. However, G_s may be assumed to be zero for elevation angles above $5°$ and for time percentages greater than 1%.

In addition, hydrometeor scatter can cause energy from one system to enter the antenna beam of another system. This situation is only of significance when the main beams of the two systems intersect within that portion of the atmosphere in which hydrometeors can exist. Such interference is not usually system limiting, and it should be possible by suitable planning of geometries to avoid scattering volumes common to the two antennas. Prediction is by means of Recommendation 452, as in Section 13.5 above. By contrast, interference from a solar-power satellite could be severe, even from harmonics of the radio frequency.

Differential attenuation on adjacent paths may produce interference problems through rain effects or scintillation effects. In the former, if two space stations operating at the same frequency are separated from each other by a small angle, situations can arise where attenuation of a wanted signal may allow interference from the neighbouring station. Recommendation P.619-1 presents an empirical formula.   However, this mechanism is not expected to produce unacceptable interference. Scintillation effects may be significant only at low elevation angles, as stated above.

13.7   Interference reduction techniques

If circumstances dictate that unacceptable interference levels at a site are to be inevitable, it may be possible to use a number of techniques to reduce their effects, e.g. the use of signal-processing, modulation and demodulation techniques, the direction of antenna-sidelobe nulls (or the suppression of antenna sidelobes), the use of filters or signal cancellation and, finally, site shielding. These were briefly reviewed within COST Project 210, but the last of these received special attention as it can be used as part of the planning of the propagation path. A radio receiving terminal may be deliberately shielded from a potentially interfering transmitter by taking advantage of hills, vegetation, buildings, pits, embankments, fences or screens.

The *site-shielding* factor is the ratio of the received interference-signal power in the absence of the obstacle to that in its presence. This is less than the diffraction loss over the top of an obstacle if there is a height-gain effect present, as is usually the case. If the height gain is more than the diffraction loss, there will be an 'obstacle gain' rather than a site-shielding loss.   However, for high-signal conditions, incident energy may be close to a plane wave, the height-gain is zero so that the site-shielding factor is the same as the diffraction loss.

Chapter 14

Basic physics of the ionosphere

H. Rishbeth

14.1 Introduction

14.1.1 What is the ionosphere?

The ionosphere is the part of the atmosphere in which free electrons are sufficiently numerous to influence the propagation of radio waves. Its lower boundary is at about 60 km. Its upper boundary may be considered as the height where the transition begins from the oxygen ions of the F region to the hydrogen ions of the overlying protonosphere, around 600–1000 km, though the limits are not well defined.

The ionosphere was, in effect, discovered by Marconi when he sent radio signals across the Atlantic in 1901. Soon afterwards, Kennelly and Heaviside suggested that free electric charges in the upper atmosphere could cause the reflection of radio waves, although the related idea of an atmospheric conducting layer, as the site of the currents that produce small daily variations of the geomagnetic field (Section 14.1.3), had originated with Gauss in 1839. Ionospheric physics really began in 1924/25 with Appleton and Barnett, and Breit and Tuve, whose experiments measured the height of the conducting layer and revealed its stratified nature. It was soon verified that the ionisation is produced by solar ionising radiation. Many facts about the ionosphere were known by 1940 but its physics and chemistry were not really established until the 1950s and 1960s.

The ionisation in the upper atmosphere is of practical importance because of its role in radiocommunication, and important scientifically because the charged particles are easier to detect experimentally than the neutral gas, and thus act as useful tracers for studying the upper atmosphere. But even in the main layers of the ionosphere, less than 1% of the air is ionised. The ionosphere is electrically neutral to a high degree of approximation, as positive- and negative-charged particles are always created and destroyed together. The electron–ion gas of the ionosphere is referred to as a 'plasma'.

14.1.2 Basic physical principles

The most important ionospheric parameter is the electron density (or concentration) N. Typical density against height '$N(h)$ profiles' for midlatitudes are shown in Figure 14.1. The variation of N with height h and time t may be computed theoretically by solving the continuity equation (eqn. 14.1) with eqns. 14.2 and 14.3, for the velocities V and temperatures T of the ions and the electrons. Corresponding equations exist for every species of charged or neutral particle.

Continuity equation (conservation of mass)

$$\partial N \,/\, \partial t = \{\text{production}\} - \{\text{loss}\} - \{\text{transport}\} \qquad (14.1)$$

Force equation (conservation of momentum)

$$dV \,/\, dt = \{\text{driving force}\} - \{\text{drag}\} - \{\text{advection}\}^* \qquad (14.2)$$

Heat equation (conservation of energy)

$$\partial T \,/\, \partial t = \{\text{heating}\} - \{\text{cooling}\} - \{\text{conduction}\} \qquad (14.3)$$

14.1.3 The geomagnetic field

The behaviour of ions and electrons in the ionosphere is largely governed by the Earth's magnetic field, which may be approximated by a dipole inclined at $12°$ to the Earth's axis. In low and middle latitudes the geomagnetic field lines are closed, but at higher latitudes they are linked to the magnetosphere, the tenuous region around the Earth that is permeated by the geomagnetic field. The high-latitude ionosphere is profoundly affected by charged particles and electric currents, originating in the magnetosphere, that cause the aurora and magnetic storms. The particles and currents mainly enter the atmosphere in the auroral ovals, rings 2000–3000 km in diameter which surround the magnetic poles (Section 14.5.5). In turn, the magnetosphere is influenced by the solar wind — the stream of charged particles emitted by the Sun — and the interplanetary magnetic field.

---

*Advection is momentum transport by viscosity etc.

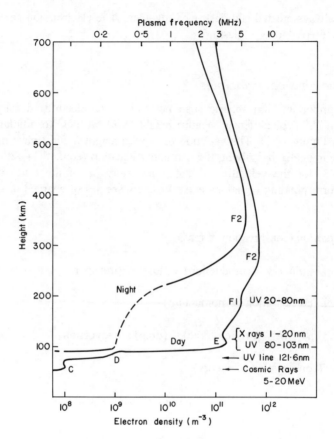

Figure 14.1   *Typical midlatitude N(h) electron-density profiles for moderate*
solar activity, showing the radiations that produce the ionospheric
layers
[*Contemp. Phys.*, 1973, **14**, 230 (Taylor & Francis, London)]

Magnetic storms tend to follow solar disturbances and the resulting
strengthening of the solar wind (Section 14.6). During storms, the electric
currents in the high-latitude ionosphere may locally perturb the geomagnetic field
by up to 5%, as compared with the quiet-day variations of 0.1% produced by
dynamo currents in the ionosphere (Section 14.4.5). The degree of geomagnetic
disturbance is characterised by 'planetary geomagnetic indices' (A_p and K_p), which
are computed for every 3 h of Universal Time from records made at several
magnetic observatories.

14.1.4   Experimental techniques for studying the ionosphere

14.1.4.1   Radio probing

Total *reflection* at plasma frequency $f_p = \sqrt{80.6N}$ (typically HF/MF/LF in daytime F/E/D layers, respectively). Used for routine sounding of ionosphere (ionosondes). Cannot observe above F2 peak, except by 'topside sounders' carried in satellites. The reflection process is influenced by the geomagnetic field.

Coherent scatter at frequencies $\gg f_p$ (HF, VHF, UHF). Scattering from irregularities of scale size $\lambda/2$ (λ = radar wavelength), which occur in the auroral oval and near the magnetic equator; also from meteor trails. Measures velocity by Doppler effect.

Incoherent scatter at frequencies $\gg f_p$ (VHF, UHF). Thomson scattering from thermal random irregularities (ion–acoustic waves) of scale size $\lambda/2$. Very weak (depends on electron cross-section, 10^{-28} m²; the total target area is ~1 mm²). Measures ion density, ion and electron temperatures, ion drift velocity, and indirectly some neutral atmosphere parameters.

14.1.4.2   Active remote sensing

(using man-made signals traversing the ionosphere)
  Radio phase and polarisation (Faraday rotation)
  LF, VLF reflection, MF partial reflection in lower ionosphere
  Radiowave absorption (signal strength)
  Radar studies of meteor echoes (neutral winds, density).

14.1.4.3   Passive remote sensing

from ground or space (using natural emissions)
  These techniques suffer from lack of height discrimination
  Absorption of cosmic radio waves; radio star refraction
  Aurora — interferometry and imaging
  Airglow — temperature and velocity of emitting gases.

14.1.4.4   In-situ measurements

(using instruments on rockets and satellites)
  Good space and time resolution but no systematic coverage
  Density, composition, temperature, velocity of ionised and neutral gas; electric and magnetic fields
  Tracking of objects and chemical trails released from spacecraft
  Analysis of satellite orbits (giving air density).

14.1.4.5   Active experiments

(e.g. RF heating, release of active chemicals).

14.2   The neutral atmosphere

14.2.1 Regions of the upper atmosphere

Figure 1.3 of Chapter 1 shows the main regions of the neutral atmosphere up to 200 km, in relation to the average vertical profile of temperature profile. They are the *troposphere* (0–12 km approximately), where the temperature decreases upwards from its ground-level value; the *stratosphere* (12–30 km), a turbulent region of fairly uniform temperature; the *mesosphere* (30–80 km), a region of higher temperature heated by the absorption of solar UV radiation by ozone; and the *thermosphere* above 80 km, strongly heated by solar extreme ultraviolet (EUV) and X-rays which dissociate and ionise the atmospheric gases and thus create the ionosphere. Above 600 km, in the *exosphere*, the air is too thin to be considered as a gas at all; the individual atoms move freely in satellite-like orbits controlled by the Earth's gravity.

Up to about 100 km the major gases, nitrogen (N_2) and oxygen (O_2), are well mixed by winds and turbulence; their concentration ratio of 4:1 does not change with height so the air has a mean molar mass $M \approx 29$. The minor constituents H_2O, CO_2, NO and O_3 are largely controlled by chemical reactions, and their distributions do not necessarily follow the same trend as the major gases.

Above 100 km, two changes occur. First, oxygen (but not nitrogen) largely becomes dissociated into atoms by the action of solar ultra-violet radiation. Secondly, turbulence ceases at the turbopause near 100 km; above this level, the gases are diffusively separated by the action of gravity so that atomic oxygen (molecular mass $M = 16$) progressively becomes more abundant than molecular nitrogen and oxygen ($M = 28$, 32). Higher still, the lightest gases, hydrogen and helium, become dominant in the exosphere, from which they slowly escape.

14.2.2 Hydrostatic equation and scale height

The vertical distribution of any gas is controlled by gravity, which balances the vertical pressure gradient. This balance is expressed by the hydrostatic or barometric eqn.

$$-dp \, / \, dh = \rho g = nmg \tag{14.4}$$

where p = pressure, ρ = density, n = concentration, m = particle mass and g = acceleration due to gravity. This equation may be combined with the perfect gas law

$$p = nkT = \rho RT \, / \, M \tag{14.5}$$

where k = Boltzmann's constant, R = gas constant and M = molar mass, and integrated with height to give the variation of pressure with height above any chosen base level h_0, at which $p = p_0$, so that

$$p = p_0 \, e^{-z} \sim p_0 \exp\{(h_0 - h)/H\} \tag{14.6}$$

where

$$z = \int(dh/H) \sim \{(h - h_0)/H\} \tag{14.7}$$

$$H = kT/mg = RT/Mg \tag{14.8}$$

H is known as the scale height of the gas and z as reduced height (measured in units of H). In eqns. 14.6 and 14.7 the simplified expressions that follow the symbol $\sim$ apply if H is constant with height. It is easily shown that the total number of particles in a vertical column above any height h_0 is $Hn(h_0)$, so H may be regarded as 'the thickness of the atmosphere'. Below 100 km the major gases have the same scale height, with a mean molar mass of 29, but in the diffusively separated atmosphere above 100 km each gas has its own scale height.

14.2.3 Upper-atmosphere temperature

In the thermosphere, the temperature varies with height. At the base, it is governed by the mesopause temperature (usually 150–180 K), but it increases rapidly upwards with a gradient dT/dh determined by the solar-heat input. Higher up, the thermal conductivity of the air becomes large and the temperature gradient flattens out. Above about 300 km, the temperature tends to a limit known as the exospheric temperature T_∞ which varies considerably with local time (LT), latitude, season and the 11-year sunspot cycle. Typically at midlatitudes T_∞ is 800 K at midday and 600 K at midnight at sunspot minimum, and 1400 K and 1100 K, respectively, at sunspot maximum. The temperature variations are mainly due to solar heating, which causes the atmosphere to expand on the daylit side of the Earth, creating a high-pressure region often referred to as the 'daytime pressure bulge'. T_∞ is raised locally in the auroral ovals because of the energy deposited by energetic particles and electric currents from the magnetosphere.

From observations of satellite orbits, which yield extensive data on air density, it is deduced that T_∞ is highest around 14 LT and lowest around 03 LT; the maxima and minima are in low latitudes near equinox but migrate seasonally, as shown, for example, in Figure 14.2 which is a simplified map of T_∞ at the June solstice. Localised heating in the auroral ovals is not shown in these maps.

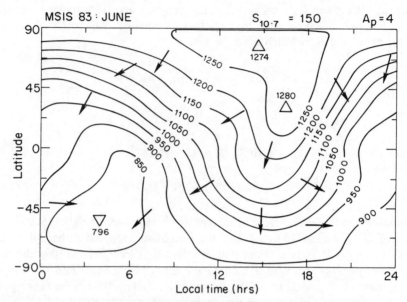

*Figure 14.2   Map of exospheric temperature at June solstice, for high solar
activity and quiet geomagnetic conditions, according to the
MSIS83 thermospheric model by A.E.Hedin, J. Geophys. Res. 1983
88, 10170, (Rutherford Appleton Laboratory)*

The chemical composition also varies with time and season. The O/N$_2$ ratio above 200 km is greater in winter than in summer, while localised variations of composition occur in auroral latitudes. These variations are caused in a complicated way by the global circulation described in Section 14.2.4.

14.2.4 Upper-atmosphere winds

The thermosphere is a vast heat engine driven by solar, auroral and interplanetary sources. The energy from these sources produces the temperature variations shown in Figure 14.2, giving rise to horizontal pressure gradients and to winds which blow in the directions shown by the arrows. The winds, together with the associated vertical air currents, form a global circulation which carries energy away from the heat sources and liberates it elsewhere. The neutral-air wind velocity U is controlled by the Coriolis force due to the Earth's rotation with angular velocity Ω, by the molecular viscosity of the air (coefficient μ), and by 'ion drag', the friction due to collisions between air molecules and ions. Ion drag exists because the ions are constrained by the geomagnetic field and cannot move freely with the wind, their velocity V being mainly controlled by electromagnetic forces (Section 14.4.4).

The equation of motion for the horizontal wind is

$$dU \,/\, dt = \mathbf{F} - 2\mathbf{\Omega} \times \mathbf{U} - KN(\mathbf{U} - \mathbf{V}) - (\mu \,/\, \rho)\nabla^2 \mathbf{U} \qquad (14.9)$$

in which

$$\mathbf{F} = -(1\,/\,\rho)(\partial \,/\, \partial x, \partial \,/\, \partial y)p \qquad (14.10)$$

is the driving force due to the horizontal pressure gradient, and K is a parameter relating to ion–neutral collisions. At great heights (small ρ) the ratio (μ/ρ) becomes very large so that $\nabla^2 \mathbf{U}$ becomes small; i.e. viscosity tends to destroy spatial variations of $\mathbf{U}$. As a result, $dU/dh \rightarrow 0$ and the wind varies little with height. Lower down, (μ/ρ) is smaller and small-scale velocity gradients are not smoothed out by viscosity, so $\mathbf{U}$ can vary appreciably with height.

The wind direction depends on the ratio of Coriolis force to ion drag. This may be illustrated by considering special steady-state cases of eqn. 14.9 in which $dU/dt = 0$ and viscosity is neglected. If Coriolis force is dominant and ion drag is small, as in the lower ionosphere, the wind blows at right angles to the pressure gradient. Then at latitude ϕ

$$\mathbf{U} = \mathbf{F} \,/\, (2\Omega \sin\phi) \quad \text{and} \quad (\mathbf{U} \perp \mathbf{F}) \qquad (14.11)$$

This is the situation familiar from weather maps for the lower atmosphere, with the wind blowing along the isobars of constant pressure. A very different situation exists in the daytime F layer, where ion drag is large and the wind is almost parallel to the pressure-gradient force, so that

$$\mathbf{U} = \mathbf{F} \,/\, (KN \sin I) \quad \text{and} \quad (\mathbf{U} \,\|\, \mathbf{F}) \qquad (14.12)$$

where I is the magnetic dip angle. In the intermediate case, both ion drag and Coriolis force are significant and the wind is inclined to $\mathbf{F}$ at the angle

$$\psi = \arctan/\left\{ (2\Omega \sin\phi) \,/\, KN \sin I \right\} \qquad (14.13)$$

and its magnitude is given by

$$\mathbf{U} = \mathbf{F} \,/\, \sqrt{\left(K^2 N^2 \sin^2 I + 4\Omega^2 \sin^2 \phi \right)} \qquad (14.14)$$

The schematic wind vectors in Figure 14.2 illustrate this. They are nearly parallel to the temperature and pressure gradient by day, but are slightly deflected by Coriolis force at night. If the ions are drifting under the influence of an electric field, they set the air into horizontal motion; i.e. the ion-drag term in eqn. 14.9

acts as a driving force. In high latitudes, such 'electrically driven' winds are strong, but in addition there are localised pressure gradients due to heating by auroral particles and electric currents. Both pressure-gradient and ion-drag terms must therefore be considered. At lower latitudes, thermal winds driven by solar heating dominate, though electric fields are important in the vicinity of the magnetic equator.

The neutral air velocity is subject to equations of continuity and energy, as well as the equation of motion (eqn. 14.9). Production and loss processes are unimportant for the major constituents of the neutral air, and the continuity equation (eqn. 14.1) reduces to

$$\partial n / \partial t = -\mathrm{div}(n\mathbf{U}) \tag{14.15}$$

The pressure distribution and the wind velocity automatically adjust themselves to satisfy eqn. 14.15. Thus any divergence (convergence) of the horizontal winds is balanced by upward (downward) winds so that div $(n\mathbf{U})$ is small. Winds that blow along isobars, as in the lower atmosphere, are almost divergence-free and so are ineffective in removing the horizontal pressure differences that drive them.

14.2.5 Atmospheric tides

Thermospheric winds may be regarded as part of a complex system of atmospheric tides, which contains many different components. Tides are 'forced' (driven) by the heating effect of the Sun or the gravitational attraction of the Sun and Moon. Unlike the case of marine tides (dominated by the Moon's gravitational attraction) the atmospheric tidal motions are controlled by solar thermal forcing by three main causes: (i) the absorption of solar EUV and X-rays in the thermosphere; (ii) the absorption of solar ultra-violet radiation in the ozone layer; (iii) the heating of the ground and lower atmosphere by visible and infra-red radiation. The resulting tidal motions comprise a complicated set of diurnal (24-hour), semidiurnal (12-hour) and other modes, plus a weak lunar gravitational (12.4-hour) tide. The tidal winds produce geomagnetic variations through the dynamo effect (Section 14.4.5).

14.2.6 Gravity waves

The atmosphere contains many kinds of waves, ranging in period from the 26-month quasibiennial oscillation to infrasonic waves of periods <1 s, some of which are associated with aurora or seismic activity. They include planetary waves, with wavelengths of order 1000 km and periods of days; tidal waves as described in Section 14.2.5; and acoustic and gravity waves. These waves represent solutions of wave-dispersion equations which are derived from the appropriate form of eqns. 14.1–14.3. Acoustic waves are oscillations of period <4 min, in which the restoring force is provided by compression and gravity is unimportant. Gravity waves are oscillations controlled by the buoyancy of the air, i.e. by gravity, and have a minimum period of about 5 min (the Brunt–Väisälä period) in the thermosphere. Unlike ocean waves which are virtually confined to the surface, gravity waves permeate the whole upper atmosphere. 'Short-period' and 'medium-period' gravity waves have periods of 5–30 min, wavelengths of 100–300 km, horizontal speeds of 100–200 m/s, and are generated in the lower atmosphere by storms, by winds blowing over mountains, and occasionally by man-made explosions. 'Long-period' gravity waves have periods of 0.5–3 h, horizontal wavelengths of 1000–4000 km, speeds of 400–700 m/s, and are generated by disturbances in the thermosphere, mainly in the auroral oval. They can travel for thousands of kilometres, and are prevalent at times of geomagnetic disturbance.

Gravity waves have a complicated phase structure. In addition to the horizontal propagation, the phase velocity has a downward component which gives the appearance of a downward travelling wave, although the group velocity and energy propagation are inclined slightly upwards. In the lower and middle atmosphere, the wave's kinetic energy density is independent of height, so the wave amplitude increases upwards as the air density decreases. This increase cannot continue indefinitely, and is limited in three ways: (i) dissipation by ion drag or viscosity; (ii) partial or total reflection, which depends on the thermal structure of the atmosphere; and (iii) nonlinear effects, when the amplitude becomes so large that the wave 'breaks' into smaller-scale motions.

14.3 Ionospheric layers

14.3.1 Ionising solar radiations

The principal ionospheric layers are produced by extreme ultra-violet (EUV) and X-rays, emitted from the Sun as spectral lines and continuum radiation. The level in the atmosphere to which any wavelength λ penetrates depends on its absorption cross-section σ, and the gases it can ionise are those for which λ is shorter than the ionisation limit (80 nm for N_2, 91 nm for O and H, 103 nm for O_2 and 134 nm for NO). The principal radiations that produce the ionospheric layers are (Figure 14.1):

F1 layer (150–180 km): EUV 20–80 nm
E layer (100–120 km): X-rays 1–20 nm, EUV 80–103 nm (especially Lyβ 102.6 nm)
D layer (70–90 km): Ly α 121.6 nm (NO), X-rays 0.1–1 nm
C layer (50–70 km): (an insignificant layer) MeV cosmic rays

No ionising photon radiation exists which has a large enough σ to be strongly absorbed above the F1 layer, so the radiation that produces the F1 layer must also produce the F2 layer, and the greater electron density of the latter is attributed to a smaller rate of loss of electrons (Section 14.4.2). Charged particles can also ionise the air, and the height to which they penetrate depend on their energy or hardness. Such ionisation is important in high latitudes, particularly in the auroral ovals.

14.3.2 *Chapman layers*

The simple Chapman theory of the production of an ionised layer starts by assuming: (*a*) monochromatic solar radiation acting on (*b*) a plane atmosphere containing (*c*) a single ionisable gas with (*d*) a constant scale height. All four simplifications (*a–d*) can be removed, but the equations then become more complicated.

Ionising radiation, initially of intensity I_∞, strikes the top of the atmosphere at zenith angle (obliquity) χ, and its intensity $I(h)$ decreases as it travels downwards, according to the equation

$$dI \,/\, I = \sigma n\, dh \sec \chi = -d\tau \tag{14.16}$$

where σ is the absorption cross-section and τ is optical depth. By integrating eqn. 14.16 along the path of the radiation, from the top of the atmosphere down to height h, and using the property that $\int n\, dh = n(h)H$ (Section 14.2.2), it is found that

$$I(h) = I_\infty \exp\left(-n\sigma H \sec \chi\right) = I_\infty\, e^{-\tau(h)} \tag{14.17}$$

From eqn. 14.17 it may be seen that, for oblique sun, attenuation begins at a higher height (as shown by dots in Figure 14.3) than for overhead sun. The production rate q is found by multiplying the intensity of the radiation (broken line) by the gas concentration n (continuous curve) and by the cross-section σ and ionising efficiency η. Thus

$$q(h) = I(h)\eta\sigma n(h) = I_\infty \eta\sigma n(h) e^{-\tau} \tag{14.18}$$

On differentiating eqn. 14.18 it is found that the peak value of q occurs at the height where $\tau = 1$ (unit optical depth). The height of peak production is given by

$$z_m = \ln \sec \chi \tag{14.19}$$

and the peak value of q is

$$q_m = q_0 \cos\chi \quad \text{where} \quad q_0 = \left(\eta I_\infty / e H\right) \tag{14.20}$$

Using the reduced height z, q is given by the Chapman formula

$$q(z, \chi) = q_0 \exp\left(1 - z - e^{-z} \sec\chi\right) \tag{14.21}$$

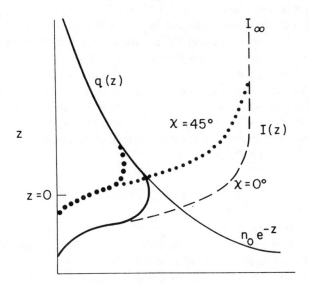

Figure 14.3   *Theoretical curves of ion production for overhead sun (full lines) and oblique sun (dotted lines), illustrating eqns. 14.17 and 14.18 (Rutherford Appleton Laboratory)*

14.3.3 Critical frequency

The 'ordinary-mode' critical frequency f_0 of an ionospheric layer is related to its peak electron density N_m by

$$f_0 = 80.6 N_m \quad \text{or} \quad f_0 \sim 9\sqrt{N_m} \tag{14.22}$$

where the numerical factors in MKS units are derived from physical constants. There is also a slightly higher extraordinary-mode critical frequency f_x, which exists because of the double refraction of radio waves caused by the Earth's magnetic field. For a simple Chapman layer in which electrons are produced according to eqn. 14.21 and recombine at a rate proportional to N^2 (Section 14.4.1), the critical frequency depends on solar zenith angle χ and the mean sunspot number R according to the formula

$$f_0 = A(1 + aR)(\cos\chi)^n \quad \text{where} \quad n = 0.25 \qquad (14.23)$$

where A and a are constants. The daytime E and F1 layers follow this formula fairly well when χ varies with time of day, season, and latitude, though the index n actually varies between about 0.2 and 0.35 (Section 14.5.2). The incident solar flux I_∞ varies with the solar cycle and so f_0 varies regularly with the sunspot number. Figure 14.4 shows the regular seasonal and solar-cycle variations of the E and F1 layers at Slough, UK, and Port Stanley, Falkland Islands, together with solar and interplanetary parameters. The normal D layer (though not possessing a measurable critical frequency) also varies quite regularly with $\cos\chi$. The F2 layer does not; careful examination of the seasonal ripple of Slough f_0F2 shows that it is out of phase with those of the F1 and E layers, which do conform to the variation of $\cos\chi$. At Port Stanley, however, the F2 layer has a predominantly semiannual variation, with maxima at the equinoxes.

14.3.4 Ionospheric models

Ionospheric models are numerical or graphical schemes which represent the variations of ionospheric parameters (usually electron density) with time, height and position, for different seasons and different levels of solar and geomagnetic activity. There are two basic types of model: 'empirical' and 'computational', though 'hybrids' between the two types have been devised.

Empirical models are descriptive; they comprise mathematical formulas or computational algorithms (e.g. sets of spherical harmonics), of which the numerical coefficients are determined by fitting to actual data. Their purpose is to give numerical values for scientific or engineering use, not to represent the physical processes of the ionosphere. Computational models (sometimes called 'first-principles' models) are derived by numerically solving the basic conservation equations (eqns. 14.1–14.3) and the equations of state for the ionospheric electrons, and often for the ions and the neutral gases too. Apart from the challenging task of setting up and solving this complex system of equations (ideally for the three-dimensional time-dependent situation) the main problems are to set realistic boundary conditions and to choose accurate values of the many parameters that are required.

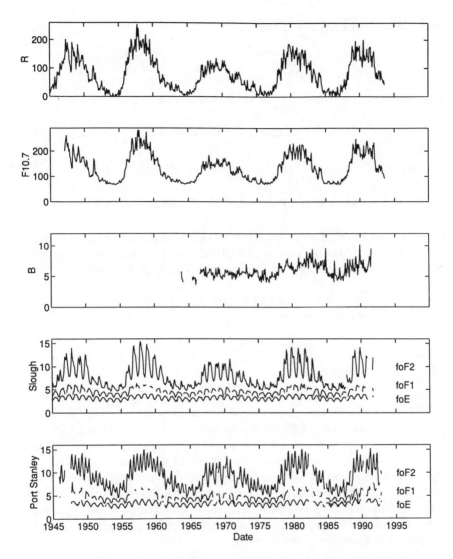

*Figure 14.4   Almost half a century of solar terrestrial data: monthly mean or
median values of Zürich sunspot number R, solar 10.7 cm flux
density, magnitude of the interplanetary magnetic field B near the
Earth, and noon critical frequencies of F2, F1 and E layers at
Slough (lat. 52N) and Port Stanley (lat. 52S) (Rutherford Appleton
Laboratory)*

14.4   Basic ionospheric processes

14.4.1 Continuity equation

The continuity equation (eqn. 14.1) for electrons and ions is

$$\partial N / \partial t = q - L(N) - \mathrm{div}(NV) \tag{14.24}$$

The loss term $L(N)$ can generally be simplified as follows:

$$\text{E and F1 layers: } L(N) = \alpha N^2 \qquad \text{F2 layer: } L(N) = \beta N \tag{14.25}$$

where α, β are the square-law and linear loss coefficients. The transport term can generally be simplified by omitting horizontal motion which contributes little to eqn. 14.24, and including only the vertical drift velocity W, in which case

$$\mathrm{div}(NV) = \partial(NW) / \partial h \tag{14.26}$$

14.4.2 Ion chemistry and loss processes

Reactions in the ionosphere have to conserve energy and momentum, besides satisfying the principles of quantum theory. One consequence is that atomic ions cannot recombine easily with electrons, unless a third body takes part in the reaction and absorbs the excess energy. This requires a three-body collision, which is only likely to occur below 100 km where the air density is sufficiently high. Molecular ions, however, recombine easily at any height, since they can dissociate to give two products. Throughout the ionosphere, reactions may leave atoms and molecules in excited states, which subsequently decay by emitting radiation. This is the origin of the upper atmosphere luminescence, the airglow. To summarise for the main layers:

D layer: The atmosphere is relatively dense, so both two-body and three-body collisions occur and complicated chemical reactions take place. Electrons may become attached to atoms or molecules to form negative ions. Negative ions are more abundant than electrons at all heights below 90 km at night but, being destroyed by visible and ultraviolet sunlight, they are only abundant below 70 km by day. Many ions, both positive and negative, become attached to water molecules to form complex clusters.

E layer: The atmosphere is more rarified and only two-body collisions occur, so atomic ions cannot recombine easily with electrons, but are readily converted to molecular ions by reactions such as eqns. 14.30–14.32 below. The overall result is that the positive ions are mostly molecular, NO^+ and O_2^+, and the loss coefficient α is of the square-law type in eqn. 14.25. Some metal atoms (Fe^+, Mg^+, Ca^+), traces of which are deposited in the atmosphere by meteors, become ionised to produce long-lived metallic ions that are the main ionic component of midlatitude sporadic E layers.

F layer: The ions are predominantly NO^+ and O_2^+ in the F1 layer, O^+ in the F2 layer.

The simplified chemistry of the E and F layers may be summarised as follows:

Production: The major gases are ionised by EUV photons (hv) of the radiations mentioned in Section 14.3.1, giving O^+, N_2^+ and some O_2^+ ions, plus electrons. For example:

$$O + hv \rightarrow O^+ + e^- \tag{14.27}$$
$$N_2 + hv \rightarrow N_2^+ + e^- \tag{14.28}$$
$$O_2 + hv \rightarrow O_2^+ + e^- \tag{14.29}$$

Charge transfer: Ions collide with neutral molecules and undergo 'charge-transfer' or 'charge-exchange' reactions, the most important of which are:

$$O^+ + O_2 \rightarrow O_2^+ + O \tag{14.30}$$
$$O^+ + N_2 \rightarrow NO^+ + N \tag{14.31}$$
$$N_2^+ + O \rightarrow NO^+ + N \tag{14.32}$$

Recombination: Ions recombine with electrons and dissociate:

$$NO^+ + e^- \rightarrow N + O \tag{14.33}$$
$$O_2^+ + e^- \rightarrow O + O \tag{14.34}$$
$$N_2^+ + e^- \rightarrow N + N \tag{14.35}$$

This photochemical scheme accounts for the general form of the $N(h)$ profiles as shown in Figure 14.5 and for the following facts:

(*a*) Despite the abundance of neutral N_2, very few N_2^+ ions exist, as they are rapidly destroyed via eqn. 14.32 as well as by eqn. 14.35.
(*b*) NO^+ ions are abundant because they are rapidly formed via eqns. 14.31 and 14.32, even though neutral NO is very scarce.

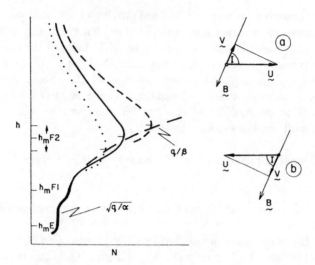

Figure 14.5   Idealised electron-density profiles in the E and F1 layers, with square-law loss coefficient; and in the F2 layer, with linear loss coefficient. [Contemp. Phys. 1973, 14, 244; (Taylor & Francis, London)]

(c) Below about 200 km molecular ions dominate; this is because the O^+ ions are rapidly converted to NO^+ and O_2^+ via eqns. 14.30 and 14.31 by the relatively abundant N_2 and O_2.

(d) Above about 200 km, the relative scarcity of N_2 and O_2 makes eqns. 14.30 and 14.31 very slow. Once formed, however, the NO^+ and O_2^+ ions rapidly recombine with electrons via eqns. 14.33 and 14.34. As a result, O^+ is the dominant ion.

(e) The loss term in eqn. 14.24 takes the form αN^2 at heights where NO^+ and O_2^+ ions dominate, i.e. in the E and F1 layers, cf. eqn. 14.25.

(f) The loss term in eqn. 14.24 takes the form βN in the F2 layer, cf. eqn. 14.25, where

$$\beta = k'n[O_2] + k''n[N_2] \qquad (14.36)$$

k' and k'' are the rate coefficients of reactions 14.30 and 14.31.

(g) The F1 layer is the transition between (e) and (f). It is a distinct feature with its own critical frequency at certain times (mainly summer and at sunspot minimum) but not at other times (e.g. winter). Its appearance may be shown to depend on the relative values of q, α and β.

(h) Above the F1 layer, the electron density increases upwards, because the loss coefficient β (which depends on the N_2 and O_2 concentrations) decreases upwards faster than the production rate q (which depends on the O concentration).

14.4.3 *Equation of motion of the charged particles*

The simplified equations of motion for singly charged positive ions and electrons are

$$m_i dV_i \,/\, dt = m_i \boldsymbol{g} + e\boldsymbol{E} + e\boldsymbol{V}_i \times \boldsymbol{B} + m_i \nu_i (\boldsymbol{U} - \boldsymbol{V}_i) - \nabla p_i \,/\, N \simeq 0 \qquad (14.37)$$

$$m_e dV_e \,/\, dt = m_e \boldsymbol{g} - e\boldsymbol{E} - e\boldsymbol{V}_e \times \boldsymbol{B} + m_e \nu_e (\boldsymbol{U} - \boldsymbol{V}_e) - \nabla p_e \,/\, N \simeq 0 \qquad (14.38)$$

The acceleration terms dV_i/dt, dV_e/dt are negligible for the large-scale motions considered here, so the sum of the forces acting on the particles can be set equal to zero. The forces are due to gravity $\boldsymbol{g}$, the electrostatic field $\boldsymbol{E}$, the geomagnetic field $\boldsymbol{B}$, frictional drag by the neutral air, and the gradient of the ions' or electrons' own partial pressure p_i, p_e. The symbol e denotes the elementary positive charge (so that $-e$ is the charge on an electron), and ν_i, ν_e are the collision frequencies of ions and electrons with neutral particles.

14.4.4 *Motion produced by winds and electric fields*

Only a few special cases of eqns. 14.37 and 14.38 need be discussed here. They concern the flow of electric currents in the E layer (Section 14.4.5) and the behaviour of the F2 layer (Section 14.4.6). A single kind of positive ion is assumed and negative ions are ignored (the equations could be generalised to remove these restrictions). Deleting the acceleration, gravitational and pressure gradient terms, eqns. 14.37 and 14.38 can be solved to give the following results:

For motion $\|\boldsymbol{B}$: if the neutral air wind has a component parallel to $\boldsymbol{B}$, it moves both ions and electrons freely in that direction. If the electric field has a component parallel to $\boldsymbol{B}$, it drives electrons and ions in opposite directions, i.e. there is a high electrical conductivity.

For motion $\perp \boldsymbol{B}$: the direction of motion depends on the ratio of the collision frequency ν (with neutral particles) to the magnetic gyrofrequency $\omega = Be/m$. Since ν decreases exponentially upwards, while ω is almost constant, the ratio ν/ω falls off rapidly with height; it is different for ions and electrons. The important cases are:

$\nu \gg \omega$ (up to 100 km for ions, 60 km for electrons): a wind drives the ions and electrons at its own speed U, and the magnetic field has no effect. An electric field moves ions parallel and electrons antiparallel to itself, but slowly because of the high collision frequency, so the conductivity is low.

$v \sim \omega$ (at about 125 km for ions, 75 km for electrons): a wind or an electric field drive ions and electrons in directions inclined to the applied force. The angle is given by arc tan (v/ω), and is thus $45°$ at the level where $v = \omega$.

$v \ll \omega$ (above 150 km for ions, 90 km for electrons): the particles move at $90°$ to the applied force. An electric field moves both ions and electrons at speed E/B in the same direction; this is the electromagnetic drift velocity (Hall drift), written vectorially as $E \times B/B^2$. Since electrons and ions move at almost exactly the same velocity, there is virtually no current. A neutral air wind U produces practically no motion of ions and electrons across the magnetic field.

An electric current flows if the positive ions and electrons move at different velocities, i.e. at heights between 60 and 150 km where the ratio v/ω is different for ions and electrons. The current density is given by 'Ohm's law for the ionosphere' which can be expressed in two ways:

$$j = Ne(V_i - V_e) = \sigma.E \qquad (14.39)$$

Equation 14.39 shows that the conductivity σ (a tensor quantity because j and E are usually not parallel) depends on the *difference* between the ion and electron velocities, as well as on the electron density N.

14.4.5 Dynamo theory

At midlatitudes, the electric current in the ionosphere is mostly produced by the atmospheric dynamo in the E layer, driven by the tidal winds described in Section 14.2.5. At this height, $v \gg \omega$ for ions but $v \ll \omega$ for electrons (Section 14.4.4). The winds accordingly move the ions across the geomagnetic lines of force, but cannot move the electrons. As the ions and electrons move differently, a current flows as given by eqn. 14.39, and charge separation takes place, building up an electric polarisation field which modifies the ion and electron motions. The current and the polarisation field constantly adjust themselves in such a way that no further charge separation occurs. Very little current flows at heights below 100 km because the electron density N is so small (Figure 14.1); the bulk of the current flows between about 100 km and 125 km (the latter height being where $v = \omega$ for the ions). At night N is small at these heights so the E layer current is weak, though some current flows in the F layer.

At the dip equator (for reasons connected with the special geometry of the magnetic field) a strong vertical polarisation field develops in the E layer. It has the effect of creating a very high electric conductivity in the east-west direction, giving rise to a strong east-west current known as the 'equatorial electrojet' that flows in a narrow belt extending about $3°$ either side of the dip equator.

In addition to the wind-driven dynamo fields and currents in the ionosphere, electric fields and currents are generated in the magnetosphere and transmitted along geomagnetic field lines to the high-latitude ionosphere, and spread to some extent to lower latitudes. They are responsible for the large-scale convection described in Section 14.5.6. Strong localised east–west currents known as the 'auroral electrojets' flow in the auroral ovals, especially during magnetic storms, and are major sources of heating at high latitudes (Section 14.2.3).

14.4.6 The F2 peak

Section 14.4.2 (*h*) explained why the electron density N increases upwards from the F1 layer, because of the upward increase of the ratio q/β. The increase of N stops when gravity eventually controls the ion distribution. The F2 peak thus occurs at the height where chemical control gives way to gravitational or diffusive control. At the F2 peak, the transport terms in the continuity equation (eqn. 14.25) are comparable with production and loss, which are roughly in balance below that height. The $N(h)$ profile then looks roughly as in Figure 14.5, in which the sketches on the right show the plasma drift V produced by a horizontal wind U in the neutral air, and B represents the geomagnetic field (dip angle I). Upward field-aligned drift is produced by a wind blowing towards the magnetic equator, as in sketch (*a*); downward drift is produced by a wind towards the magnetic pole, as in sketch (*b*).

Introducing the diffusion coefficient D for the ions, which is inversely proportional to the ion–neutral-collision frequency, the F2 peak is governed by the relations (in which the suffix *m* denotes values at the peak height h_mF2):

$$\beta_m = D_m / H^2 \tag{14.40}$$

$$N_m \sim q_m / \beta_m \sim I_\infty n[\text{O}] / \{k'n[\text{O}_2] + k''n[\text{N}_2]\} \tag{14.41}$$

Thus N_m depends on the atomic/molecular ratio of the neutral air. The height h_mF2 tends to lie at a fixed pressure level in the atmosphere, i.e. a fixed value of the reduced height z defined by eqn. 14.7. The height of the peak can be shifted by a neutral air wind or an electric field. A horizontal wind blowing towards the magnetic equator drives the ionisation up magnetic-field lines, raising the peak and increasing N_m. Opposite effects are produced by a poleward wind (Figure 14.5). Equatorward winds tend to occur at night, poleward winds by day. Eastward/westward electric fields produce upward/downward electromagnetic drift, i.e. the vertical component of the drift velocity $E \times B/B^2$ (Section 14.4.4), though the effect of this drift on the midlatitude F2 layer may be shown to be greatly reduced by the reaction of the neutral air (the 'ion-drag effect').

14.4.7 Topside ionosphere

At heights well above the F2 peak, the ion and electron distribution is diffusively controlled, gravity being balanced by the pressure-gradient terms in eqns. 14.37 and 14.38, much as is the neutral air, cf. eqn. 14.4 in Section 14.2.2. Above about 700 km there is a gradual transition from the O^+ ions of the F2 layer to the H^+ ions of the protonosphere, some He^+ ions being present also. An important feature of the topside F2 layer is the flux of plasma along the geomagnetic field lines to or from the overlying protonosphere, which involves the charge-exchange reaction

$$O^+ + H \Leftrightarrow H^+ + O \tag{14.42}$$

By day the F2 layer supplies ions to the protonosphere through the 'forward' reaction, but at night the 'reverse' reaction, accompanied by downward flow, plays an important part in maintaining the O^+ content of the midlatitude F2 layer, so the protonosphere acts as a reservoir of plasma. At higher latitudes, where the geomagnetic-field lines are not closed, the ionosphere (particularly at auroral latitudes) acts as a source of energetic O^+ ions to the magnetosphere.

14.5 Ionospheric phenomena

14.5.1 D layer

This chapter does not survey the complex chemistry of the lower ionosphere, and only a few salient points are mentioned. From the radio-propagation point of view, the D layer plays an important part in LF/VLF propagation; at MF and HF it is important because of its absorbing properties, which stem from the relatively high air density and consequent large collision frequency between electron and neutral molecules. Increases of D-layer electron density therefore affect radio propagation. Daytime-absorption measurements show a fairly regular dependence of D-layer ionisation on cos χ. The attachment of electrons to form negative ions at sunset (Section 14.4.2) and their detachment at sunrise cause changes of LF/VLF reflection height. On some winter days the electron density is abnormally high, apparently because of enhanced concentrations of nitric oxide (ionised by solar Lyman α, Section 14.3.1); this anomaly may be connected with dynamical phenomena such as planetary waves.

Some D-layer processes are linked with meteorological phenomena at lower heights. The temperature increases (stratospheric warmings) that occur in winter at moderately high latitudes, and represent major changes in the large-scale circulation, are accompanied by increases of radio-wave absorption in the D layer. The auroral D layer is often enhanced by particle precipitation (Section 14.5.5).

14.5.2 E layer

The normal E layer shows slight departures from idealised Chapman-layer behaviour (Section 14.3.3), owing to complications such as the scale/height gradient dH/dh and drifts due to tidal electric fields. In consequence, the index n in eqn. 14.23 usually exceeds 0.25.

In the past, many attempts were made to determine q and α from observations during solar eclipses. They proved unsuccessful because much of the ionising radiation originates in the solar corona and is not fully cut off by the eclipsing Moon.   More reliable data have come from laboratory measurements of recombination coefficients and spacecraft measurements of solar-radiation fluxes.

14.5.3 Sporadic E

Sporadic E or Es is a generic term for thin layers of enhanced ionisation that occur at around 100–120 km, generally in a rather irregular fashion that is not predictable in detail. They are important in practice because they are often dense enough to affect radio propagation seriously.   Twelve types of Es have been defined, distinguished by their appearance on ionograms, but in reality there may be only three main physical types:

(*a*) Midlatitude Es layers, sometimes much denser than the normal E layer or even the F2 layer, are thought to be produced by wind shears (small-scale gradients of wind velocity).   The sheared wind interacts with the geomagnetic field in such a way as to compress the long-lived metal ions into thin layers, typically 1 km thick but several hundred km in horizontal extent. The wind shears may be associated with large-scale tidal motions, which have a certain regularity, or with smaller-scale winds or gravity waves which have little or no predictability.
(*b*) Equatorial Es is a plasma instability caused by the high electron-drift velocity (i.e. the large current density) in the daytime equatorial electrojet (Section 14.4.5).
(*c*) Auroral Es layers are produced by precipitation of keV electrons, particularly during auroral substorms (Section 14.5.5).

14.5.4 F2-layer behaviour

The F2 layer is much less regular than the E layer, and shows many features that differ from those of a purely solar-controlled layer. Examples may be seen in the world map presented in Figure 17.3 of Chapter 17, and in Figure 14.6, which shows the F2-layer variations at a midlatitude station throughout a complete year, the data being arranged in lines of 27 days to display any recurrences attributable to the solar 27-day rotation period.   Storm phenomena are dealt with in Section 14.6.3.

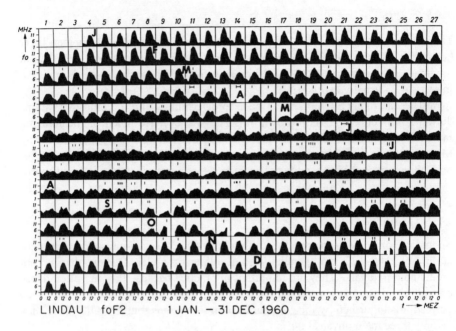

Figure 14.6   *'Lindau mountain diagram' showing the critical frequency* f_0F2 *at Lindau (Latitude 52°N) throughout the post-sunspot-maximum year 1960*

Each box represents 00–24 LT and 1–16 MHz; the letters mark the first day of calendar months [Max-Planck-Institut für Aeronomie, Katlenburg-Lindau, Germany]

The shapes of the day/night variations of N_mF2 and h_mF2 are largely due to neutral air winds, and to a lesser extent electric fields. Around sunset, the meridional component of the wind changes from equatorward to poleward and the resulting F2-layer drift changes from downward to upward (see Figure 14.5). This is a probable cause of the commonly observed increase of N_mF2 in the late evening. The common predawn decrease of N_mF2 is probably connected with the poleward-to-equatorward wind reversal around that time. In detail, these phenomena depend on the geometry of the geomagnetic field. For example, when sunrise at the magnetically conjugate point occurs before local sunrise, the inflow of photoelectrons from the opposite hemisphere can produce increases of electron temperature and consequent effects on the electron density.

At night the F2 layer is maintained (*a*) by being raised by winds (Figure 14.5) to heights where the loss coefficient β is small; (*b*) by inflow from the protonosphere (Section 14.4.7); (*c*) by weak EUV fluxes from the night sky; and perhaps (*d*) by energetic particles.

The anomalous seasonal variation (Figure 14.4) is attributable to changes of the $[O/N_2]$ ratio of the neutral air produced by the global circulation (Section 14.2.3), which affect the loss coefficient β and to some extent the production rate q. There is also an annual variation, largely because the variation of Earth–Sun distance modulates the flux of ionising radiation, and a semiannual variation, most pronounced in the Southern Hemisphere (see Figure 14.4), that is not well understood but may be due to thermospheric winds and compositional changes.

Near the geomagnetic equator, the ionosphere shows very anomalous behaviour. The main feature of interest in the F2 layer is the daytime 'equatorial trough' or 'Appleton anomaly' (as shown in the map of Figure 17.3 of Chapter 17). It is produced by dynamo electric fields, which cause a drift away from the magnetic equator, and disappears at night.

14.5.5 Auroral ovals

At high latitudes, ionospheric structure is related to the auroral ovals and the magnetosphere, to which the ionosphere is linked by geomagnetic field lines (Section 14.1.3). Electric currents and energetic particles can flow into the ionosphere, causing heating and consequent expansion of the air, which modifies the chemical composition and the thermospheric wind system, with consequent effects on the F layer. The auroral ovals are the site of many active phenomena, especially particle precipitation which causes enhanced ionisation in all the ionospheric layers. There are two main types of precipitation: the intense impulsive 'splashes' of relatively soft keV particles associated with substorms and the auroral oval, and the more persistent background 'drizzle' of harder MeV particles at rather lower latitudes. The consequences include the following phenomena:

Splash events	*Drizzle events*
Discrete auroral forms	Steady diffuse aurora
Auroral absorption events	Slowly varying absorption
Auroral sporadic E	Low diffuse sporadic E
Rapidly fading VHF scatter	VHF forward scatter
Impulsive pulsations (P_i)	Continuous P_c pulsations
Bursts of VLF auroral hiss	VLF chorus emissions
Bursts of keV electrons	Sustained electron fluxes
Bremsstrahlung X-ray bursts	Long duration hard X-rays
Spread F echoes	
Negative magnetic bays	

Figure 14.7 is a schematic plan of the northern auroral oval and features linked to it. The oval can be fairly well represented by a circular ring, with its centre offset from the geomagnetic (dip) pole by $2 - 4°$ towards the midnight side. The outer ring represents the plasmasphere (midlatitude ionosphere) which corotates with the Earth. The stippled ring represents the approximate location of the hard 'drizzle' precipitation. The inner shaded ring is the auroral oval, showing the approximate locations of the 'cusp' in the noon sector and the Harang discontinuity (HD) near midnight. The fine lines represent typical flow lines of the plasma convection pattern (some curves are left incomplete to reduce congestion). The diameter of the oval depends on the level of geomagnetic activity, which in turn depends on solar activity and the interplanetary magnetic field (IMF). In principle this dependence on solar and interplanetary parameters can be exploited for short-term forecasting and longer-term predictions. It must be emphasised that the pattern shown in Figure 14.7 should be regarded as a statistical average, and does not necessarily represent the real physical situation at any actual instant.

14.5.6 High-latitude convection, the polar cap and trough

From early days in ionospheric physics, it was a puzzle as to how the ionosphere was maintained throughout the polar winter in the absence of solar ionising radiation. Since the polar caps (within the auroral ovals) are not normally locations of strong particle precipitation, and the night sky provides only weak sources of ionising radiation, horizontal transport of ionisation seemed the most likely solution.

From many observations of plasma motions, by satellite-borne instruments and by coherent and incoherent scatter radar, it is known that the magnetic linkage between the solar wind, as it sweeps past the Earth, and the geomagnetic field sets up a system of electric fields throughout the high latitude ionosphere. The resulting $E \times B/B^2$ drift, shown in Figure 14.7, causes a fast day-to-night flow of plasma across the polar cap. This flow maintains the F2 layer in the dark side of the polar ionosphere, though there is a competing loss process because the rapid drift also increases the decay rate of the plasma by increasing the coefficient k'' in eqn. 14.36. The drift tends to be faster at times when the south-to-north (B_z) component of the IMF is directed southward, which facilitates linkage between the geomagnetic field and the IMF, and has a more complex pattern when B_z is directed northward.

Figure 14.7 shows that the day-to-night drift in the polar cap is part of a huge high-latitude circulation or 'convection'. On the low-latitude side of the auroral oval, this convection gives rise to the 'ionospheric trough', a region of low plasma density originally discovered by satellites. The combination of the Earth's rotation and the convection velocity causes plasma to become almost stationary in the evening sector of the trough. If this 'region of stagnation' is in darkness, the ion density can decay to very low levels. The 'drizzle' precipitation often produces small-scale irregular structure and strong airglow emission in the trough region.

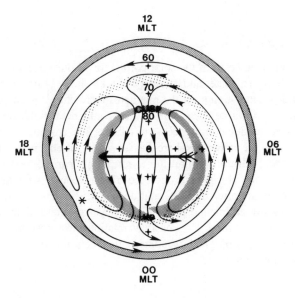

Figure 14.7   *The auroral oval and associated phenomena*

View of the north polar region, centred on the magnetic pole. The bottom of the diagram represents magnetic midnight, 00 MLT; the top represents magnetic noon, 12 MLT; the outer boundary is at 50° magnetic latitude, the crosses being at magnetic latitudes 80°, 70° and 60°. The diagram corresponds roughly to moderate magnetic activity ($Kp = 3$) with a southward B_z component of the interplanetary magnetic field (Rutherford Appleton Laboratory)

The equatorward boundary of the trough is linked to the plasmapause, the boundary between the inner magnetosphere which is magnetically linked to the ionosphere at middle and low latitudes and rotates with the Earth, and the tenuous outer magnetosphere which does not rotate with the Earth.

14.6 Solar–terrestrial relationships and ionospheric storms

14.6.1 Solar flares and sudden ionospheric disturbances

The intense X-rays from solar flares produce strong enhancements of electron density in the D layer, giving rise to 'sudden ionospheric disturbances' (SID).

The term SID embraces several phenomena, including radio short-wave 'fadeouts', effects on LF/VLF propagation, and magnetic perturbations, which occur only in the sunlit hemisphere and typically last for half an hour. Soon afterwards MeV particles arrive, causing enhanced radio absorption at high latitudes (polar cap absorption, PCA).

14.6.2 Geomagnetic storms

Streams of plasma ejected from the solar corona travel outwards in the solar wind (which thereby becomes faster and denser) and take about 24 h to reach the Earth. The impact of a plasma stream on the magnetosphere causes a 'magnetic storm', which often starts with a compression of the geomagnetic field, observed at the ground as a 'storm sudden commencement' (SSC). The ensuing interaction of the plasma stream and the magnetosphere has several consequences, including the setting up of a 'ring current' around the Earth mainly carried by keV ions. The ring current causes the storm-time decrease of the geomagnetic field, which is greatest at 12–24 hours after the SSC and decays in 2–3 days. The auroral oval expands equatorward during storms and the plasmapause and trough move to lower latitudes than normal.   Possible relationships between geomagnetic disturbances and meteorological weather have been found but their physical mechanisms are not established.

14.6.3 Ionospheric storms

During a geomagnetic storm, particle precipitation at high latitudes produces disturbances at D-layer heights, probably chemical in nature (e.g. through the production of extra NO). The effects are propagated to the midlatitude D layer, where they may last for ten days or so. The F2 layer is often profoundly affected during magnetic storms, with severe effects on radio propagation. At midlatitudes the F2-layer electron density initially increases, then often decreases during the storm's main phase, and recovers in 2–3 days. Examples of these 'negative storm effects' can be seen in Figure 14.6. In winter, however, there is often a 'positive' effect, i.e. an increase of F2-layer density in the main phase. 'Positive' F2-layer effects are quite common at low latitudes.

These storm effects, which vary in time and location in a complicated manner, seem to originate from the energy inputs (joule heating and particles) at high latitudes, which generate winds and waves that spread globally. Thermospheric composition changes, electric fields and particle precipitation all play a part. The mechanisms involved are not understood in detail, but there must be a 'storm circulation' driven by the high-latitude heating, somewhat as sketched in Figure 14.8, which is superimposed on the quiet day wind system described in Section 14.2.4. The equatorward winds of the storm circulation cause some lifting of the

F2 layer (white arrows in Figure 14.8), which affects N_mF2 in the manner shown
in Figure 14.5, sketch (*a*). The return circulation, envisaged to be below 150 km,
is speculative, as are the winds blowing into the polar cap

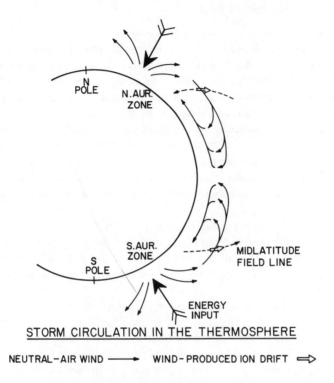

STORM CIRCULATION IN THE THERMOSPHERE

NEUTRAL−AIR WIND ⟶   WIND−PRODUCED ION DRIFT ⟹

Figure 14.8   Idealised sketch of the thermospheric storm circulation (Radio Sci.,
*1974, **9**, 185, American Geophysical Union)*

The 'negative' storm effects at midlatitudes are due to neutral composition
changes produced by the 'storm circulation'. Air of enhanced $[N_2/O]$ ratio is
brought by the 'storm circulation' from high latitudes to midlatitudes, resulting in
increased loss coefficient β and hence decreased N_mF2, by eqn. 14.36. These
effects might be augmented by localised particle precipitation, which is known to
produce depletions of N_mF2 and strong airglow emission in the trough region.
The 'positive' F2-layer storm effects in winter, and at low latitudes, may be due to
decreases of $[N_2/O]$ ratio or to drifts produced by winds or electric fields.

14.7   Conclusion

14.7.1   Progress in recent years

Ionospheric science nowadays has several aspects. First, it is an old but still active branch of the domain of solar–terrestrial physics; experimental and theoretical studies of the   ionosphere have contributed enormously to understanding of the Earth's environment. Secondly, it contributes to the science and technology of radiocommunication; the practical use of the ionosphere for communication is still great, and could be improved by further progress in the challenging field of 'predicting and forecasting'.   Thirdly, the mathematical subject of wave propagation in the ionosphere deals with waves of many types; besides man-made radio transmissions, a vast range of electromagnetic emissions, magnetic pulsations and other phenomena originate in the ionosphere, or travel through it. Fourthly, the ionosphere offers great scope for plasma physics, both experimental and theoretical.

The real problems remain at the boundaries, both above and below; i.e. the interactions with the lower atmosphere and the magnetosphere.   The latter embraces the whole question of the effect of geomagnetic disturbance and of the interplanetary magnetic field, primarily at high latitudes, but also how the effects spread worldwide.

Influences of the lower atmosphere on the ionosphere remain a subject for much present and future study.   That the lower atmosphere, with its huge mass and energy density, should affect the upper atmosphere is hardly surprising.   It has long been known that some tidal components and oscillations (Section 14.2.5) are excited low in the atmosphere, and that some gravity waves (Section 14.2.6) originate in storms and explosions, and possibly earthquakes. The large electric fields generated in thunderstorms can affect the electrodynamics of the ionosphere.   There have been suggestions of topographic influences (e.g. of mountain ranges) on the ionosphere.

This chapter has concentrated on the large-scale structure, but the ionosphere has smaller structure with a wide range of scale sizes.   The small-scale irregularities at auroral and equatorial latitudes are examples of plasma-instability phenomena, and even at midlatitudes there is irregular structure that causes radiowave fading, scintillation of radio signals traversing the ionosphere, of which spread F and sporadic E are examples. Medium-scale structure, 10–1000 km in scale, is widely observed but not well explained.

As a natural plasma, often nonlinear and nonthermal, the ionosphere is a site for active experiments with radiowaves, powerful enough to produce observable 'ionospheric modifications' (heating and plasma instabilities), and with chemical releases and particle-beam experiments. Such experiments provide novel physical and chemical insights, though they often raise as many questions as they solve.

14.7.2 Global change and the ionosphere

The concentrations of carbon dioxide and methane gases in the lower and middle atmosphere are increasing and are expected to lead to 'global warming'. Increased concentrations of these gases in the mesosphere at 60–80 km are, however, expected to *cool* the mesosphere and thermosphere, because of the enhanced loss of heat to space by infra-red radiation. If the mesospheric concentrations of CO_2 and CH_4 are doubled, as expected within about the next fifty years, it is estimated that:

(*a*) the mean tropospheric temperature will increase by 1–2 K;
(*b*) the mean mesospheric temperature will decrease by 5 K;
(*c*) the mean exospheric temperature T_∞ will decrease by 40 K.

The resulting contraction of the thermosphere will decrease the air density at 300 km by about 40%, which will prolong the lifetime of artificial satellites. It will also lower the ionospheric layers, decreasing h_mE by about 2 km and h_mF2 by 20 km, with some change of MUF. This drop of h_mF2 should be noticeable even in the presence of the large daily and solar-cycle changes (a drop of 8 km in the last 33 years may, indeed, have been detected already). Other possible long-term changes include a possible drastic change in upper-atmosphere tides, resulting from the depletion of the ozone layer where much of the 'tidal forcing' originates; and chemical changes due to contamination of the ionosphere by spacecraft launches.

14.7.3 The frontiers: major areas of active research

This simplified account of ionospheric physics avoids many questions and complexities. The basic mechanisms of the ionosphere seem to be known, but new surprises are always possible in any active area of science, and many questions remain to be solved. The frontiers include:

(i)   the detailed analysis of the thermospheric circulation;
(ii)   influences of the magnetosphere, IMF and the solar wind;
(iii)   relationships between the ionosphere and magnetosphere and the weather;
(iv)   influences of oceans, mountains, earthquakes, thunderstorms;
(v)   mechanisms of ionospheric storms, and their predictability;
(vi)   the cause of irregular structure of a wide variety of scales;
(vii)   active experiments: the ionosphere as a physics/chemistry laboratory;
(viii) predictability of the ionosphere on various time scales; and
(ix)   global change in the upper atmosphere.

Future experimental facilities, such as the new high-latitude radars, will bring new opportunities and new puzzles to solve and give rise to demanding theoretical projects, but the basic everyday monitoring of this important part of the environment — spanning five solar cycles since 1931 — will be as vital for progress in the future as it has been in the past.

14.8 Bibliography

RISHBETH, H., and GARRIOTT, O. K.: 'Introduction to ionospheric physics' (Academic Press, New York, 1969)

RATCLIFFE, J. A.: 'Sun, earth and radio' (World University Library, 1970)

BANKS, P. R. and KOCKARTS, G.: 'Aeronomy, 2 vols' (Academic Press, New York, 1973)

GIRAUD, A. and PETIT, M.: 'Ionospheric techniques and phenomena' (Reidel, Dordrecht, 1978)

JURSA, A. S. (Ed.): 'Handbook of geophysics and the space environment' Air Force Geophysics Laboratory, Bedford, Mass., 1985

TASCIONE, T. F.: 'Introduction to the space environment' (Orbit, 1988)

PHILLIPS, K. J. H.: 'Guide to the Sun' (Cambridge University Press, 1992)

HARGREAVES, J. K.: 'The upper atmosphere and solar-terrestrial environment' (Cambridge University Press, 1992)

RUSSELL, C. T. and KIVELSON, M.: 'Introduction to space physics' (Cambridge University Press, 1995)

A longer version of this chapter was published in *J. Inst. Electron. Radio Eng.,* 1988, **58**, (S207).

Chapter 15

Surface waves, and sky waves below 2 MHz

J.D. Milsom

15.1 Introduction

The principal modes of radiowave propagation at frequencies below 2 MHz are the surface wave and sky wave. In this chapter these two modes are introduced and described. Rather than concentrate on the details of elaborate path-loss prediction theories, they are merely introduced and the discussion then concentrates on their application by the planning engineer.

The sky-wave propagation prediction methods described in this chapter are only applicable at frequencies below 2 MHz. However, the surface-wave models are based on more general theories and can also be applied in the HF band.

Antenna and external-noise aspects of system planning are discussed.

Throughout, readers are directed towards relevant data sources, prediction procedures and computer programs so that they might apply the planning methods described.

15.2 Applications

The radio spectrum below 2 MHz is allocated to a variety of radio systems:

Aeronautical radionavigation
Amateur
Fixed services
Land mobile
Maritime mobile coastal radiotelegraphy
Maritime radionavigation (radio beacons)
Mobile (distress and calling)
Radiolocation
Radionavigation

Sound broadcasting
Standard-frequency and time services

Some of these applications are discussed in more detail in Chapter 1.

15.3 Surface-wave propagation

15.3.1 What is the 'surface wave'?

Consider the case of a transmitting antenna, T, above a perfectly conducting flat
ground as depicted in Figure 15.1. The voltage, V, induced in the receiving
antenna, at an arbitrary receiving position, R, might be expressed as a vector sum
of direct and ground-reflected components:

$$V = QI\left\{ Q_1 \frac{\exp(-jkr_1)}{r_1} + Q_2 R \frac{\exp(-jkr_2)}{r_2} \right\} \qquad (15.1)$$

where:
I is the current in the transmitting antenna;
Q is a constant;
Q_1 and Q_2 take account of the transmitting- and receiving-antenna polar diagrams;
R is the appropriate reflection coefficient (see Chapter 4).
Other terms are defined in Figure 15.1.

  In many cases, especially where the radiated frequency is in the VHF or higher
frequency bands, the above calculation will give a perfectly acceptable result for
practical applications. However, it transpires that a complete description of the
field at R requires an additional contribution to the resultant:

$$V = QI\left\{ Q_1 \frac{\exp(-jkr_1)}{r_1} + Q_2 R \frac{\exp(-jkr_2)}{r_2} + S \frac{\exp(-jkr_2)}{r_2} \right\} \qquad (15.2)$$

where S is a complicated factor which depends on the electrical properties of the
ground, transmitted polarisation, frequency and the terminal locations.

  When introduced in this way, as an apparent afterthought to make up the
numbers and satisfy Maxwell's equations, one is tempted to regard it as a minor
contribution of interest primarily to the mathematical physicist. In fact this third
term represents the 'surface wave' and it is a propagation mode of great practical
value to radio systems operating in the HF and lower frequency bands. We will
see later that, when the points T and R are close to the ground, the direct and
ground-reflected waves act to cancel each other, leaving only the surface wave.

Therefore, for example, during the daytime when ionospheric absorption smothers MF sky-wave modes the surface wave is the carrier of all the signals which occupy the medium-wave broadcasting band. Surface waves also support the operations of LF broadcasting, VLF/LF communication and navigation systems, HF short-range communication and some classes of HF radar.

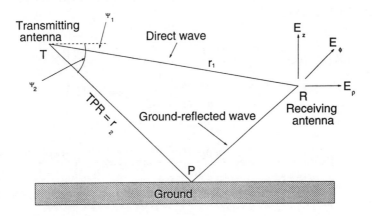

Figure 15.1   Geometry of direct and ground-reflected waves

The surface wave propagates by virtue of currents which flow in the ground and does not depend for its existence on the atmosphere. Horizontally polarised surface waves are very heavily attenuated and have little or no practical worth. All the applications mentioned earlier utilise vertically polarised surfaces waves.

At this stage it is appropriate to introduce some new terms and clarify the relationship between surface waves, space waves and ground waves:

ground wave = direct wave + reflected wave + surface wave

space wave

Unlike ionospherically propagated signals, the surface wave suffers negligible dispersion so that, in principle, wideband signals can be transmitted when the surface wave alone is active.

Fading only occurs when there is some temporal variation in the propagation path. Overland ground waves are stable signals. Oversea ground wave can be subject to slow fading due to changing tidal effects and shifts in the seawave characteristics.

15.3.2 Theory for a homogeneous smooth Earth

15.3.2.1 Plane finitely conducting Earth

Sommerfeld–Norton flat-Earth theory: After Sommerfeld [169], Norton [138, 139] derived expressions for the ground-wave field-strength components above a finitely conducting plane Earth due to a short vertical current element. In its full form eqn. 15.2 becomes

$$E_z = j30kIdl\left[\left\{\cos^2\psi_1\,\frac{\exp(-jkr_1)}{r_1} + \cos^2\psi_2 R_v\,\frac{\exp(-jkr_2)}{r_2}\right\} + \left\{(1-R_v)(1-u^2+u^4\cos^2\psi_2)F\,\frac{\exp(-jkr_2)}{r_2}\right\}\right]$$

(15.3)

$$E_\rho = -j30kIdl\left[\sin\psi_1\cos\psi_1\,\frac{\exp(-jkr_1)}{r_1} + \sin\psi_2\cos\psi_2 R_v\,\frac{\exp(-jkr_2)}{r_2} - \right.$$
$$\left.\cos\psi_2(1-R_v)u\sqrt{(1-u^2\cos^2\psi_2)}\left\{1-\frac{u^2}{2}(1-u^2\cos^2\psi_2)+\frac{\sin^2\psi_2}{2}\right\}F\,\frac{\exp(-jkr_2)}{r_2}\right]$$

(15.4)

where

ψ_1 and ψ_2 are defined in Figure 15.1,
j is the square root of -1,
k is the radio wavenumber $= 2\pi/\lambda$,
Idl is the product of source current and length — the 'dipole moment',
R_v is the plane-wave Fresnel reflection coefficient for vertical polarisation,
F is an attenuation function which depends on ground type and path length. It is given by the expression

$$F = \left[1 - j\sqrt{\pi w}\exp(-w)\left\{\mathrm{erfc}\left(j\sqrt{w}\right)\right\}\right]$$

(15.5)

erfc denotes the complementary error function (Abramowitz and Stegun [47]) and

$$w = \frac{-j2kr_2u^2(1-u^2\cos^2\psi_2)}{(1-R_v)^2}$$

(15.6)

$$u^2 = \frac{1}{(\varepsilon_r - jx)}$$

(15.7)

and

$$x = \frac{\sigma}{(\omega\varepsilon_0)} = 1.8 \times 10^4 \frac{\sigma}{f_{MHz}} \qquad (15.8)$$

σ is the conductivity of the earth in S/m, $\varepsilon_r = \varepsilon/\varepsilon_0$ is the relative permittivity of the earth and f_{MHz} is the frequency in MHz.

Note that eqns. 15.3 and 15.4 above represent field components in the vertical and radial directions of a cylindrical co-ordinate system.

Special case of ground-based terminals: When the points T and R are both at the ground so that $R_v = -1$ (see Chapter 4) and $\psi_1 = \psi_2 = 0$, the direct and ground-reflected waves act in opposition and sum to zero. Such circumstances will prevail in many practical applications. When this happens the surface wave dominates and may be described by somewhat simplified forms of eqns. 15.3 and 15.4, thus:

$$E_z = j60kIdl(1 - u^2 + u^4)F\frac{\exp(-jkr)}{r} \qquad (15.9)$$

$$E_\rho = j30kIdl\left\{u\sqrt{1-u^2}\left(2 - u^2 + u^4\right)\right\}F\frac{\exp(-jkr)}{r} \qquad (15.10)$$

In the pure surface wave the vertical and radial components of the electric field are still present. In physical terms this means that the propagating wavefront is tilted. The radial component given by eqn. 15.10 is small relative to the vertical component described by eqn. 15.9. The phase relationship is such that the modest wavefront tilt is in the direction of propagation. The degree of tilt depends on ground conductivity and frequency. Measurements of wave tilt can be used to infer the electrical properties of the local ground. Because E_ρ is finite and the magnetic-field component is horizontal, there exists a downward component of the Poynting vector and energy is lost from the horizontally propagating wave. In this way attenuation occurs in addition to that due to ordinary inverse-square-law spreading. Responsibility for describing this extra attenuation within the Sommerfeld–Norton theory falls to the term F which appears in the earlier expressions.

For ground-based terminals the 'attenuation factor', F, introduced in eqn. 15.5 still appears, but w simplifies to become

$$w = \frac{-jkru^2}{2}(1 - u^2) \qquad (15.11)$$

Having survived this short excursion into the realm of the theoretical physicist it is time to retreat and attempt an engineering interpretation of the above results.

Interpretation and key results: Radio coverage predictions are almost invariably conducted in terms of electric-field strengths. This approach also prevails in LF and MF broadcasting even though most domestic receivers now incorporate ferrite-rod antennas which are sensitive to the radio-frequency magnetic field. The propagating surface wave contains a horizontal magnetic component, H_ϕ, which is approximately related to the major electric component via the expression

$$H_\phi = \frac{-E_z}{Z_0} \qquad (15.12)$$

where Z_0 is the intrinsic impedance of free space ($377\,\Omega$). It is sufficient therefore to design in terms of the electric-field strength.

Attenuation of the surface wave arises through the forward tilt of its electric field. The rate of attenuation becomes more marked as the tilt angle increases. By combining eqns. 15.9 and 15.10, it is possible to show the ratio of electric-field components to be simply related by

$$\frac{E_p}{E_z} \approx u = \frac{1}{\sqrt{K_r}} \qquad (15.13)$$

K_r is the complex dielectric permittivity of the ground. It varies with frequency and the electrical properties of the ground.

Some representative values are presented in Table 15.1

Table 15.1   Typical values of complex dielectric permittivity for different ground types and frequencies

Ground type	Frequency (kHz)	
	200 (LF)	1000 (MF)
Sea ($\sigma=5$ S/m, $\varepsilon_r=70$)	$70-j450\,000$	$70-j90\,000$
Good ground ($\sigma=10^{-2}$ S/m, $\varepsilon_r=10$)	$10-j900$	$10-j180$
Poor ground ($\sigma=10^{-3}$ S/m, $\varepsilon_r=4$)	$4-j90$	$4-j18$

Large values of K_r correspond, according to eqn. 15.13, to low degrees of forward tilt and therefore attenuation. We can conclude that surface-wave attenuation is greatest over ground of low conductivity and at high radio frequencies. Sea water has an outstandingly high conductivity and the surface wave, with a near-vertical electric field, propagates over it with relatively low attenuation. This conclusion will be shown more graphically in Section 15.3.4.

In eqn. 15.9 the factor $(1 - u^2 + u^4)$ is close to unity for all practical situations. The amplitude of the vertical component of electric field is therefore given by

$$|E_z| = \frac{300}{r} \sqrt{P} \, |F| \qquad (15.14)$$

where P is the total radiated power from the Hertzian-dipole current element, expressed in kW, r is the path length in km and E_z is the electric field strength in mV/m.

All the interesting effects are associated with $|F|$. Within a few wavelengths of the signal source $|F|$ is approximately unity. In this regime the field strength varies as $1/r$, i.e. in inverse-distance fashion. At sufficiently large distances $|F|$ makes a transition to become inversely proportional to distance so that the field strength varies as $1/r^2$. This long-range behaviour will persist for as long as the Sommerfeld assumption of a plane Earth remains valid.

Usually it is necessary to compute fields from more practical antennas. This merely entails substituting a different constant on the right-hand side of eqn. 15.14. Chapter 3 gives some useful factors.

15.3.2.2   *Spherical finitely conducting Earth*
The next stage of refinement in our efforts to devise a realistic propagation model involves substituting a spherical Earth shape for Sommerfeld's plane model. At short ranges Sommerfeld's ground-wave model can be applied without adaption. At longer ranges, and when the two terminals are beyond line of sight, it is necessary to compute fields with proper regard for diffraction over the curved Earth [140, 177,178,179,180].

It would serve no useful purpose to detail the associated theory in this overview of surface-wave propagation. In any case the mathematics is extremely complicated and not easily understood. Bremmer [63] gives a good account of the theory.

It transpires that the curved Earth introduces a third range regime, normally beyond that where inverse-square-law field variation occurs, in which the decrease in field strength becomes exponential. The starting distance of this exponential behaviour can be estimated by the expression

$$\frac{80}{\sqrt{f_{MHz}}} \quad \text{km} \qquad (15.15)$$

Apart from this new far-range behaviour, most of the other characteristics of the surface wave above a spherical Earth are identical to those deduced from Sommerfeld's plane-Earth model.

15.3.3 Atmospheric effects

Terrestrial surface waves would propagate in the total absence of an atmosphere. All the theoretical work of Sommerfeld, Norton, Van der Pol and Bremmer reported above ignored atmospheric effects. They assumed that a wave propagating in space above the ground would travel in a straight line. In practice, the Earth's atmosphere is stratified and possesses a refractive index which normally decreases with height. On average the height variation is exponential. Near the ground it is sometimes sufficient to assume a linearly decreasing height profile.

In any atmosphere where the refractive index decreases with height, a radiowave will be refracted towards the ground. If the profile is linear, a remarkably simple change to the Van der Pol/Bremmer/Norton theory enables the atmospheric refraction to be accommodated. It is only necessary to increase artificially the Earth radius above its true value of some 6371 km. Such a trick is commonly applied in the VHF and higher-frequency bands (see Chapter 4). An effective Earth radius should only be used when both terminals are near to the ground and at frequencies greater than 10 MHz.

Rotheram [159,160,161] has explored the behaviour of the effective Earth radius multiplying factor in the frequency band where surface waves are of practical importance. Above 30 MHz a factor of 1.374 is appropriate. At frequencies below 10 kHz the atmosphere has negligible effect and the factor tends to unity. In the neighbourhood of the MF broadcasting band the factor lies in the range 1.20–1.25 for most classes of ground. Rotheram's results have been computed for average atmospheric conditions. During times of abnormal atmospheric conditions, effective Earth-radius factors outside the range 1–1.374 may be required to simulate the prevailing propagation effects.

When problems arise where one or both terminals are elevated, energy propagating between the two encounters a refractivity/height profile which is approximately exponential and clearly nonlinear. Such paths cannot be modelled using an effective Earth-radius factor and an atmosphere-free propagation theory. The nonlinear refractivity profile becomes significant at all frequencies when a terminal is elevated above 1 km and at frequencies below 10 MHz even for terminals on the ground [161].

Happily, these doubts about the validity of atmosphere-free theories and the Earth-radius factor have been overcome by the work of Rotheram. Rotheram has developed a general-purpose ground-wave prediction method and an associated computer program. The method incorporates an exponential atmospheric-refractivity profile. It is now recommended by the ITU-R for system planning and has been adopted by many agencies. Because of its practical worth, Rotheram's program, GRWAVE, is described in fair detail in Section 15.3.4.

15.3.4 ITU-R recommended prediction method

Rotheram describes three methods of predicting space-wave and surface-wave fields over a smooth homogeneous Earth surrounded by a uniform atmosphere which exhibits an exponential refractivity/height profile. The methods cater for elevated terminals and a very wide frequency band. No single method is effective for all path geometries but, using appropriate numerical techniques, it is possible to establish whether or not one method is working and, if it is not, switch to a better approach. The methods and their approximate regions of validity are:

(i)   Residue series (mode summation): Used at the farthest distances, for elevated terminals this is beyond the radio horizon. For terminals near the Earth's surface it is for distances greater than approximately $10 \, \lambda^{1/3}$ km where λ is the radio wavelength in metres.

(ii)   Extended Sommerfeld flat-Earth theory: An extended Sommerfeld theory can be applied at short ranges and small heights. These restrictions turn out to be ranges less than approximately $10 \, \lambda^{1/3}$ km and heights below $35 \, \lambda^{2/3}$ m.

(iii)   Geometrical optics (ray theory): This final method is applied within the radio horizon when the terminal heights are above that which can be handled by the Sommerfeld approach. It involves calculating the phase and amplitude of the direct and ground-reflected paths with due regard for the atmospheric refraction.

Fortunately, the three methods are able to deal with all reasonable geometries so that inelegant interpolation between two inappropriate results is unnecessary. Under circumstances where two methods are simultaneously valid the results are found to be in good agreement.

Rotheram has written a Fortran computer program to compute ground-wave fields using these theories. This program, GRWAVE, has been used by the ITU-R to produce a series of curves which show how vertically polarised electrical-field strength varies as a function on range, ground type and frequency (10 kHz to 30 MHz). In doing this the ITU-R has elected to adopt a global average refractivity profile given by

$$n = 1 + (n_s - 1)\exp(-h/h_s) \qquad (15.16)$$

where n_s = surface refractivity = 1.000315 and h_s = refractivity profile scale height = 7.35 km.

These curves are to be found in ITU-R Recommendation P.368-7 [7].

While the ITU-R curves are comprehensive, it is sometimes useful to use GRWAVE program itself.

Figure 15.2 shows some example ITU-R ground-wave curves for sea, 'land' (σ = 0.03 S/m) and 'very dry ground' (σ = 0.0001 S/m). The latter two represent some extreme ground conditions. Sea is best regarded as being in a class of its own. For these curves both terminals are on the ground so that, in their computation, the geometrical-optics parts of GRWAVE will not have been invoked. A close inspection of Figure 15.2 will reveal the inverse-distance, inverse-square-distance and exponential range attenuation regimes which were inferred in earlier Sections. Also note the very strong frequency and ground-conductivity dependence of surface-wave attenuation over dry ground.

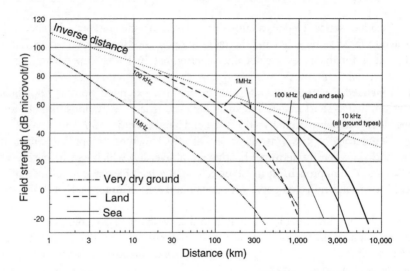

Figure 15.2   ITU-R surface-wave curves

Figure 15.2 shows the relatively low attenuation experienced by the surface wave over sea. In the frequency bands below approximately 3 MHz the wave passes over the first 100 km of sea with an inverse-distance attenuation rate, i.e. as if the ground was perfectly conducting. However, by a range of approximately 400 km even a VLF transmission at 10 kHz begins to suffer losses beyond that given by the inverse-distance line.

In applying these curves for system planning purposes it is essential to have a clear understanding of the reference radiator used in their calculation. For the ITU-R curves the transmitting antenna is a Hertzian vertical dipole with a current–length product (dipole moment) of $5\lambda/2\pi$. This moment has been carefully selected so that the characteristics of the dipole are identical to those of a short vertical monopole radiating 1 kW, a configuration which is easier to visualise.

Such a monopole, located over a perfectly conducting plane, will establish a field of 300 mV/m at a distance of 1 km along the plane. This factor appears explicitly in eqn. 15.14 where the same reference antenna was used. It is a trivial matter to adjust the curves so that actual radiated power and antenna gain, relative to the 1 kW short monopole, are incorporated.

For some applications it is convenient (or conventional) to work in terms of transmission loss rather than field strength. Radar systems are one such case. Transmission losses are defined in terms of ratios of power transmitted to power received, but there are several ways of constructing this type of ratio and care is required (see Chapter 3).

When the transmitting and receiving terminals are on the ground, as assumed in Figure 15.2, it is easy to relate ITU-R field-strength curves to basic transmission loss:

$$L_b = 142.0 + 20\log_{10} f_{MHz} - E \qquad (15.17)$$

where E is the field strength, in dB (μV/m) for the ITU-R reference radiator.

The frequency term in the above equation arises because the collecting aperture of a receiving antenna, and therefore the power available, depends on radio frequency.

ITU-R Recommendation P.341-4 [6] and Chapter 3 discuss the use of L_b in system design calculations. This will not be reiterated here. It is, however, important to address one key point and this relates to the antenna-gain definition which must be used when L_b is involved. In the system calculation, actual antenna gains must be introduced as relative to an 'isotropic' antenna *at the same location.* This is not the convention commonly used by antenna manufacturers in their sales literature. They tend to use dB relative to a truly isotropic antenna in free space.

This apparently trivial matter of antenna gain in ground and surface-wave path-loss calculations can lead to great confusion and, worse, an incorrect system design. The problem is compounded by the existence of several different (but internally consistent) sets of definitions, and papers in the open literature which present vague or incorrect accounts of antenna gain.

15.3.5 Ground conductivity maps

In practice, when the propagation engineer has been asked to compute the viability of a link or the coverage of a broadcasting station, one of the most difficult stages is to acquire a proper description of the ground conductivity along the path or in the area of interest. A useful source of data lies in ITU-R Recommendation P.832 [25]. Here, geographic maps are presented on which areas of differing electrical properties are delineated. The maps are a conglomerate of information collected over many years, and the degree of detail varies from region to region.

Once the ITU-R maps have been consulted, it is often worth seeking supplementary local information about the ground types present. Broadcasting authorities sometimes have details of ground conductivity in their areas which may not have found its way into Recommendation P.832.

When the ITU-R and local information sources both prove to be inadequate, it is sometimes instructive to consult geological maps of the district and attempt an association of mineral type with electrical properties A useful association is given in ITU-R Report 879-1 [38].

15.3.6 Smooth Earth of mixed conductivity

So far we have discussed increasingly elaborate models of surface-wave propagation and the ways in which they might be applied. However, all of the models only treat the case of a homogeneous smooth Earth. In practice, it is often necessary to solve planning problems which involve changes in ground type along the propagation path.

Suppose we have a situation where a surface-wave link must be established between two terminals which are located on ground of different electrical properties. At some point along the smooth propagation path a transition occurs between the two ground types. The upper half of Figure 15.3 shows the situation in schematic form. How do we compute the electric field strength at R due to the transmission from T?

Eckersley [85] suggested that this might be done by using sections of the surface-wave attenuation curves appropriate to the radio frequency and different ground types. Figure 15.3 shows the idea in graphical form. The Eckersley construction can be made using the curves for homogeneous ground such as those published by the ITU-R. Intuitively, this method appears correct. However, it produces results in poor agreement with experiment and, furthermore, violates the need to have reciprocity on the path. Reciprocity demands that, if the transmitter and receiver were transposed, so that the wave encountered the two types of ground in reverse order, then the field strength at the receiver would be unchanged.

Millington [135], in a classic paper on ground-wave propagation, presented a simple but effective method of solving the problem depicted in Figure 15.3. His work was done at a time when reliable measurements of field-strength changes in the neighbourhood of a conductivity transition were scarce. Others were developing analytic solutions to the problem but their results were complicated and unsuitable for practical application. Millington's approach was a blend of known theory and physical intuition. The argument proceeded along the following lines.

Suppose T in Figure 15.3 is well removed from the conductivity transition at X. In this case the surface wave which is launched in the direction of R will have a rate of attenuation with horizontal distance and a variation with height above ground which is characteristic of ground type 1. The field can be computed using a model for the homogeneous Earth with little loss of accuracy. Similarly, if R is well removed from X then the surface-wave field strength will vary with horizontal distance and height in a way which is largely dictated by the ground type 2. In effect, it might appear as if the signal had travelled from T to R over homogeneous ground of type 2. The only residual evidence of its passage over the type 1 ground will be a shift in the absolute signal level. Homogeneous-Earth propagation models cannot be applied directly to compute the absolute field strength on the receiver side of X.

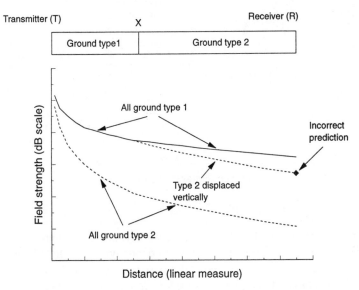

Figure 15.3   *Smooth Earth of mixed conductivity. Eckersley's prediction method*

Consider now what might be happening close to the conductivity transition point X. Millington argued that some sort of distortion or disturbance of the surface wave must occur. On approaching the transition from T the field will probably be affected even before X itself is encountered. However, this must be modest in comparison to the trauma undergone by the field on the receiver side of the transition.

Millington then elaborated the discussion and sought to establish a prediction method which would give a consistent change in the height profile of the field strength near the ground-type transition. An essential constraint on the form of the method was that it must satisfy the reciprocity requirement. A proposal was made, without proof or mathematical rigour, that the field strength might be estimated by a double application of Eckersley's method, followed by an averaging of the two results (expressed in dB). One application is made in the forward direction and then another in the reverse direction, as if the placement of T and R were reversed. Formation of an average forces the solution to satisfy reciprocity. Figure 15.4 shows how the 'Millington method' should be applied.

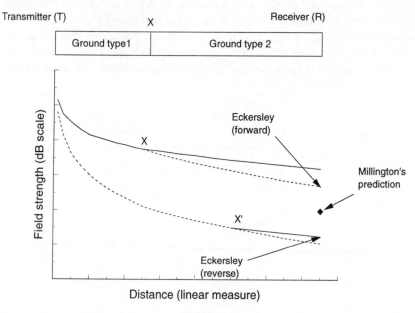

Figure 15.4   Millington's prediction method

When the method is applied to the situation of signals propagating from ground of high conductivity to ground of a lower conductivity, the disturbance at the boundary appears as a somewhat abrupt decrease in field strength. Eventually, the field variation takes on the character of the low-conductivity ground.

When the method is applied to signals propagating from ground of low conductivity to ground of high conductivity, a much more remarkable effect is predicted. The phenomenon is most marked at a land–sea boundary. The field strength undergoes an abrupt increase with range, immediately on crossing the coast. Millington sought to explain the unexpected recovery in terms of a redistribution of signal energy from elevated portions of the wavefront down to lower levels.

In 1950 Millington and Isted [136], in a carefully executed experiment, measured the recovery effect at 3 MHz and 75 MHz and demonstrated an excellent agreement with the new prediction method. Figure 15.5 shows the degree of prediction accuracy at 3 MHz.

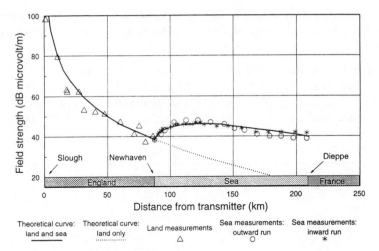

Figure 15.5   Calculated curve and experimental observation of the land–sea recovery at 3 MHz [136]

The method can be applied to paths with more than two ground sections by exactly the same procedure. Eckersley's method is used for forward and reverse routes and then the average of the two results is computed.

Since 1950, Millington's method has been very widely used and is still recommended by the ITU-R for surface-wave planning where ground-conductivity changes occur along the propagation path.

15.4   Sky-wave propagation below 2 MHz

15.4.1   What is the 'sky wave'?—hops and modes

The sky wave is that part of the total received signal which relies on the presence of the ionosphere for its existence. The sky wave is a more easily understood concept than the surface wave.

In the frequency bands below 2 MHz there are essentially three methods of estimating the sky-wave field strength:

(i)   A theoretical 'waveguide mode' method by which propagation is analysed as the sum of waves corresponding to modes in the waveguide formed by the Earth and lower ionosphere.

(ii)   A theoretical method called 'wave hop' in which the signals are modelled as one or more geometrical raypaths reflected from the lower ionosphere. This approach is similar, in principle, to that used in HF sky-wave prediction methods.

(iii)   An empirical method has been developed; see ITU-R Recommendation P.114-7 [5] for the planning of sound-broadcasting services in the LF and MF bands.

To a degree, the three methods are complementary.   The choice of method depends mainly on the combination of frequency and ground ranges of interest. Figure 15.6 is a rough guide to the regions of applicability of each method.

The three methods are outlined in Sections 15.4.2, 15.4.3 and 15.4.4, respectively.

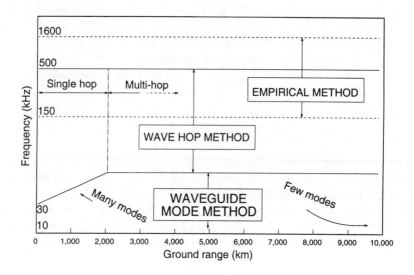

Figure 15.6   Sky-wave prediction methods. Approximate domains of applicability

15.4.2 Waveguide-mode field-strength prediction theory

At frequencies below about 60 kHz the distance between the Earth and ionosphere is less than a few wavelengths and the cavity tends to act as a waveguide.   For example, at the main Omega frequency of 10.2 kHz, the separation is less than approximately three wavelengths. Surface wave and sky wave therefore cannot be considered independently, except at short ranges.

Propagation losses for ELF/VLF and lower LF signals in the Earth–ionosphere waveguide are very modest.   However, as discussed in Section 15.5, poor transmitting antenna efficiency can be a severe constraint.   Communication services suffer from unavoidably narrow bandwidths and therefore low data rates. On the other hand, the excellent phase stability of continuous-wave signals makes these bands ideal for long-range hyperbolic navigation services.

Figure 15.7, derived from Morfitt *et al.* [137], shows a typical VLF field-strength/distance curve measured over a sea path at night at about 15.5 kHz. It can be seen that the measured curve agrees closely with the ITU-R surface wave curve for distances up to 1200 km. Beyond this the influence of the ionosphere is clearly apparent.

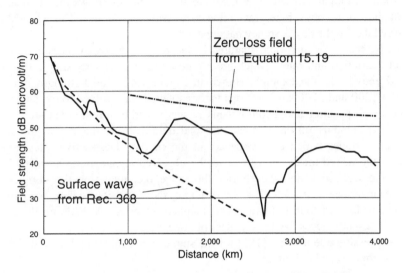

Figure 15.7   Measured field strength at night on 15.567 kHz [137]

It is interesting to compare the measured field strength with that predicted by a very simple prediction model. Imagine the Earth–ionosphere waveguide to be loss free and that the radiated power P is distributed uniformly over the wavefront. If the Earth were flat the area of the cylindrical wavefront would be $2\pi hd$, where d is the distance from the transmitter and h is the height of the ionosphere ($d \gg h$). The power flux over the wavefront would therefore be $P/2\pi hd$, which is also equal to E^2/Z_0, where E is the field strength. The field strength, in mV/m, according to this simple model is thus given by

$$E = 245\sqrt{\frac{P}{hd}} \qquad (15.18)$$

where P is in kW and h and d are in km.

Equation 15.18 shows that the field strength would decrease as the square root of the distance; this is of course less than the inverse square law spreading in free space. The rate of attenuation is decreased still further by Earth curvature; if this is taken into account the expression for E becomes

$$E = 245\sqrt{\frac{P}{ha\sin(d/a)}} \qquad (15.19)$$

where a is the radius of the Earth in km.

Figure 15.7 also shows the field strength calculated using eqn. 15.19, for $h = 90$ km. Comparison with the measured field strengths shows the attenuation within the Earth–ionosphere waveguide to be small at night. The attenuation measured during the day is somewhat greater.

Clearly, there are important features in the relation between measured field-strength variation and range which are not described by the very simple theories offered above. In the example shown in Figure 15.7, signal-strength minima occur at 1200 and 2700 km. Various theories have been evolved to explain low-frequency propagation and a comprehensive survey can be found in ITU-R Report 895-2 [39]. One useful theory treats the total field as being the sum of the main waveguide modes which can propagate in the cavity. Destructive interference between the active modes gives rise to the observed signal-strength nulls. The higher the frequency and the shorter the range the greater is the number of significant modes. At 15.5 kHz, for example, the Earth–ionosphere waveguide can support at least four TM (transverse magnetic) modes. At ELF, on the other hand, it is usually necessary to consider only one mode.

An expression due to Wait [182] for the sum of waveguide modes excited by a short monopole can be written:

$$E = 300\sqrt{\frac{P}{a\sin(d/a)}}\frac{\sqrt{\lambda}}{h}\exp\{-j(kd + \pi/4)\}\sum_n \Lambda_n \exp(-jkS_n d) \qquad (15.20)$$

where λ is the wavelength, in km, k is the free-space wavenumber $= 2\pi/\lambda$, Λ_n is the excitation factor of the nth mode and kS_n is the propagation coefficient of the nth mode.

The terms Λ_n and S_n are complex. The excitation factors give the relative amplitude and phase of the various modes excited in the waveguide by the source. The real part of the propagation coefficient kS_n contains the phase information of each mode while the imaginary part gives its attenuation rate. These factors depend on wavelength, ionospheric height, electrical properties of the ground and the reflection coefficients of the ionosphere.

ITU-R Report 895 describes in more detail how the reflection coefficients of the ionosphere may be computed and used in the above waveguide-mode prediction method. In addition, the Report describes a more advanced form of waveguide-mode theory which provides a 'full-wave' solution. Various workers have developed computer programs to evaluate waveguide-mode methods. The most complete implementation currently available is the program suite due to Ferguson [88].

15.4.3 Wave-hop field-strength prediction theory

At frequencies above about 60 kHz (wavelengths shorter than 5 km), and at lower frequencies when the path length is less than approximately 2000 km, it is no longer appropriate to model the propagation mechanism as a waveguide because of the large number of significant modes. Instead it is more straightforward to use the ray theory to compute the sky-wave field strength and combine this with the surface-wave field strength derived separately using, say, the method introduced in Section 15.3.4.

Figure 15.8 shows a measured variation of field variation with range, together with the theoretical curve for the surface wave only [61]. The oscillation in measured field is due to interference between the surface wave and one-hop sky-wave modes. The range of oscillation is small at short ranges because the sky wave is small compared with the surface wave. At ranges beyond about 1500 km the sky wave is dominant and the oscillations decay. The interference nulls are generally known as the Hollingworth pattern. The field-strength nulls change location in response to the diurnal changes in sky-wave reflection height.

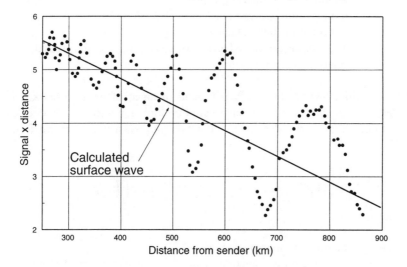

Figure 15.8   Hollingworth interference pattern measured at 85 kHz [61]

Sky-wave reflection takes place at heights near 70 km during the day and 90 km at night. In these lower reaches of the ionosphere the variation of electron concentration with height is significant within the wavelength of an LF signal. Under such circumstances the magnetoionic theory applied so successfully at HF is not valid. Waves are returned to the Earth by partial, rather than total, reflection.

The reflection coefficients used in a wave-hop prediction procedure may be empirical or based on a theoretical model.

LF sky waves propagate strongly at night and may be almost as strong during the winter days in temperate latitudes. Solar-cycle variations appear to be small.

A complete account of a wave-hop field-strength prediction method is presented by the ITU-R in Recommendation P.684-1 [24]. The basic formula for predicting the effective sky-wave field strength E_s when reception is via a small in-plane loop antenna is

$$E_s = 300\sqrt{P}\frac{2}{L}\cos\psi \, {}_\parallel R_\parallel \, DF_t F_r \tag{15.21}$$

where

P	=	radiated power (kW)
L	=	sky-wave (slant) path length (km)
$\|R\|$	=	ionospheric reflection coefficient which gives the ratio of electric-field components parallel to the plane of incidence
D	=	an ionospheric focusing factor
F_t	=	a transmitting-antenna factor
F_r	=	a receiving-antenna factor
Ψ	=	angle of departure and arrival of the sky wave at the ground, relative to the horizontal.

For propagation beyond the one-hop ground distance of about 2000 km it is necessary to compute the signal strength after multiple ionospheric reflections and the intermediate ground reflection(s). For this purpose a more general form of eqn. 15.21 is

$$E_s = 300\sqrt{P}\frac{2}{L}\cos\psi \, {}_\parallel R_{g\parallel}{}^{(n-1)} D^n D_g{}^{(n-1)} F_t F_r \prod_{i=1}^{n} {}_\parallel R_i \| \tag{15.22}$$

where

n	=	number of ionospheric hops
$\|R_g\|$	=	effective reflection coefficient of finitely conducting ground
D_g	=	divergence factor caused by the spherical Earth (approximately equal to $1/D$).

Recommendation 684 gives graphs of the above factors. An unusual feature of the method is its consideration of negative elevation angles in evaluating antenna factors. A negative elevation angle path corresponds to a geometry where the signal propagates, at each end of a zero-elevation sky-wave hop, by sections of diffraction around the bulge of the spherical Earth.

15.4.4 An empirical field-strength prediction theory

In the planning of sound broadcasting systems in the LF and MF bands, the ITU-R recommends the use of an empirical sky-wave field-strength prediction method. A complete description of the method is given in ITU-R Recommendation P.114-7 [5]. The salient points of the method are presented here. Sky-wave propagation in these bands differs from that at lower frequencies because the radio waves return to Earth by ionospheric refraction rather than partial reflection. Therefore the magnetoionic theory may be applied.

One of the principal characteristics of the frequency band above 150 kHz is that sky waves propagate efficiently at night but are greatly attenuated during the day. This attenuation occurs mainly in the D region of the ionosphere. At sunset the D-region ionisation decays rapidly, and waves reflected from the higher E or F regions become significant.

For LF and MF broadcasting the surface wave is the most important propagation mode because it provides a stable signal at all times. Nevertheless the sky-wave mode is important because it can provide a greatly extended, though inferior quality, night-time coverage. It can also give rise to troublesome night-time interference between stations which function independently with the surface-wave mode during the day.

15.4.4.1 General features

As the sky-wave propagates from transmitter to receiver it is subject to various losses. These losses are considered in more detail in the following Sections. In practice most transmitting and receiving antennas used in the LF and MF broadcasting bands use vertical polarisation, and this is assumed here.

The main ITU-R prediction formula for the annual median sky-wave field strength is

$$E = (P + G_v + G_h) + G_s - L_p + A - 20\log p - 10^{-3}k_r p - L_t \qquad (15.23)$$

where

E = annual median of half-hourly field strengths in dB (μV/m)

P = *radiated* power in dB (kW)

G_v = a transmitting antenna gain factor due to vertical directivity (dB relative to the maximum gain of a small monopole)

G_h = a transmitting antenna gain factor due to horizontal directivity (dB); G_h=0 for an azimuthally omnidirectional antenna

G_s = a 'sea gain' associated with sea near the transmitting and/or receiving antenna (dB)

L_p = a 'polarisation-coupling' loss associated with magnetoionic effects (dB)

A = an empirical factor which depends on geomagnetic latitude (dB)

p =   slant propagation distance (km)
k_r =   an empirical loss factor in which is bundled ionospheric absorption, focusing, terminal losses and intermediate ground reflection losses
L_t =   an empirical 'hourly' loss factor (dB)

The prediction equation is applicable between 150 and 1600 kHz and path lengths up to 12 000 km, but should be used 'with caution' for geomagnetic latitudes outside $\pm 60°$.

15.4.4.2   Terminal losses and sea gain

The strength of the transmitted wave, and voltage induced in the receiving antenna, are both influenced by ground loss, which would be zero only if the ground were flat and perfectly conducting near the antennas. With flat but finitely conducting ground the interaction of direct and ground-reflected wave gives rise to a ground loss at each terminal of

$$L_g = 6 - 20\log\left|1 + R(\Psi)\right| \quad \text{dB} \tag{15.24}$$

where R is the Fresnel plane-wave reflection coefficient for vertically polarised waves at elevation angle Ψ. As the elevation angle approaches grazing, R tends to -1 and the ground loss becomes infinite. When the Earth's curvature is taken into account the losses remain large but finite.

Terminal loss factors due to imperfect ground do not appear explicitly in the Recommendation 435 prediction method. Instead they are implicitly included for average ground as part of the empirical term k_r.

Sea water has a much higher conductivity than land, with the result that the ground losses for terminals located within a few tens of kilometres of the sea can be much less than those for the average ground included in k_r.

For a terminal located on the coast, a correction factor, known as sea gain G_s, can be computed using reflection coefficients appropriate to first average ground (say a conductivity of 10 mS/m) and then sea. The difference in loss for the two ground types is the sea gain appropriate to one terminal. Figure 15.9 shows the result of such a calculation. Note that sea gain has maxima at ground ranges which are multiples of 2000 km owing to the presence of low-angle signals. At 2000 km the low-angle one-hop E-region reflection is dominant, at 4000 km the low-angle two-hop mode dominates and so on. When a terminal is located inland, or the sea only occupies a narrow channel, then the sea gain for the terminal will be reduced. An algorithm to deal with such complications is presented in Recommendation 435.

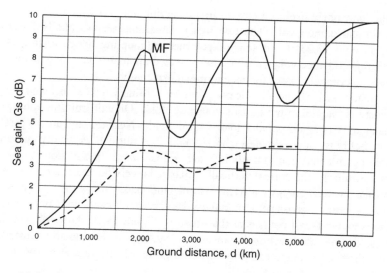

Figure 15.9   Sea gain G_s for a single terminal on the coast [5]

When using the recommended prediction method to compute the coverage of an LF or MF broadcasting station it is relatively straightforward to allow for sea gain at the transmitter. Evaluating sea gain at the receiver is computationally intensive and is only really practical when a digital coastline database is available. In practice G_s at the receiver is often taken to be zero.

15.4.4.3   Polarisation coupling loss

The ionosphere, a plasma with an embedded magnetic field, is a birefringent medium. A linearly polarised radio wave incident on the bottom of the ionosphere splits into two waves which propagate independently. The division of power between the two waves depends on polarisation of the incident wave relative to the local geomagnetic field. One of these waves, the 'extraordinary', is more heavily attenuated than its partner, the 'ordinary' wave. In general, on exit from the ionosphere the two waves have elliptical polarisation which may or may not be well orientated to excite a current in the receiving antenna.

The Earth's gyromagnetic frequency varies between 800 kHz near the equator and 1600 kHz near the poles, and therefore lies in the MF broadcasting band. At the gyromagnetic frequency the extraordinary wave is almost completely attenuated so that only the ordinary wave contributes to the received signal. This effect also prevails over a band of frequencies either side of the gyrofrequency and the extinction of the extraordinary wave can be assumed in all MF planning. In the LF broadcasting band the effect can be ignored.

When the incident wave excites an extraordinary wave which is subsequently absorbed, this is a loss mechanism, the so-called polarisation-coupling loss. A further loss occurs when the elliptically polarised ordinary wave which emerges from the ionosphere excites a current in a receiving antenna which responds only to the vertical component. The two loss mechanisms are essentially the same. The L_p in eqn. 15.23 is the sum of polarisation-coupling losses arising at the transmitting and receiving ends of the propagation path. On long paths it is necessary to compute losses for each terminal separately.

The major axis of the elliptically-polarised ordinary wave accepted by the ionosphere is parallel to the Earth's magnetic field. On exit from the ionosphere the major axis is again parallel to the local field. Near the equator, where the Earth's field is approximately horizontal, the normal vertically polarised transmitting antenna couples badly into the ordinary mode in directions to the east or west. Similarly, the normal vertically polarised receiving antenna is orthogonal to ordinary-wave energy arriving from the east or west. Thus the polarisation-coupling loss per terminal can be very significant and is most extreme on east–west/west–east paths near the equator. Figure 15.10 shows the predicted coverage of an omnidirectional transmitting station on the equator. Instead of roughly circular coverage contours, the foreshortened coverage east and west is due to polarisation coupling losses.

15.4.4.4 Temporal variability

Sky-wave field strength in the LF and MF bands varies on timescales ranging from minutes to years.

Short-period variations, usually Rayleigh distributed and measured in minutes, arise due to continuous turbulence in the ionosphere. Occasionally, when only two sky-wave modes are present, the fading may be more severe.

Short-term median field strength measured on one day will generally differ from an equivalent measurement on the following day. A sequence of such day-to-day measurements can often be approximated by a log-normal distribution.

The diurnal variation in field strength is considerable. The sky-wave field strength is largest late at night and is weak or insignificant during the day. A large number of measurements under a variety of circumstances has permitted the estimation of an hourly average loss factor L_t, the form of which is shown in Figure 15.11. For multihop paths (>2000 km) the times of sunset and sunrise are taken to be those at a point 750 km from the terminal where the Sun sets last or rises first, because the remainder of the path is then in darkness.

Seasonal variations in field strength also arise. In the MF band equinoxial months are associated with higher field strengths than others. The overall seasonal variation can be as much as 15 dB at the lower frequencies in the MF band, but is only about 3 dB at 1600 kHz. In the LF band different variations are observed, for example a pronounced summer maximum.

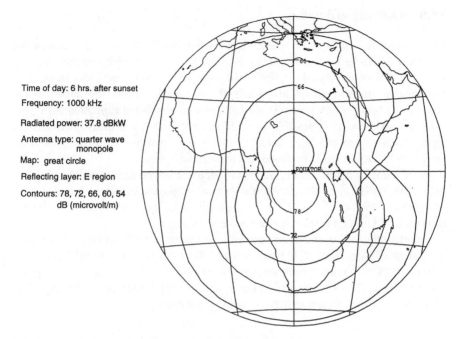

Time of day: 6 hrs. after sunset

Frequency: 1000 kHz

Radiated power: 37.8 dBkW

Antenna type: quarter wave
                    monopole

Map: great circle

Reflecting layer: E region

Contours: 78, 72, 66, 60, 54
                 dB (microvolt/m)

Figure 15.10   Coverage of an MF transmitter on the equator

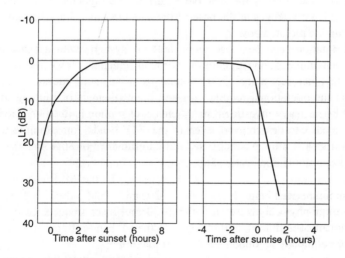

Figure 15.11   Hourly loss factor L_t [5]

15.5   Antenna efficiency

Radio transmissions in the frequency band below 2 MHz are normally made from a vertical mast, the transmitter being connected between the base of the structure and a radial-wire ground screen. The emission from such an antenna is azimuthally omnidirectional and vertically polarised. Some MF broadcasting stations use two or more masts, located and phased to achieve azimuthal gain.

In the MF band it is possible to build masts which are a quarter-wavelength high. Such an antenna is resonant, with a purely resistive input impedance of $36\,\Omega$ which can be readily matched to a transmitter.

At frequencies below about 250 kHz it becomes impractical to build masts $\lambda/4$ high. The radiation resistance R_r of a vertical monopole of height h is given by $40\pi^2(h/\lambda)^2$. In the ELF and VLF bands this ratio is small and the radiation resistance of realisable structures can be very modest.

Unfortunately, all of the power output by a transmitter cannot be radiated from the antenna. A portion of the transmitter power is dissipated in other resistive elements of the antenna system such as the series-dielectric resistance R_{sd}, copper-loss resistance R_c, load-coil resistance R_i and ground-system resistance R_g. The efficiency e of the antenna system is given by the ratio:

$$e = R_r / \left(R_r + R_{sd} + R_c + R_i + R_g \right) \qquad (15.25)$$

A second consideration is the reactive nature of the antenna impedance. This gives rise to large voltages at the base of the antenna and voltage breakdown can set a limit on the radiated power.

These limitations can be eased using antenna systems with a capacitive top loading to maximise R_r, and an extensive ground screen and low-loss components to minimise other resistances. The extent to which such measures are taken depends on the operational requirement and the economics of running the radio station. When a heavy investment in antenna construction is justified, moderately high efficiences can be achieved, even in the VLF band. For example the VLF antenna system at Cutler, Maine, USA, has an efficiency of 86% at 20 kHz. The structure is enormous, see Watt [186]:

Maximum mast height:	298 m
Average physical height:	201 m
Number of masts:	26
Area of the top loading:	$2.25\,\text{km}^2$
Ground screen:	3300 km of copper wire

The radiation resistance of this antenna is a mere $0.15\,\Omega$.

15.6   Surface-wave/sky-wave interactions

At short ranges from a transmitter the surface-wave mode dominates. At long ranges the sky-wave mode will normally deliver the strongest signal.   At intermediate ranges the surface wave and sky wave may be comparable and the interference between the two will lead to a region of signal peaks and troughs. Unlike the sky wave, the surface wave has a stable phase. Diurnal and short-term variations in the height and structure of the lower ionosphere cause the signal peaks and troughs to shift position.   Gross variations in the sky-wave signal strength causes the region of interplay between the two modes to move towards or away from the transmitter.

For the long wavelengths of the VLF/LF bands the surface-wave/sky-wave interaction gives rise to the relatively stable Hollingworth pattern referred to in Section 15.4.3. The systems required to operate in the interference region must be designed to function within the minima of the pattern.

In the MF band the region of interaction is called the night-fading zone (NFZ). The location of the NFZ depends on any factor which affects the relative amplitude of the surface and sky waves, e.g. antenna vertical-radiation pattern, time of night, ground conductivity and polarisation. Other factors such as radiated power have no effect. The fading is especially disruptive to broadcast reception and in planning such systems it is prudent to estimate the NFZ location. A fading zone which, night after night, resides over an important urban area will soon cause listeners to retune their receivers. The NFZ is usually taken to be the region in which the difference between surface- and sky-wave field strengths is 8 dB or less. When designing a broadcasting station so that the NFZ is not too troublesome, it must be borne in mind that night-time interference from other cochannel transmitters may, in fact, have a more significant effect on the quality of reception.

15.7   Background noise

The performance of any radio system is determined, in part, by the level of background noise with which the wanted signal must compete. This can be generated within the receiver or arrive at the receiver input terminals via the antenna. The four potential sources of background noise originate: in the receiver itself, the galaxy, atmospheric lightning discharges and man-made electrical equipment (see Chapter 3). Galactic noise will normally be reflected by the top side of the ionosphere at the frequencies considered here and can be ignored. For most applications the level of receiver noise is also insignificant compared with the atmospheric and man-made contributions.   An exception is in submarine communication where the atmospheric and above-surface man-made noise contributions may be both heavily attenuated by the sea so that the receiver-noise contribution becomes dominant. Chapter 3 also discusses the characteristics of radio noise.

15.8   Acknowledgments

Parts of the text in this chapter concerning sky-wave propagation are based on material originally written by Dr. Philip Knight [116] for the previous edition of this book. Figures 15.9 and 15.11 are used with the kind permission of the ITU as copyright holder.

Chapter 16

Propagation of radiowaves in the ionosphere

P.A. Bradley

16.1 Propagation in an ionised medium: magnetoionic theory

16.1.1 Introduction

Radiowaves propagating in the ionosphere set the charged particles into oscillation causing them to radiate secondary wavelets in all directions. In the forward direction their path lengths are equal and so the combined secondary wave is strong. This forward-going scattered wave travels in the same direction as the original wave and combines with it. It experiences a phase advance of $\pi/2$ so that the resultant phase also is slightly advanced with respect to the original wave. The reradiated wave is more intense and the combined wave is further advanced in phase the greater the concentration of charged particles. The change of speed is larger for electrons, which can more easily be set into oscillation than for heavier ions. For a wave obliquely incident on the ionosphere from below with an increase in the electron concentration with height, the speed of the resultant wave increases upwards. Separate parts of the wavefront find themselves in places where the charge concentration is different; the top travels more rapidly than the bottom, and so the wave is refracted earthwards. Reflection occurs when the group ceases to increase height.

During oscillations the charges may collide with the neutral air particles. The regular oscillations are interrupted and energy has to be fed in from the main wave. So the wave becomes weaker, or is absorbed as it progresses. When a charged particle travels in a magnetic field, it follows a spiral path, simultaneously moving along the field line and rotating around it. The speed of rotation depends on the charge and mass of the particle and on the strength of the field. In the Earth's magnetic field electron rotation rates are around 10^6 times per second. The refraction and the absorption of a wave is affected by this circular motion. The importance of the rotation depends on wave frequency. Greatest effect arises when the rotation rate matches the wave frequency.

The theory of wave propagation in an ionised medium in the presence of a magnetic field was first developed by Lorentz to explain light passage through crystals. However, when applied to radiowave propagation in the ionosphere, this failed to explain some observed features. The modified form developed by Appleton and Hartree is now known as the magnetoionic theory [66, 155].

16.1.2 The Appleton–Hartree equations

When a linearly polarised wave passes through an assembly of charged particles in the presence of a magnetic field it causes them to spiral around the field lines. They then reradiate wavelets in which the electric field rotates. The composite wave which results when these wavelets add to the original wave has its electric field rotating so that the polarisation is different from that of the original wave. If the original wave has its electric field rotating in a certain way it makes the charges rotate in the same way, they reradiate wavelets with the same kind of rotation, and when these are added to the original wave they produce a composite wave whose field also rotates in the original way. In this case the polarisation has not been changed, and the wave is a characteristic ordinary (O) wave. A second characteristic extraordinary (X) wave is possible in which the rotation is in the opposite sense.

The Appleton–Hartree theory applies for a medium which is electrically neutral with no resultant space charge and equal numbers of electrons and positive ions. A uniform magnetic field is assumed and the effect of positive ions on the wave is neglected. Steady-state solutions for characteristic waves of plane polarisation are generated. The complex refractive index n at angular frequency ω is given [155] as

$$n^2 = (\mu - i\chi)^2 = 1 - \frac{X}{1 - iZ - \dfrac{Y_T^2}{2(1 - X - iZ)} \pm \left\{ \dfrac{Y_T^4}{4(1 - X - iZ)^2} + Y_L^2 \right\}^{\frac{1}{2}}} \quad (16.1)$$

where

$$X = \frac{Ne^2}{\varepsilon_0 m\omega^2} \qquad Y_L = \frac{e B_L}{m\omega} \qquad Y_T = \frac{e B_T}{m\omega} \qquad Z = \nu / \omega$$

N is the electron concentration, e and m are the electronic charge and mass and ε_0 is the permittivity of free space. ν is the electron collision frequency. The subscripts T and L refer to the transverse and longitudinal components, respectively, of the Earth's magnetic field B with reference to the direction of the wave normal. In particular, the refractive indices of the ordinary (upper-sign) and extraordinary (lower-sign) waves differ. The corresponding wave polarisations R are

$$R = \frac{i}{2Y_L}\left[\frac{Y_T^{\,2}}{1-X-iZ} \mp \left\{\frac{Y_T^{\,4}}{(1-X-iZ)^2} + 4Y_L^{\,2}\right\}^{1/2}\right] \qquad (16.2)$$

Equation 16.1 shows that:

(i)   below the ionosphere refractive index is unity. For a given wave frequency it decreases with increasing electron concentration and for a given electron concentration it decreases with increasing wave frequency. If N is sufficiently large then, ignoring the magnetic field and collisions, μ will become zero and $X=1$ at the height of reflection at vertical incidence; otherwise the wave traverses the whole ionosphere and escapes. Hence for a frequency f at reflection N satisfies

$$f^2 = \frac{Ne^2}{4\pi^2\varepsilon_o m} \qquad (16.3)$$

(ii)   in the presence of a magnetic field the ordinary wave is reflected as if the field were absent but the extraordinary wave at HF is reflected from a lower height where

$$X = 1 - Y \qquad (16.4)$$

with $\quad Y^2 = Y_L^2 + Y_T^2$.

The wave polarisations given from eqn. 16.2 for the ordinary and extraordinary waves indicate the amplitude ratio and phase difference between the component electric vectors in the wavefront plane lying parallel to and normal to the projection of the magnetic field. In general wave polarisation is elliptical with the ordinary and extraordinary waves having equal axial ratios but opposite senses of vector rotation.   In the case of no collisions ($Z=0$), R_0 and $R_x = 1$ and the two waves have orthogonal major axes.   With longitudinal propagation ($Y_T^4 \gg 4(1-X)^2 Y_L^2$) the two magnetoionic waves are circularly polarised. With transverse propagation ($Y_T^4 \ll 4(1-X)^2 Y_L^2$) the ordinary wave is linearly polarised with its electric vector parallel to the imposed magnetic field.

For a wave travelling in the z direction

$$E = E_0 \exp\left(-\frac{\omega}{c}\chi z\right)\exp i\left(\omega t - \frac{\omega}{c}\mu z\right) \qquad (16.5)$$

The quantity $(\omega\chi/c)$ is a measure of the decay of amplitude per unit distance and is called the absorption coefficient k:

$$k = \frac{\omega\chi}{c} \qquad (16.6)$$

In the absence of a magnetic field the absorption in nepers per metre (1 neper = 8.69 dB) is given as

$$k = \frac{\omega XZ}{2c\mu(1+Z^2)} = \frac{e^2}{2\varepsilon_0 mc\mu}\frac{N\nu}{\omega^2 + \nu^2} \qquad (16.7)$$

When N is small $\mu = 1$ and eqn. 16.7 gives

$$k = \frac{e^2 N\nu}{2\varepsilon_0 mc\omega^2} \qquad (16.8)$$

This is called 'nondeviative' absorption and arises primarily in the D region. Near reflection when μ becomes small

$$k = \frac{\nu}{2c}\left(\frac{1}{\mu} - \mu\right) \qquad (16.9)$$

and the absorption is 'deviative' since it occurs in a region where considerable ray deviation takes place.

In the presence of collisions, the wave polarisations (eqn. 16.2) are complex. This means that the major axes of the polarisation ellipses of the ordinary and extraordinary waves are no longer orthogonal. The ellipses each rotate from the no-collision case by the same amount in opposite directions, such that each ellipse is the reflection of the other in the plane making an angle of 45° with the magnetic meridian.

The assumption that ν is independent of electron velocity is one of the major limitations of the Appleton–Hartree theory. In 1960 Sen and Wyller [165] generalised the magnetoionic theory to include the energy dependence of the electrons. The use of the generalised expressions is particularly important in considering VLF and LF propagation, and in calculating the absorption of HF waves in the D region.

16.1.3 Phase and group velocity

The phase velocity v is

$$v = \frac{c}{\mu} = c\left(1 - \frac{Ne^2}{m\varepsilon_0\omega^2}\right)^{-1/2} \tag{16.10}$$

for propagation with no collisions and no magnetic field. This indicates that the phase velocity in the medium is greater than the velocity of light and the wavelength in the medium is greater than in free space

$$\lambda = \lambda_0\left(\frac{v}{c}\right) \tag{16.11}$$

If the phase velocity of a wave in a medium varies as a function of the wave frequency, it is said to be dispersive. Two waves with slightly different frequencies will therefore travel with slightly different velocities. It is the interference pattern between two such waves which determines where, and with what velocity, the energy of the composite wave will travel. For a wave $\cos(kz-\omega t)$ the group velocity u is given by

$$u = \frac{\delta\omega}{\delta k} \tag{16.12}$$

For a nondispersive medium in which ω/k is constant, $u = v$.

The group refractive index μ' may be defined as

$$\mu' = \frac{c}{u} = c\frac{dk}{d\omega} = c\frac{d}{d\omega}\left(\frac{2\pi}{\lambda}\right) = \frac{d}{d\omega}(\mu\omega)$$

$$= \mu + \omega\frac{d\mu}{d\omega} = \mu + f\frac{d\mu}{df} \tag{16.13}$$

For the no field situation where $\mu^2 = 1 - (f_N/f)^2$ we have that

$$\mu' = \frac{d}{df}(\mu f) = \frac{1}{\mu} \tag{16.14}$$

16.1.4 Propagation in an anisotropic medium

A medium is said to be isotropic if the phase velocity of a wave propagating within it is independent of direction. This is not the case within a magnetoionic medium, where refractive index depends on direction of propagation relative to the field. In general the directions of the phase and ray paths then differ. It can be shown that the angle α between the wave normal and the ray direction is

$$\tan\alpha = -\frac{1}{v}\frac{dv}{d\theta} = +\frac{1}{\mu}\frac{d\mu}{d\theta} \tag{16.15}$$

where θ is the angle at which the wave normal direction cuts a reference axis. The phase path in an anisotropic medium is

$$P = \int_s \mu \cos\alpha\, ds \tag{16.16}$$

integrated over the raypath s. The corresponding group path is

$$P' = \int_s \mu \cos\alpha\, ds \tag{16.17}$$

16.2 Propagation-path determination

16.2.1 Ray-tracing techniques

16.2.1.1 General

Ray tracing is the process of determining the path of an electromagnetic signal by the successive application of ray theory over a series of thin homogenous slabs of medium. It requires that the wave parameters such as polarisation and refractive index do not change appreciably within a wavelength, that the division of energy between the ordinary and extraordinary waves be determined at the place of entry to the ionosphere and that thereafter these two waves propagate independently. Ray tracing requires a knowledge and representation of the state of the ionosphere through which the rays pass.

The most common application of ray tracing is to find the position at which a ray launched into the ionosphere returns to earth. Homing techniques can be applied to find the launch direction which give propagation to a selected reception point. Ray tracing may also be applied to determine the total phase and group paths and the ionospheric absorption. The path attenuation is given by the change in cross-section area of a small bundle of rays.

There are a number of ray-tracing techniques of varying complexity and accuracy. In any application the simplest method giving adequate accuracy should always be used, because ray tracing is expensive and time consuming, even with currently available high-speed computers. The introduction of the effects of the Earth's magnetic field is an appreciable complication. It is not usual to take account of the modifying influence of electron collisions on ray paths but to consider collisions only as responsible for absorption. At VHF and higher frequencies adequate determinations of the raypath parameters can usually be made with approximate equations in terms of the total electron content which is the number of electrons in a vertical column of ionosphere of unit cross-section.

16.2.1.2 Specific procedures
Concentric model ionosphere with no magnetic field: Bouger's law may be applied to trace rays via a succession of thin concentric slabs of ionosphere. It enables the angle of incidence at a slab of refractive index μ and height h to be determined for a ray launched with elevation angle Δ_u relative to the Earth of radius R (Figure 16.1). The law gives that

$$R\cos\Delta_u = \mu(R+h)\sin i \qquad (16.18)$$

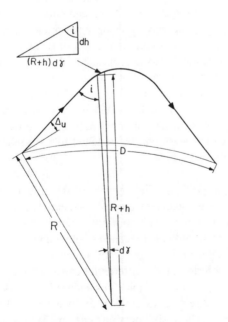

Figure 16.1   Raypath geometry for spherically stratified ionosphere

Application of eqns. 16.16 and 16.17 yields

$$P = \int_0^h \frac{\mu \, dh}{\cos i} = \int_0^h \frac{\mu^2 \, dh}{\sqrt{\mu^2 - \left(\dfrac{R}{R+h}\right)^2 \cos^2 \Delta_u}} \tag{16.19}$$

$$P' = \int_0^h \frac{dh}{\sqrt{\mu^2 - \left(\dfrac{R}{R+h}\right)^2 \cos^2 \Delta_u}} \tag{16.20}$$

and

$$D = R \int_0^\gamma d\gamma = R \int_0^h \frac{\tan i \, dh}{R+h} = R^2 \cos \Delta_u \int_0^h \frac{dh}{(R+h)^2 \sqrt{u^2 - \left(\dfrac{R}{R+h}\right)^2 \cos^2 \Delta_u}} \tag{16.21}$$

Raypaths in nonconcentric layers with no magnetic field: The real ionosphere is not concentric and contains horizontal gradients, particularly at times of sunrise/sunset and for paths at high latitudes or those across the magnetic equator. The equations based on Bouger's law for segments of a concentric model ionosphere can be applied successively using appropriate electron-concentration data derived in accordance with a cumulative record of ray position. Step sizes are determined from operational tests appropriate to the models being investigated and are chosen such that further reductions of step size lead to consistent results within the desired accuracy. Small steps are needed near a reflection position but usually much larger steps are adequate elsewhere over the path. Techniques can be devised to reduce the step size automatically in accordance with an error criterion. Thereby segment sizes can be matched to the ionospheric conditions and the calculations optimised.

Ray tracing with a magnetic field: The raypaths of the ordinary and extraordinary waves differ and these are usually displaced oppositely of each other from the corresponding raypath if there were no field. Differences from the no-field raypaths become significant only at frequencies below about 5–6 MHz. A widely used method of ray tracing in the presence of a magnetic field, produced by the Hazelgroves, involves solution of six differential equations based on Hamiltonian optics. These give three position variables in standard polar-co-ordinate form and three components of the direction variable of the wave normal with respect to a local-co-ordinate system. These six equations enable changes in the positions of the rays and in the directions of the wave normals to be deduced by suitable numerical-integration techniques.

16.2.2 Raypaths at VHF and higher frequencies

Propagation is nearly always quasilongitudinal to the magnetic field. Hence from eqn. 16.1 ignoring electron collisions and setting $X \ll 1$ gives

$$\mu \sim 1 - \frac{X}{2}\left(1 \mp Y_L\right) \tag{16.22}$$

where the $-$ and $+$ signs refer, respectively, to the ordinary and extraordinary waves. Further, under conditions when the magnetoionic splitting of the rays may be ignored (putting $Y_L = 0$),

$$\mu = 1 - \frac{X}{2} \tag{16.23}$$

Thus, the change in phase path due to the presence of the ionosphere is

$$\Delta P = \int_s (\mu - 1) ds = -\frac{e^2}{8\pi^2 \varepsilon_0 m f^2} \int_s N ds \tag{16.24}$$

The ordinary and the extraordinary waves at these frequencies are essentially circularly polarised and they suffer no change of polarisation during propagation. However, because they have different phase velocities, the plane of polarisation of their resultant, which is a linearly polarised wave, rotates gradually. This phenomenon is known as Faraday rotation.

The difference in phase paths of the ordinary and extraordinary waves is

$$P_0 - P_x = \int_s (\mu_0 - \mu_x) ds = \int_s X Y_L ds = \frac{\left|\varepsilon^3\right|}{8\pi^3 \varepsilon_0 m^2 f^3} \int_s B \cos\theta N ds \tag{16.25}$$

Often results of adequate accuracy are obtained using a mean value of $B\cos\theta$ over the raypath. Hence Ω, the angle through which the plane of polarisation of the resultant wave rotates during ionospheric passage, is

$$\Omega = \frac{\pi f}{c}\left(P_0 - P_x\right) = \frac{\left|e^3\right| B \cos\theta}{8\pi^2 \varepsilon_0 m^2 c f^2} \int N ds \tag{16.26}$$

Note that both change in phase path and resultant wave rotation are proportional to total electron content and inversely proportional to frequency squared.

The change in group path due to the presence of the ionosphere is

$$\Delta P' = \int_s (\mu' - 1)ds \tag{16.27}$$

Substituting $\mu' = \dfrac{1}{\mu}$ from eqn. 16.14 in the expression of eqn. 16.23 for μ leads to

$$\mu' = 1 + \frac{X}{2} \quad \text{or} \quad \mu' - 1 = 1 - \mu \tag{16.28}$$

Hence, from eqn. 16.24,

$$\Delta P' = -\Delta P \tag{16.29}$$

The refractive effects of the ionosphere may be considered in terms of the angular bending β of the raypath from the line-of-sight direction. β is the integral over the raypath of the change in i, the angle of incidence, i.e. $\beta = \int di$. It is given from Snell's law as

$$\beta = \int_s \tan i \, \frac{d\mu}{\mu} \tag{16.30}$$

Analytical expressions are available for β for certain ionospheric models. As with the other path parameters considered above, β is also proportional to total electron content and inversely proportional to frequency squared.

16.3   Propagation effects at HF

Figure 16.2 shows the raypaths via the same single-layer model ionosphere for rays at three separate frequencies launched with a series of different elevation angles from a ground-based transmitter. A number of features are apparent:

(i)   for the lowest frequency there is sufficient ionisation present to reflect the waves at all elevation angles, including the vertical; at the higher frequencies, rays launched with an elevation greater than some critical value escape;

(ii)   waves launched more obliquely in most cases travel to greater ranges;

(iii)   waves suffer more refraction at the greater heights;

(iv)   waves of higher frequency are reflected from a greater height;

(v)   waves launched more obliquely are reflected from a lower height.

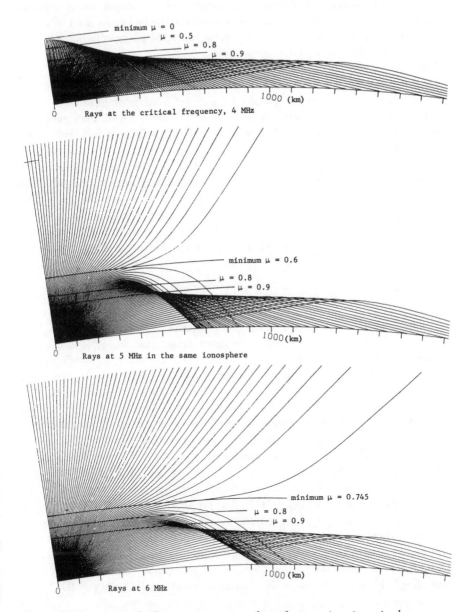

Figure 16.2   *Raypaths for propagation at three frequencies via a single Chapman model ionosphere of critical frequency 4 MHz, height of maximum electron concentration 300 km and scale height 100 km (from Croft [77])*

The maximum range attainable after one ionospheric reflection arises for rays launched at grazing incidence and this depends primarily on the height of maximum electron concentration. For typical E, F1 and F2 layers, the maximum range is 2000, 3400 and 4000 km, respectively.

For a given ionosphere there will be some limiting upper frequency reflected vertically at the height of maximum electron concentration. At frequencies above this critical frequency there is a ground distance out from the transmitter at points along which illumination is not possible by waves reflected from the ionosphere. This distance is known as the skip distance. The skip distance increases as the wave frequency increases and in the limit for a very high frequency can extend to the maximum ground range possible for rays launched at grazing incidence; in that case all rays escape into space. It follows, for a fixed point of reception, that there is some maximum frequency at which the waves can be reflected to it. This is the frequency making the distance from the transmitter to the point equal to the skip distance. The frequency is known as the basic maximum usable frequency (BMUF). The BMUF is defined in [10] as the highest frequency that can propagate between ground-based terminals on a specified occasion by ionospheric refraction alone.

The BMUF increases with ground distance and depends also on the amount of ionisation present. It depends too on the height of the ionosphere since the determining factor as to whether reflection or transmission occurs is the angle of incidence at the layer. The greater the layer height, the steeper the angle of incidence to achieve propagation to a fixed range, and therefore the lower the BMUF. This means that although the critical frequency of the E layer is less than that of the F1 layer, which in turn is less than that of the F2 layer, sometimes the E-BMUF can be the greatest of the three separate layer BMUFs. This is most likely to be the case in the summer daytime at low solar epochs (when the ratio of E to F2 critical frequencies is greatest) over path ranges of 1000–2000 km.

Since the Earth's field leads to the production of O and X waves which follow different raypaths, these waves also have differing BMUFs. The O wave is refracted less than the X wave, becomes reflected from a greater height and so has a lower critical frequency and BMUF. For propagation between a pair of fixed terminals the path BMUF is the greatest of the individual BMUFs for reflection from the different layers. This frequency undergoes systematic variations with time of day, season and solar epoch as the electron concentration and layer heights vary; there are also large day-to-day changes which create problems for modelling. Figure 16.3 shows the maximum observed frequency (MOF) on a sample path recorded in a single month using an oblique sounder and, for comparison, the estimated monthly median basic BMUF determined by conventional modelling techniques.

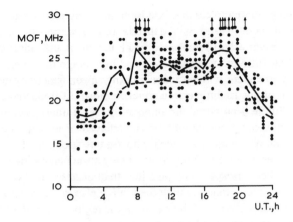

Figure 16.3   *Maximum observed frequencies for Cyprus–Slough path in July 1969 (from Bradley and Howard [62])*

-    ●   daily values
-    —   monthly median values
-    ---   predicted monthly median values (from [11])

Now consider propagation to some point beyond the skip distance. Figure 16.2 shows that, as the elevation angle is increased at a fixed frequency, rays travel to shorter ground ranges until the skip distance is reached. Rays of slightly larger elevation angle do not penetrate the ionosphere into space because, contrary to a popular misconception, ray apogee at the BMUF is below the height of maximum electron concentration, except in the limiting case of vertical incidence. These larger elevation rays are then reflected from a greater height, and they travel back to ground at increased range by virtue of having a significant length of near-horizontal path close to apogee. In principle, such so-called high-angle or Pedersen rays can exist out to a limiting ground range where ionospheric reflection is from the layer maximum. This limiting range can exceed that of the low-angle ray and may well be in excess of 7500 km in temperate regions and 10 000 km in equatorial regions [37]. The band of elevation angles providing high-angle rays is usually only a few degrees. There is then a range of ground distances at all points along which there are both low- and high-angle rays. The path length through the ionosphere of the high-angle ray exceeds that of the low-angle ray by an amount which increases when moving out from the skip distance. Therefore the strength of the high-angle ray tends to be less that of the low-angle ray both because of increased spatial attenuation and also, particularly in the case of reflection from the E-layer, because of increased ionospheric absorption. In practice signal-strength considerations determine the effective upper ground-range limit of the high-angle ray. Conversely, for propagation to a fixed ground range, there is a band of frequencies below the BMUF over which the high-angle ray has appreciable amplitude. As the frequency is reduced from the BMUF, so the excess path length and group-path length of the high-angle ray relative to the low-angle ray increase, while at the same time the differential absorption also rises.

The presence of two rays with different group-path lengths is a disadvantage for it gives rise to signal distortions. Since the low- and high-angle rays merge at the BMUF, this frequency is sometimes alternatively known as the junction frequency JF. Both the O and X waves have their own separate families of high-angle rays and associated JFs. Figure 16.4 shows an oblique-incidence ionogram recorded over a 6700 km path in which propagation time is displayed as a function of wave frequency. The separate traces are associated with signals successively reflected twice, three and four times from the F2 region and being sustained by intermediate ground reflections. The corresponding junction frequencies, labelled 2F2JF, 3F2JF and 4F2JF, respectively, together with the high-angle rays, can be seen. In this example there is some smearing of the record in the region of the JFs which is attributed to ionisation gradients along the path.

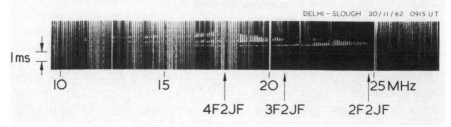

Figure 16.4   Sample oblique-incidence ionogram with classical 'noses' (from Kift et al. [114])

Aside from signal-strength considerations, for a particular mode to be present, the wave frequency must be below the BMUF and, for F modes, also the lower ionosphere must not screen or blanket it. Screening of the 1F2 mode, but not of the 2F2 mode because of the lesser path obliquity, is a common summer-daytime occurrence at certain frequencies. The strongest or dominant mode on a long path is usually the lowest-possible order F2 mode unless the antennas discriminate against this. Higher-order F2 modes traverse the ionosphere a greater number of times to become more absorbed and also experience more ground reflections, so that they tend to be weaker. A given range can be spanned by fewer F than E hops. Modes involving more than two reflections from the E layer are rarely of importance. Reflections from the F1 layer arise only under restricted conditions and the 1F1 mode is less common than the 1E and 1F2 modes. The 1F1 mode can be important at ranges of 2000–3400 km, particularly at high latitudes. Multiple-hop F1 modes are very rare in practice because the necessary ionospheric conditions to support an F1-layer reflection do not occur simultaneously at separated positions.

Geographical changes in ionisation cause so-called mixed modes with successive reflection from different layers. Mixed modes are a common feature of transequatorial paths and east–west paths across a daylight–darkness boundary. Other more complex examples of mixed modes are those involving upwards reflection from the E layer between two F reflections, known as M modes. Changes in ionisation of a smaller-scale size influence raypaths on single hops. These are variously referred to as ionisation gradients, horizontal gradients or ionospheric tilts. They cause the upwards and downwards legs of a hop to differ in length and direction.

Longitudinal tilts produce differences in the elevation angles on the two legs; lateral tilts create off-great-circle paths. Longitudinal tilts are usually the more important because they can give rise to changes in propagation modes. Lateral deviations are generally small in comparison with antenna beamwidths. An exception, even for propagation via the regular modes, where ionospheric tilts lead to marked departures from the great-circle path, arises when the transmitter and receiver are almost antipodal. Simultaneous propagation may then take place in several directions and the dominant-mode direction may vary with time of day, season and frequency. An effective tilt may result from geographical changes in either electron concentration or layer height. It follows that longitudinal tilts modify the BMUF over a fixed path length. On long paths with low elevation angles these longitudinal tilts can give rise to modes involving multiple reflection from the ionosphere without intermediate ground reflection. In such cases, if ray perigee at the middle of the path is within the ionosphere and above the D and lower E regions, there is little resulting absorption so that received signals are relatively strong. These so-called perigee modes can be particularly important across the equator and at high latitudes where significant ionisation gradients commonly exist. Associated with perigee modes are ground dead zones, additional to the skip zone, for which raypath illumination is not possible.

As well as propagation modes resulting from ionospheric reflections, there are others associated with scattering and ducting. Various mechanisms are believed to be involved and so it is not surprising that there are uncertainties in the interpretation of particular observational data and therefore in assessing the relative importance of the different phenomena. These create modelling difficulties. Signals are scattered by ionospheric irregularities in the D, E and F regions — patches of varying electron concentration such as those which give rise to the phenomena observed on vertical-incidence ionograms known as sporadic-E and spread-F. The scattering may result in onwards propagation (forward scatter), deviation out of the great circle (sidescatter) or return along the same path (backscatter). Ionospheric scatter modes are usually weaker than the corresponding reflected modes and they tend to fade more. However, they are important at the higher frequencies of the HF band since they enhance the practical (operational) MUF (referred to simply as the MUF [10]) so that it exceeds the basic MUF. Their geographical and temporal occurrence is governed by the incidence of the irregularities.   Sporadic-E is most prevalent at low latitudes in the daytime and at auroral latitudes by night. It tends to be opaque to the lower HF waves and partially reflecting at the higher frequencies. F-region irregularities can exist simultaneously over a wide range of heights. They are found at all latitudes, but are particularly common at low latitudes in the evenings where their occurrence is related to rapid changes in the height of the F region. Hence forward-scatter modes associated with spread-F are important on long transequatorial paths. F-region irregularities are field aligned and sidescatter from these has been observed on paths at high and low latitudes; in some instances the received signals were incident simultaneously from a range of directions.

Normal ground terrain is sufficiently rough that it too scatters significant signal power out of the great-circle direction. Ground sidescatter and backscatter result. Since sidescatter paths are longer than the more direct routes, they tend to have correspondingly greater MUFs. There is some practical evidence supporting a dependence of signal intensity on scattering angle and whether sea or land is involved. The backscatter mechanism is of value in providing a means of remote probing (e.g. studying the state of the sea) or for monitoring ionospheric conditions.   Special backscatter sounders can be used to determine the skip distance and deployed in support of systems operation. It is believed that another mechanism for wave propagation in the ionosphere concerns channelling as in a waveguide. This waveguide may be formed within the F layer and have an upper but no lower boundary, being sustained by the concave ionosphere, or it may be a double-walled duct in the electron-concentration minimum between the E and F regions. The waveguide is sometimes known as a whispering gallery. Signal coupling into the waveguide is assumed to involve ionospheric tilts like those which develop in the twilight periods or to be caused by the existence of ionisation irregularities such as Es or those responsible for spread-F. A further ducted type of signal propagation occurs along columns of field-aligned ionisation.

Mention has been made of ionospheric absorption in Section 16.1.2.   For propagation along the direction of the Earth's magnetic field the absorption in decibels $L(f_v)$ at vertical incidence in traversing a height region h at a wave frequency f_v is given as

$$L(f_v) = K \int_h \frac{N\nu}{\mu} \frac{dh}{(f_v \pm f_L) + \frac{\nu^2}{4\pi^2}}$$
(16.31)

where K is a constant of proportionality. This equation applies approximately over a considerable range of wave directions with f_L taken as the electron gyrofrequency about the component of the Earth's magnetic field along the direction of propagation.   The positive sign applies for the O wave and the negative sign for the X wave. For ground-based reflection the limits of integration are from the base of the ionosphere to the height of wave reflection. For propagation at oblique incidence the absorption is proportionately increased because of the greater lengths of path traversed. Inspection of eqn. 16.31 shows:

(i)   that the absorption in a given slab of ionosphere is proportional to the product of electron concentration and collision frequency. Electron concentration increases with increase of height whereas the collision frequency for electrons, which is proportional to the atmospheric pressure, decreases.   Hence the absorption reaches a maximum in the lower E region with most of the contribution to the total absorption occurring in the D region;

(ii)   large amounts of additional deviative absorption arise near the height of reflection where μ is small;

(iii)   absorption decreases with increase of frequency; and

(iv)   the O-wave absorption is less than that of the X wave and differences are accentuated the lower the frequency, provided that the first term of the denominator of eqn. 16.31 remains dominant.

The absorption is low at night time because of the reduced D and E region ionisation. The nondeviative absorption reaches a maximum around local noon in the summer, but the influence of deviative absorption can modify the resultant seasonal variation. Ionospheric absorption is one of the most important factors influencing received skywave signal strengths at MF and HF so that accurate methods of modelling it are needed.   There are particular difficulties at MF because raypath-reflection heights of around 85–90 km are common and much of the absorption is deviative absorption occurring within 2–3 km of ray apogee. Such electron-concentration data as exist for these heights display considerable irregular variations.

When signals are propagated between terminals via multiple paths, whether these involve different modes, low- and high-angle rays or O and X waves, there exists a difference in the group paths of the separate components. Hence there is a spread in time of the received signals.   Multipath time dispersions can limit system performance just as can an inadequate signal/noise-power ratio.   Large time spreads are often associated with scatter propagation. There are also large variations in the angles of elevation of the incident signals.

If the ionosphere were unchanging, the signal amplitude over a fixed path would be constant.   In practice, however, fading arises as a consequence of variations in propagation path, brought about by movements or fluctuations in ionisation. The principal causes of fading are: (i) variations in absorption, (ii) movements of irregularities producing focusing and defocusing, (iii) changes of path length among component signals propagated via multiple paths, and (iv) changes of polarisation, such as for example due to Faraday rotation.   These various causes lead to different depths of fading and a range of fading rates. The slowest fades are usually those due to absorption changes which have a period of about 10 min. The deepest and most rapid fading occurs from the beating between two signal components of comparable amplitude propagated along different paths. A regularly reflected signal together with a signal scattered from spread-F irregularities can give rise to so-called 'flutter' fading, with fading rates of about 10 Hz. Fading is discussed in detail in Chapters 3.9 and 3.10

Amplitude fading is accompanied by associated fluctuations in group path and phase path, giving rise to time- and frequency-dispersed signals. When either the transmitter or receiver is moving, or there are systematic ionospheric movements, the received signal is also Doppler-frequency shifted. Signals propagated simultaneously via different ionospheric paths are usually received with differing frequency shifts. Frequency shifts for reflections from the regular layers are usually less than 1 Hz, but shifts of up to 20 Hz have been reported for scatter-mode signals at low latitudes. Frequency spreads associated with individual modes are usually a few tenths of a hertz.

The effect of ionospheric propagation on a radio signal may therefore be expressed in terms of a corresponding channel-scattering function (Figure 16.5) in which each mode has its own attenuation due to transmission loss and its own time and frequency offsets and dispersions. As a caution, it must however be noted that even this representation is an oversimplification. Particularly for transequatorial and auroral paths, the modes coalesce because the spread associated with each is so great. Time spreads of several milliseconds and frequency spreads in excess of 10 Hz have been reported under such conditions.

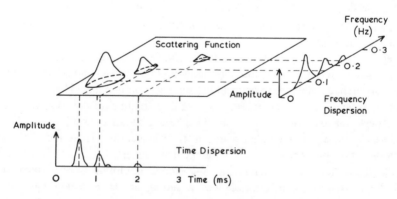

Figure 16.5   Channel-scattering function for three-moded ionospheric signal propagation (from Vincent et al. [181])

16.4 Propagation effects at VHF and higher frequencies

Radio waves at VHF and beyond traverse the whole ionosphere but, as noted in Section 16.2.2, are subject to refraction, phase and group delay and Faraday rotation. In the presence of ionospheric irregularities, particularly at low and high latitudes, scintillations also arise. Ionospheric effects generally decrease the greater the frequency, but can still be important to the operation of Earth–space communication, navigation and surveillance systems.

Chapter 17

HF applications and prediction

P.A. Bradley

17.1 Services relying on HF

At HF there are a multitude of frequency sub-bands for the different services with many of these being shared among services. Some subbands are common to the land-mobile and maritime-mobile services; others are separate. Of the 28 MHz of available spectrum this is estimated as being occupied approximately as follows:

Fixed service	55%
Land and maritime mobile services	15%
Sound broadcasting service	15%
Aeronautical mobile service	10%

The remaining 5% of spectrum is used by the amateur service, the standard frequency and space-research services.

Maritime elements include coast stations, ship-station radiotelephone working to coast stations and intership working. Aeronautical systems include single-sideband HF radiotelephone links between aircraft and the ground in the aeronautical mobile channels of the 2–22 MHz band and radio-teletype links between ground terminals in the aeronautical fixed channels in the 2.5–30 MHz band. Variations in frequency allocations between different geographical regions arise principally from changes in operational requirements, rather than from propagation effects. However, the tropical region has been defined to allow specifically for the differing propagation phenomena and increased background noise from thunderstorms at low latitudes. Sound broadcasting in the tropical zone is permitted between 2300 and 2495 kHz — as well as in three other special sub-bands in the lower part of the HF band.

The majority of long-distance fixed radio circuits nowadays rely on satellites and cables for both civilian and military applications. Likewise, there has in recent years been a growth in the use of satellites for aeronautical, maritime and land-mobile communications. Nonetheless, increases in the numbers involved mean that there are actually more HF circuits than, say, 20 years ago. Particularly for military purposes, HF systems are regarded as providing a necessary backup service to fixed links primarily established by other means. High frequencies create a useful alternative to satellites in establishing communications with small isolated communities in such places as the Arctic and Middle Eastern desert areas.

17.2 Requirements for propagation predictions

17.2.1 System design

Long-term predictions based on estimates of propagation conditions are needed for radio-circuit design. Ray-path launch- and arrival-angle data are of value for optimum antenna determinations. Studies of the relationships between transmitter power and received field strengths at a range of frequencies enable the necessary size of transmitter and its frequency coverage to be determined, when also the noise background intensities are known. There is no major restriction on the permissible amount of calculation or the speed with which the results are needed. Accuracy is the prime consideration.

17.2.2 Service planning

To date, relatively little effort has been applied on a world-wide basis to the optimisation of the different radio services; most of these have grown in a haphazard fashion. Frequency sharing is a useful means of optimising spectrum utilisation but any changes to current practice need very careful review before being introduced. There is plenty of scope for further studies based on long-term predictions to determine the ideal service-planning strategies.

17.2.3 Frequency management

Frequency management may be defined as the selection of the frequency to use on a particular occasion from those assigned and available. Given a realistic set of assigned frequencies, frequency management in principle could be aided by some form of short-term prediction procedure. It is evident that any short-term method adopted needs to be capable of rapid evaluation and requires on-line data links to a mainframe computer, or local use of a microcomputer. Such an approach must be seen in perspective in comparison with alternative techniques such as path sounding and real-time channel evaluation and in the light of existing operating practices which differ appreciably for the separate radio services. Military applications for reliable short-term predictions can be envisaged but most civilian systems work satisfactorily with the user selecting the best of a number of simultaneous transmissions. In all cases storm predictions would be of particular value.

17.3   Principles of long-term prediction and available techniques

17.3.1 Introduction

Prediction procedures have been devised to give estimates of median values of the basic and operational maximum usable frequencies (BMUF and MUF), received signal strength, background noise and lowest usable frequency (LUF), and to indicate their diurnal, seasonal and solar-cycle variations. The techniques adopted have usually involved the following stages: (i) determination of a representative model of the electron concentration over the propagation path, taken as being along the great circle between transmitter and receiver, (ii) some kind of ray assessment leading to an estimate of the modes present, (iii) calculation of the received signal intensity in terms of the various separate transmission-loss factors judged to be significant, (iv) estimation of the intensities of atmospheric noise and background man-made noise arising from unintended emissions, (v) statistical quantification of the random day-to-day variability of signal and noise intensities, and (vi) choice of some reference required signal/noise ratio and achievement probability to yield an acceptable grade of service.

17.3.2 Model of the ionosphere

A first requirement for accurate predictions must be a model of the vertical distribution of electron concentration in the E and F regions. This needs to take account of the known large geographic and temporal variations in the ionosphere. The most extensive ionospheric database is that derived from the world network of ionosondes. Hence models of the vertical distributions are produced with the parameters of the models given by empirical equations in terms of the ionospheric characteristics which are scaled on a routine basis at all ionosonde stations. The International Union of Radio Science (URSI) has developed such a model [57] which is available in either smoothed or segmental options where the various segments correspond to D, E, F1 and F2 layers of parabolic or polynomial function form. The model (Figure 17.1) is 'anchored' by the different layer peak heights and maximum electron concentrations, together with other empirical relationships.

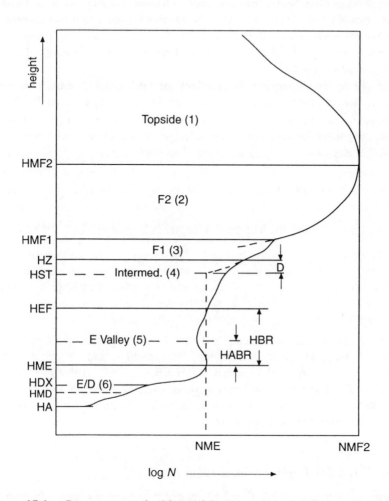

Figure 17.1 *Diagrammatic build-up of the International Reference Ionosphere (IRI) electron-concentration profile*
Segment 'anchor' points are labelled but not defined here

17.3.3 Numerical representation of the ionospheric characteristics

To generate the model for a given place and time, predicted values of the ionospheric characteristics are used. Numerical representations have been applied to past measured vertical-incidence ionosonde data from many locations throughout the world where standardised recordings are made each hour of every day to provide predicted values of these ionospheric characteristics. The ITU Radiocommunication Sector has produced reference data [11] giving monthly median formulations. foE and foF1 are obtained from empirical expressions which assume a variation with latitude, time-of-day and season depending on the solar-zenith angle χ. A solar-activity dependence is included in terms of R_{12} the smoothed sunspot number.

There are separate computer formulations for foF2 and M(3000)F2 for every month of two reference years with an assumed linear dependence on R_{12} for intermediate solar epochs. Each consists of orthogonal polynomial expressions in terms of geographic latitude λ, geographic longitude θ and Universal Time T. The general characteristic $\Omega(\lambda, \theta, T)$ is expressed as a time series

$$\Omega(\lambda,\theta,T)=\sum_j \left\{ a_j (\lambda,\theta)\cos jT + b_j (\lambda,\theta)\sin jT \right\} \tag{17.1}$$

where the *a*s and *b*s give the latitude and longitude variations, being defined as:

$$a_j (\lambda,\theta)=\sum_k U_{2j,k} \, G_k (\lambda,\theta)$$
$$b_j (\lambda,\theta)=\sum_k U_{2j-1,k} \, G_k (\lambda,\theta) \tag{17.2}$$

The *U*s are numerical coefficients and the *G*s are trigonometric functions of geographic longitude and a combined geographic- and magnetic-latitude parameter. Several tens of thousands of coefficients are involved in defining foF2, M(3000)F2 and the other ionospheric characteristics which are represented in this same way. Figure 17.2 gives an example of a prediction map for foF2 based on 988 numerical coefficients *U*.

17.3.4 Basic MUF, operational MUF and FOT

Propagation by means of F2, E and F1 modes is allowed for, depending on the path length. The path basic MUF is taken as the highest basic MUF (BMUF) of any mode reflected from the different layers, so that it is necessary to first determine the separate F2, E and F1 BMUFs depending on the path length. Basic MUFs may be evaluated by ray-tracing procedures. However, a simpler alternative approach recommended by the ITU [11] takes them as the product of critical frequency and an 'M' factor given from empirical equations as a function of path length and reflecting layer height.

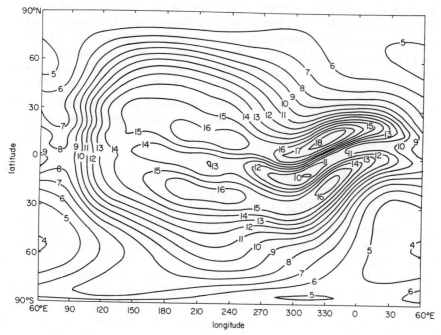

Figure 17.2   Predicted median foF2, MHz for 00 h UT in March 1958 (from [11])

For path lengths beyond the limit of single-hop F2-mode propagation, itself a variable depending on ionospheric layer height, the so-called two-control-point procedure is used in which the path BMUF is taken as the lower of the two BMUFs for the maximum-length single hop along the great circle closest to the transmitter and closest to the receiver. Although there is no rigorous basis for this approach and opinions are divided on its merits, its use is considered justifiable in many applications.

The operational MUF is the highest frequency that would permit acceptable operation of a radio service between given terminals at a given time under specified working conditions [10]. It depends, among other factors, upon the types of antenna used, the transmitter power, class of emission, information rate and required signal/noise ratio. The differences between the operational MUF and the basic MUF can be explained by various ionospheric phenomena, such as scattering in the E and F regions, off-great-circle propagation and propagation by unusual modes when ionisation irregularities exist; also spread-F may be an important factor. Empirical relationships between the operational MUF and the basic MUF are available [11].

The frequency of optimum traffic (FOT), known alternatively as the optimum working frequency, is defined as the highest frequency that is likely to propagate at a given time between a specified pair of terminals via any ionospheric mode for 90% of the days [10]. It is given in terms of the monthly median predicted path operational MUF from a knowledge of day-to-day ionospheric variability. The E and F1 layers experience relatively little variability from one day to another, and when these control the path MUF the FOT is taken as 0.95 of the MUF. For F2 modes F_1, the ratio of the lower decile to median MUF, has been evaluated from a wide range of past signal measurements and tabulated as a function of solar epoch, season, local time and geographic latitude to provide a reference set of values [11].

17.3.5 Oblique ray paths

The assessment of the active modes and their elevation angles is based on a representation of the ray paths by undeviated propagation between the ground and mirror-reflecting points in the ionosphere. The heights of the mirroring points are taken as the virtual heights of reflection of waves of 'equivalent' frequency at vertical incidence. Raypaths are assumed to follow the great circle and in some predictions are deduced from a single model of the vertical distribution of electron concentration taken as applying over the whole path. The values of the parameters of this model are given in terms of the average of the predicted ionospheric characteristics at defined positions, depending on path length.

Empirical relationships have been developed based on a quasiparabolic segment approximation to the ionospheric model discussed in Section 17.3.2 giving mirror-reflection height as a function of wave frequency, time, location and path length [11]. Other techniques involve a different model ionosphere for each hop of the path. Oblique ray paths at a given wave frequency to a particular ground range can be determined by an iterative process for rays launched in different directions. Longitudinal tilts in the neighbourhood of ray reflection can be estimated from changes in virtual heights and allowed for in terms of a tilted plane–mirror mechanism. E-layer screening of F modes is of importance and needs to be taken into account as a function of path obliquety and the prevailing E-region ionisation. In the ITU recommended HF field-strength prediction method [18] it is given on paths up to 2000 km range in terms of foE at the midpath and on paths up to 9000 km in terms of the foE values at two control points located 1000 km from each end of the path.

17.3.6 Signal strength

17.3.6.1 General
Two different approaches to sky-wave signal-intensity prediction are possible. One is to fit empirical equations to measured data for different paths, times and

frequencies. The other is to estimate intensity in terms of a number of separate factors known to influence the signals. These factors may be given by expressions which have been deduced either from theory or measurement. Unfortunately, both approaches have limitations. The former is likely to be simpler but, unless a large database exists, trends must be inferred and are liable to error. The latter approach is conceptually more elegant and enables variations to be specified in a physically meaningful manner. However, there remains the possibility of error due to failure to allow for a significant term or to an inexact allowance. There is also a likelihood of devising a method which is over complex and for which the accuracy achieved does not merit some of the complications that have been introduced. Existing models differ in regard to what factors to include and what allowances to use for these. The ITU signal prediction method [18] does not attempt to estimate propagation modes separately and their losses for path lengths exceeding 9000 km since raypath errors increase with range, making that technique particularly unsuitable at long ranges.

Monthly median values of mean available receiver power are determined in terms of transmitter radiated power, transmitting and receiving antenna gains and the basic transmission loss (see also Chapter 3.6):

$$P_r = P_t + G_t + G_r - L_b \tag{17.3}$$

where P_t =   transmitter power (dBW)
  P_r =   received power (dBW)
  G_t =   transmitting-antenna gain (decibels relative to an isotropic antenna)
  G_r =   receiving-antenna gain (decibels relative to an isotropic antenna)
  L_b =   basic transmission loss (decibels).

The corresponding RMS sky-wave field strengths E (dB> 1mV/m) are given in terms of P_r by

$$E = P_r + 20 \log_{10} f + 107.2 \tag{17.4}$$

where f is the wave frequency in megahertz.

17.3.6.2 Antenna gain
Antenna gains are those appropriate to the raypath launch and arrival angles. Because of uncertainties in determining these angles, models of antenna performance which include sharp nulls should be avoided. Instead, use of smoothed reference antenna patterns with nearest equivalence for nonstandard types is recommended, such as those now being used for broadcast planning [33].

These give radiation patterns over imperfect earth assuming sinusoidal current distributions and determine absolute gain by integrating over a hemisphere.

17.3.6.3 Spatial attenuation and focusing

For the estimation of basic transmission loss the spatial attenuation is taken to be that which would arise in free space at a distance equal to the mirror-reflection slant-path total length. Ray-path convergence focusing may be allowed for by means of empirical equations derived from raypath calculations for sample ionospheric conditions. In [43] horizon focusing, which arises principally on low-elevation paths, is given separately for E and F modes as a function of elevation angle. It is taken as having a maximum value at grazing incidence determined by ionospheric roughness of 9 dB. In [18], however, focusing is incorporated indirectly via an empirical excess loss-adjustment factor.

17.3.6.4 Ionospheric absorption

Equations for the normal ionospheric absorption arising at low and middle latitudes may be based principally on measured vertical-incidence data and the results of ray calculations for sample model ionospheres or on oblique-path measurements. Note that the absorption experienced in traversing a thin slab of ionisation is directly proportional to the product of the electron concentration, the collision frequency and the slab thickness, and inversely proportional to the refractive index. The important advantages of one such procedure [43] based on vertical-incidence data are that:

(i)   the variation with frequency includes an allowance for the change in height of reflection and for the different refractive indices at different heights, as well as for the way these depend on path obliquity;
(ii)   latitude and seasonal variations indicated by the measurements are included independently from the diurnal variation (Figure 17.3). In other prediction methods (e.g. [18]) position and time changes are combined via an assumed solar-zenith-angle dependence; and
(iii)   finite absorption is predicted at night-time.

Explicit allowances may be included for auroral absorption arising at high latitudes from precipitating-particle-induced ionisation. The absorption in [43] is taken as resulting from two separate sources of particles (see also Chapter 14). For each there is a gaussian variation with latitude and time of day about the maximum value. Longitudinal and seasonal dependencies are included. Important solar-cycle changes in the intensities, positions and widths of the auroral absorption zones are also modelled in the representation. By contrast, in [98] auroral absorption is included at the higher latitudes together with other losses as a variable empirical correction factor taken to be dependent on geomagnetic latitude, season, midpath local time and path length.

17.3.6.5   Polarisation-coupling loss

When an upgoing wave is incident on the ionosphere, it leads to the excitation of an ordinary (O) and an extraordinary (X) wave. These two waves have different but related polarisations which change as they progress, may be regarded as propagating independently within the ionosphere, and are subject to different amounts of absorption. The polarisation of a wave radiated from a transmitting antenna depends on the antenna configuration and the wave direction and frequency; likewise for the wave polarisation to which a receiving antenna responds (see also Section 15.4.4.3). Waves travel through free space with unchanged polarisation but the power coupling between incident or emergent waves and the O and X waves at the base of the ionosphere depends on their relative polarisations. This coupling may be calculated explicitly using the magnetoionic expressions for wave polarisation. In particular, these require a knowledge of the wave and Earth's magnetic-field directions. The X-wave absorption may also be estimated and the resultant received power from the O and X waves thereby deduced.

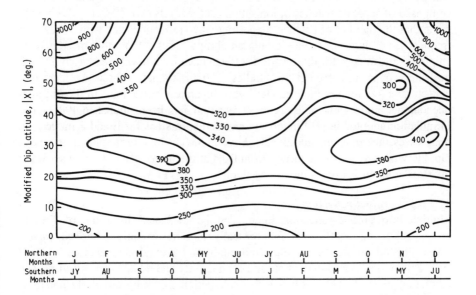

Figure 17.3   *Absorption factor A_T shown for the different months as a function of modified latitude $|X|$, for an overhead sun and smoothed sunspot number of zero (from [43])*

$X = \arctan \{I/\sqrt{\cos \lambda}\}$
with I = magnetic dip in radians and λ = geographic latitude

17.3.6.6   Sporadic-E losses

Improved understanding of the properties of sporadic-E ionisation now permits the inclusion of allowances for reflection from and transmission through this layer [43]. These allowances are based on oblique path measurements at HF and VHF. Es modes are assumed to be mirror reflected from a height of 110 km and the resulting reflection loss is given as an empirical function of distance, mode order and the ratio of the wave frequency to foEs. Other equations give the obscuration loss of transmitted waves in terms of this ratio and elevation angle. Sporadic-E obscuration losses suffered by F modes should be calculated separately for each leg of each hop.

17.3.6.7   Above-the-MUF loss

Strong signals are often received at frequencies above the predicted basic MUF, not just because of prediction errors. The predicted values are monthly median figures so that for half the days the ionosphere can support higher frequencies. Other reasons are that significant signal contributions arise via sidescatter paths and from sporadic-E modes. It has also been suggested that the regular F layer is composed of separate patches of ionisation each with its own BMUF. This would mean that the number of patches supporting wave reflection falls with increase of frequency, no single frequency giving an abrupt cutoff. A number of composite empirical allowances for these various effects have been developed based on both measured data and theoretical studies. International efforts are currently concentrated on attempting to model losses separately attributable to the different phenomena. A review of the current position has been published [98]. The formulation adopted in [18] involves separate expressions for E and F modes. In both cases, there is an additional above-the-MUF-loss term which increases rapidly with increase of frequency from 0 dB at the basic MUF. It attains a value of 20 dB at about 1.4 times the basic MUF.

17.3.6.8   Ground-reflection loss

Multiple-hop ground-reflection losses can be evaluated in terms of Fresnel ground-reflection coefficients for vertically and horizontally polarised waves. These depend on frequency, elevation angle and ground constants as deduced from a numerical world map of ground conductivity and relative permittivity. The prediction method [43] involves a full calculation assuming circularly polarised incident waves, but a constant loss of 2 dB per hop irrespective of terrain, frequency or angle is adopted in [18].

17.3.6.9   Path lengths exceeding 9000 km range

An empirical formulation based on the fit to oblique-path-measurement data collected over many paths for several decades to receiving sites located in Germany in place of a full modal treatment is used in [18]. This gives that the field strength depends on path distance and is a variable function of frequency, rising to a peak intermediate between reference lower and upper cutoff frequencies. These frequencies are given by empirical expressions in terms of the basic MUF and the LUF. Transmitting-antenna gain is taken as the largest value along the great-circle path with an elevation below 8° and an expression for antipodal focusing is also incorporated. For path lengths between 7000 and 9000 km there is an interpolation procedure with the short-path predicted values.

17.3.6.10 Excess-system loss and prediction accuracy
For the purposes of standardisation and testing the accuracy of HF-signal-prediction models, the ITU has established a databank of past measurements and has formulated reference procedures for the collection, tabulation and analysis of future data [32]. A representative sample of the data already deposited for 16 paths with ranges of 450–16 200 km is available. The measurements have been normalised to give the corresponding monthly median values of RMS sky-wave field strength for 1 kW radiation from an isotropic transmitting antenna. A further fixed factor can be included in predictions based on such tests to make the median error zero. When this is done, typically for 90% of the paths and hours, the RMS difference is less than 20 dB. Current work is aimed at the development of a new databank with measured values expressed in terms of available receiver power. In this way, improved prediction-method-standardisation factors, avoiding errors arising from certain present assumptions, are hoped for, thereby giving predictions of improved accuracy.

17.3.7 System performance

HF prediction methods usually yield monthly medians of hourly smoothed field strengths and available receiver powers and their statistical day-to-day variations about these values. The distributions of daily BMUF are assumed to follow a given law, so that it is possible to determine, for each wave frequency and examined mode, the fraction of days for which that mode can exist over the path. This is known as the availability. Reference values exist for the upper and lower standard deviations of the day-to-day signal variability to permit the estimation of the signal strengths exceeded for different fractions of the month. The optimum working frequency or frequency of optimum traffic (FOT) is defined as the lower decile of the daily values of operational MUF [10].

The type and quantity of information to be conveyed over a proposed radio circuit determine the modulation system and necessary receiver bandwidth. The next step in the circuit design is to specify the wanted signal/noise-power ratio at the receiver. Reference minimum signal/noise ratios judged to give satisfactory reception for different services are available [3]. Likewise, the ITU has reference models of composite atmospheric, man-made and galactic noise at HF [9], as discussed in Chapter 3.

An important monthly median system-performance parameter is the LUF or lowest usable frequency for which the monthly median signal/noise ratio equals that which is wanted. The LUF may be specified for a particular propagation mode, or for the circuit as a whole via any mode. Two other important parameters quantifying system performance are the reliability and the service probability. Again these may relate to a single mode or to the circuit as a whole. The reliability of a mode is given as the probability that this mode shall be present and that its signal/noise-power ratio equals or exceeds the wanted value. The day-to-day distribution of signal/noise ratio is estimated by combining the day-to-day variabilities of the signals and noise, assuming these to be uncorrelated, and by again assuming that some known distribution holds. Thereby the probability is given that a specified signal/noise ratio will be equalled or exceeded. On the assumption that the day-to-day ionospheric changes influencing mode support are not correlated with those giving rise to changes in signal/noise ratio, the mode reliability is then taken as the product of the mode availability and the probability that the mode provides a specified signal/noise ratio.

Various other types of reliability have been defined and methods of their determination formulated [31] depending on whether a single circuit is involved, whether transmission of the same information is made at more than one frequency and whether reception is at a point or over an area (see Chapter 3). The reliability is referred to as basic reliability when the wanted signals are competing against the natural noise background, and an overall reliability (or simply reliability) when interfering signals are considered. The case of multiple adjacent channel interferers has recently been taken into account in [31] for the determination of circuit reliability.

All parameters used in the reliability predictions are somewhat uncertain, and a standard error may be ascribed to each. The terms involved include the uncertainty in the predictions of the monthly median noise and signal powers, and of the standard deviations of the noise and signal day-to-day variations. The total uncertainty variance, found by adding the appropriate individual uncertainty variances, may be used to define an uncertainty distribution giving the probability that a required reliability is achieved. This is known as the service probability.

By combining predictions for different modes, the probability of multipath can be estimated. Multipath is defined as existing when two or more modes are jointly present having a difference in signal powers of less than some specified amount and a difference in group-path times exceeding a given figure. Predictions of multipath involve a simple extension of the procedures described.

17.4 Prediction procedures

With the general availability of microcomputers, there has been a proliferation of different implementations of prediction procedures incorporating variations depending on input/output requirements and the personal viewpoints of the originators. Here examples of procedures developed and available from the ITU are presented. Software for specific circuit diagnostics involving ray tracing through near-real time model ionospheres is outside the scope of this Chapter.

A list of computer programs available from the ITU associated with ionospheric propagation data is given in [41]. As an example of one available program, HRMNTH based on procedures described in [11], Figure 17.4 gives a tabulation for any mapped ionospheric characteristic (median or decile values) for a specified location in different hours, months and levels of solar activity.

```
    MEDIAN foF2      USING ITU-R COEFFICIENTS(OSLO)                DISK SET B

    LATITUDE    51.5       LONGITUDE    359.4     SLOUGH,UK                1988


        JAN   FEB   MAR   APR   MAY  JUNE  JULY   AUG  SEPT   OCT   NOV   DEC

R12    55.0  59.0  63.0  69.0  75.0  81.0  88.0  94.0 101.0 108.0 114.0 120.0

UT
 2.0    2.8   3.1   3.6   3.9   4.5   4.8   4.9   4.3   4.3   4.4   3.7   3.6
 4.0    2.5   2.6   3.0   3.6   4.5   4.9   4.6   4.1   3.6   3.4   3.0   3.3
 6.0    2.1   2.7   3.7   4.5   5.2   5.5   5.3   5.2   4.7   4.5   3.4   2.8
 8.0    4.7   5.4   5.8   5.7   6.2   6.1   6.0   6.4   6.5   7.9   7.4   6.7
10.0    6.8   6.9   7.1   6.7   6.6   6.3   6.5   6.8   7.5   9.9  10.3  10.1
12.0    7.5   7.6   7.7   7.1   6.7   6.3   6.4   6.7   7.9  10.7  11.4  11.3
14.0    7.3   7.6   7.6   7.1   6.8   6.2   6.3   6.6   7.8  10.6  11.1  11.1
16.0    6.0   7.1   7.5   7.1   6.6   6.0   6.1   6.5   7.7  10.0   9.5   9.2
18.0    4.1   5.5   6.8   7.1   6.9   6.3   6.5   7.0   7.6   8.1   6.4   6.2
20.0    3.0   3.9   5.3   6.3   6.8   6.6   6.7   7.0   6.4   5.7   4.1   4.2
22.0    2.7   3.1   4.2   5.0   5.9   6.1   6.1   5.9   5.2   4.7   3.6   3.5
24.0    2.8   3.1   3.7   4.4   5.2   5.4   5.6   5.1   4.7   4.5   3.6   3.5
```

Figure 17.4   Example of output of program HRMNTH

Figure 17.5 shows a printout from program MUFFY. This program, also following [11], gives path basic MUF, operational MUF and FOT at different UT hours for a specified month and level of solar activity between given transmitter and receiver terminals. Either the short or long great-circle paths may be selected. The propagation modes considered are 1E, 2E, 1F1, single and multiple-hop F2, depending on path length.

Evaluation of the current version of the HF field-strength prediction procedure [18] using program REC 533A is illustrated in Figure 17.6. For a series of UT hours and specified frequencies this gives the propagation mode with strongest signals and its associated elevation angle, the median strength of the resultant signal, the median resultant signal/noise ratio and the probability that a specified signal/noise ratio will be achieved. Other optional program outputs are available.

This particular propagation program embodies a gain package based on [33] for a range of common antennas and antenna arrays deployed at both transmitter and receiver. With the antenna type selected it is necessary to specify element numbers, dimensions and dispositions, as well as ground terrain characteristics. A free-standing version of that program HFANT provides both tabular gain and graphical outputs. As well as standard vertical and azimuthal pattern displays the novel Sanson–Flamstead projection showing both front- and back-lobe azimuth–elevation angle patterns is available (Figure 17.7).

PROGRAM 'MUFFY'

BASIC MUF,OPERATIONAL MUF AND FOT DETERMINATION - CCIR REPORT 340-6

OSLO FOF2 COEFFICIENTS

MAY   1990   SUNSPOT NO. 152.0

SYDNEY			TO ADELAIDE			AZIMUTHS		MILES	KM.
33.93S	151.17E		34.92S	138.58E		261.01	88.15	720.4	1159.3

SHORT PATH   TX PWR   1.0 KW

UT	BMUF	MUF	FOT	UT	BMUF	MUF	FOT	UT	BMUF	MUF	FOT
01	19.8	21.7	18.5	09	12.6	15.1	12.9	17	7.5	9.0	7.6
02	19.9	21.9	18.6	10	10.7	12.8	10.9	18	6.7	8.0	6.8
03	19.9	21.9	18.6	11	9.2	11.1	9.4	19	5.8	7.0	6.0
04	19.7	21.6	18.4	12	8.1	9.8	8.3	20	6.4	7.7	6.5
05	19.4	21.4	18.2	13	7.7	9.2	7.8	21	9.6	10.5	9.0
06	19.0	20.9	17.8	14	7.7	9.2	7.8	22	14.4	15.9	13.5
07	17.6	19.3	16.4	15	7.8	9.4	8.0	23	18.2	20.0	17.0
08	15.1	16.6	14.1	16	7.8	9.4	8.0	24	19.6	21.5	18.3

Figure 17.5   Example of output of program MUFFY

17.5 Short-term prediction and real-time channel sounding

Short-term models for frequency management are directed towards assessing the best frequency to use with an existing system in the light of the prevailing ionospheric conditions of the time. Therefore the requirement is to use system-performance predictions of the form already described, but with the forecast ionosphere replaced by a more accurate representation. The greatest fractional variations in the E and F regions arise in foF2. Hence a useful improvement in modelling capability would be achieved if it were possible to use near-real-time values of foF2 and to retain monthly median estimates of the other ionospheric characteristics.

Procedures involving vertical-incidence sounders at one of the path terminals to measure foF2 directly have only limited use because, typically, the correlation of daily departures from the median falls to a value of 0.7 in a distance of about 2000 km for E–W paths and 1000 km for N–S paths. Rush and Gibbs [162] have compared the accuracy of forecasts of daily foF2 in terms of observed monthly median models with those deduced from weighted means of the preceding past days. Figure 17.8 is an example from their published results. They found that, on average, for a series of locations and times, the use of a previous-five-day-period value gives estimates which are comparable with or better than those from the observed monthly median. This means they are to be preferred to using predicted monthly median figures. Nevertheless, an uncertainty of the order of 0.5 MHz exists at all times and, when extrapolating to remote locations where measured data are not available, errors are likely to be prohibitive.

Interest centres on the identification of precursors of solar disturbances responsible for changes in the ionosphere and in the Earth's magnetic field. Optical, X-ray and radio emissions from the Sun are observed daily at a number of ground-based sites and also aboard satellites. Ionospheric disturbances following solar flares occur either in close time succession and last for several hours, or begin 24–36 hours later and last for several days. The former arise from enhanced X-ray, ultra-violet and high-energy particle radiation, while the latter are associated with lower-energy particles.

Various attempts have been made to correlate daily foF2 values with indices of solar and magnetic activity. During magnetically quiet periods, daily and 60-day-average 10.7 cm solar-flux values are equally good in predicting hourly foF2 a day ahead. Ionospheric-disturbance forecasts and short-term prediction services are currently offered by the National Oceanic and Atmospheric Administration, Boulder, Colorado, USA, and the Marconi Research Centre, Great Baddow, Chelmsford, UK.

There is much attraction and current interest in developing intelligent receivers with embedded microcomputers which incorporate a crude propagation prediction for frequency management, real-time spectrum-occupancy measurement of assigned channels and real-time examination of reception quality for potentially good channels to identify the optimum. It is hoped that the next few years will see the emergence of a variety of systems available for purchase.

```
           JUL    1977            SSN =  29.
  NORFOLK ANT=103        LUECHOW VM NO SCREEN   AZIMUTHS    SP   N. MI.        KM
  36.80 N     76.50 W    52.98 N    11.22 E     43.91  292.73   3612.6    6690.0
  MIN ANG   3.0 DEG         TBEAR  53.0         RBEAR    0.0     PWR    1.00 KW
  TX AHR/4/3/0.5/SO/10     7.0- 14.0 MHz   RX VM  8/0/0/0        2.0- 13.1 MHz
       S/N% 50           -157 NOISE DBW        3000 HZ RX BDWTH  REQ S/N  8 DB

UT  MUF                                                             LUF  FOT OPMUF

 2 12.0  5.0   7.2   8.1 10.9 12.5 16.4 20.0  0.0   0.0   0.0 FREQ 7.0 11.7 13.7
    3F2   3F2   4F2   3F2  3F2  3F2  3F2  3F2    0     0     0 MODE
     14     5    14     8   13   14   15   16    0     0     0 ANGL
     26  -106    19    22   26   17 -110 -117 -999  -999  -999 DBU
     32  -107    19    23   31   24 -201 -207 -999  -999  -999 S/N
    .93   .01   .81   .88  .92  .85  .01  .01  .00   .00   .00 FS/N

 6  9.9  5.0   7.2   8.1 10.9 12.5 16.4 20.0  0.0   0.0   0.0 FREQ 7.0 10.5 12.4
    3F2   6F2   4F2   4F2  3F2  3F2  3F2  3F2    0     0     0 MODE
     13    29    19    19   13   14   14   15    0     0     0 ANGL
     24  -115    15    17   14    6 -117 -123 -999  -999  -999 DBU
     29  -108    21    22   19   13 -208 -213 -999  -999  -999 S/N
    .90   .01   .82   .82  .78  .64  .01  .01  .00   .00   .00 FS/N

10 13.7  5.0   7.2   8.1 10.9 12.5 16.4 20.0  0.0   0.0   0.0 FREQ 11.9 13.4 15.7
    3F2   3F2   6F2   5F2  4F2  3F2  3F2  3F2    0     0     0 MODE
     11    11    27    22   17   11   11   11    0     0     0 ANGL
     12  -245   -37   -22   -2   10 -114 -119 -999  -999  -999 DBU
    -80  -239   -29   -14    6   17 -206 -209 -999  -999  -999 S/N
    .01   .01   .01   .01  .39  .74  .01  .01  .00   .00   .00 FS/N

14 16.6  5.0   7.2   8.1 10.9 12.5 16.4 20.0  0.0   0.0   0.0 FREQ 12.7 15.6 18.3
    2F2   2F2   5F2   5F2  4F2  3F2  2F2  2F2    0     0     0 MODE
      6     6    26    26   20   14    6    6    0     0     0 ANGL
   -100  -245   -56   -45  -17    0 -100 -111 -999  -999  -999 DBU
   -191  -239   -48   -37   -9    7 -192 -201 -999  -999  -999 S/N
    .01   .01   .01   .01  .05  .45  .01  .01  .00   .00   .00 FS/N

18 16.5  5.0   7.2   8.1 10.9 12.5 16.4 20.0  0.0   0.0   0.0 FREQ 11.9 15.4 18.2
    3F2   3F2   6F2   5F2  4F2  3F2  3F2  3F2    0     0     0 MODE
     14    14    30    25   20   14   14   14    0     0     0 ANGL
   -102  -245   -52   -33  -10    5 -102 -113 -999  -999  -999 DBU
   -194  -240   -46   -28   -5   10 -194 -203 -999  -999  -999 S/N
    .01   .01   .01   .01  .11  .58  .01  .01  .00   .00   .00 FS/N

22 17.1  5.0   7.2   8.1 10.9 12.5 16.4 20.0  0.0   0.0   0.0 FREQ 8.3 16.0 18.8
    3F2   8F2   5F2   5F2  3F2  3F2  3F2  3F2    0     0     0 MODE
     11    30    19    20   10   11   11   12    0     0     0 ANGL
    -91  -145     1     5   21   23  -90 -103 -999  -999  -999 DBU
   -182  -144     2     7   24   28 -182 -194 -999  -999  -999 S/N
    .01   .01   .26   .42  .89  .94  .01  .01  .00   .00   .00 FS/N
```

Figure 17.6   Example of output of program REC 533A

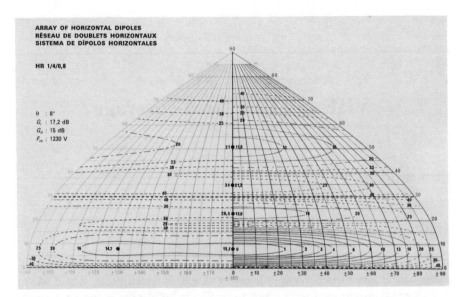

Figure 17.7   Example of Sanson–Flamstead projection graphical output for program HFANT

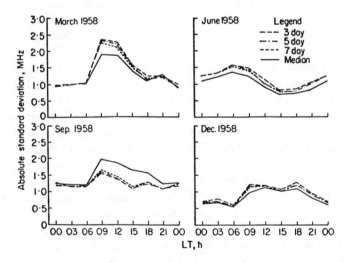

Figure 17.8   Errors in estimates of daily values of foF2 at Slough, UK, using monthly median and weighted means of past-days measurements (from Rush and Gibbs [162])

Chapter 18

VHF and UHF area-coverage

D.F. Bacon

18.1 Introduction

An area-coverage radio system provides a service which should be available throughout a given space. A well known example is broadcasting, where one or more fixed transmitters serve a given area. The radio planner in this case knows where the transmitting antenna is, but not where each receiving antenna will be. With television broadcasting, the planner may assume typical locations for receiving antennas, such as at roof-top level. For sound broadcasting many of the receivers will be mobile, and often at street level or inside buildings. The other main example of area-coverage is mobile radio, most of which operates on the same principle as broadcasting, with a fixed station serving a given area. However, somewhat different terminology is in use. The fixed equipment is called the 'base station', which communicates with the 'mobiles', and of course the radio link is bidirectional. Broadcast-planning methods have traditionally treated 10 m as the normal receiving antenna height, although for radio much lower heights are often in use. Mobile services tend to take 1.5 m for the receiver height, and are generally interested in shorter minimum distances than the broadcasters.

For convenience this Chapter will tend to use broadcasting terminology except when specifically referring to mobile-radio systems. Most of the principles discussed, however, apply to all types of area-coverage.

Terrestrial area-coverage services cannot guarantee line-of-sight radio paths, since mobile-radio users and broadcast receivers will often be inside buildings, or in the radio shadow of buildings and other obstacles. This makes the VHF and UHF bands the favoured range of frequencies; at shorter wavelengths diffraction loss, in particular, becomes a serious problem.

Broadcasting services in the UK, particularly UHF television coverage, are planned to well defined criteria. Private mobile-radio systems, such as typically used by a taxi company, often have service areas defined fairly informally. For cellular radio, however, where continuous geographical coverage is required, the accurate determination of service area is important.

Area-coverage systems are in distinct contrast with fixed radio links, where the antennas at both ends of the path are in defined positions, and the complete link can be engineered to achieve a given performance. In area-coverage, there is always an element of uncertainty as to one end of each radio path. Although the principles of propagation apply equally to both types of service, their planning has developed somewhat differently, and to some extent with different terminology. The following points are worth noting:

(i)   Broadcasting coverage, and sometimes mobile, is normally planned on the basis of field strength, whereas point-to-point services use the concept of transmission loss.

(ii)   To cover a given area, the statistics of location variability must be considered. For instance, a radio system might be required to be available over 95% of a service area. This principle does not apply to point-to-point links.

(iii)   Both point-to-point and area-coverage planning must consider temporal variations, since radio signals vary with time. Area-coverage services, with a crowded spectrum and serious limitations owing to interference from unwanted transmissions, are typically planned to be available for up to 99% of the time. Point-to-point links, with more variables under the control of the planner and with greater demands upon their reliability, are sometimes engineered for only 0.001% nonavailability due to propagation conditions.

iv)   Assessment of the area covered by a given transmitter or base station requires a large number of radio paths to be evaluated. Thus there has been a tendency to use simple methods which can be implemented quickly, although with improving computer technology this is becoming less of an issue.

18.2   Service-area planning

18.2.1   General

The starting point is to define a required grade of service. This is a measure of quality as perceived by the user, and might include such factors as the quality of a television picture, or the level of background noise or interference in a sound channel. The grade of service is then translated into radio-frequency conditions which must exist in order to achieve it. This is a question which concerns the characteristics of the radio system and the equipment which will be used. The two most important radio parameters are the signal/noise (S/N) and signal/interference ratios at the input of a receiver. (A small point to note about the various 'ratios' used in radio planning is that they are ratios before being expressed in dB; they are usually quoted in dB and added or subtracted.)

18.2.2 Noise and interference limitations

For planning purposes, the level of natural and man-made noise in which the system will operate, including receiver noise, defines a minimum level of field strength below which the system cannot deliver the required grade of service. This is the minimum usable, or 'reference' field strength, which must exceed the total noise by at least the required S/N ratio. In the absence of interference, the distance at which a transmitter's field strength falls to the minimum reference value will determine the boundary of a service area. This is illustrated in Figure 18.1, where the field strength from a transmitter falls with distance until it drops to the minimum usable level, in this case at 40 km.

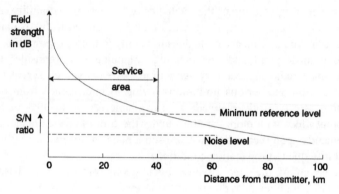

Figure 18.1   Noise-limited service area

In the VHF and UHF bands, minimum usable field strength is usually determined by man-made radio noise, which will vary with location.   ITU Recommendation 412-3 [2] gives the following values for VHF FM sound broadcasting:

Location	*Monophonic*	*Stereophonic*
Rural areas	48 dB (μV/m)	54 dB (μV/m)
Urban areas	60 dB (μV/m)	66 dB (μV/m)
Large cities	70 dB (μV/m)	74 dB (μV/m)

In today's crowded spectrum, service areas are rarely limited by noise. Frequencies are reused as densely as possible, and it is normal for service areas to be interference limited.   Area-coverage planning is usually based on two important quantities:

(i)   *Protection ratio*

This specifies the extent to which an unwanted field strength must be less than a wanted signal. Protection-ratio values depend on the type of service for both the wanted and unwanted signals, and also on whether the interfering signal is at the same frequency as the wanted service, or at a given frequency difference. Protection ratios are normally positive, indicating that the wanted signal must be greater than the interference. In general, values fall for increasing frequency difference, and for large offsets negative values are possible.

(ii)   *Minimum protected field strength*

This is the lowest value of wanted field strength for which protection can be claimed. In other words, at all locations where the wanted service achieves the protected field strength or more, all unwanted fields strengths must be less than the wanted field strength minus the protection ratio.

These concepts are illustrated in Figure 18.2, which shows two transmitters separated by 100 km. The field strengths from the transmitters, plotted as solid curves, fall with distance as before. The broken lines represent the maximum interference field strengths which each service can tolerate, the vertical distance between the full and broken lines being everywhere equal to the protection ratio in dB. The separation of the two transmitters is such that, at the point where each wanted signal has fallen to the minimum protected field strength, the unwanted signal has just become lower by the required protection ratio. It is clear that for positive protection ratios a gap must exist between adjacent service areas.

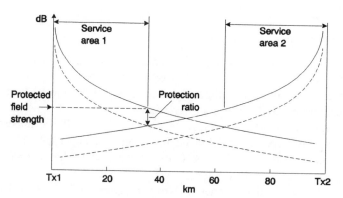

Figure 18.2   Interference-limited service areas

Note that if both transmitters in Figure 18.2 increased power by the same amount, neither would gain a larger service area. The curves would simply move upwards on the dB scale but not change relative to each other. Thus, for the most effective and economical use of available frequencies, neighbouring services should be interference limited, but with the field strength at the edge of a service area only just above the minimum reference field strength. This means that no more transmitter power than necessary is used to achieve the service areas.

However, the above arguments are based on field strengths which fall smoothly with distance. In practice field strengths vary irregularly with distance, owing to the shadowing effect of hills, buildings, vegetation etc. Service-area boundaries thus tend also to be irregular, often with detached 'islands' of reception and 'holes' within the area.

Interference calculations normally consider each possible interference signal individually, testing each to ensure that it is at least the protection ratio below the wanted signal level. For this reason protection ratios often incorporate an allowance for multiple interferers. The value of this extra safety factor depends on the expected number of interferers likely to be close to the limit, which can be inferred from the type of service in view and expected operational conditions.

A further point illustrated in Figures 18.1 and 18.2 is that, for locations well inside the border of a service area, the field strength is likely to be far higher than necessary for an adequate service, and for large service areas this may be by a very large margin. High-power transmissions from antennas well above ground level are more likely to cause interference at long distances, and from this point of view it is desirable to plan area-coverage using a large number of relatively small service areas. Against this, however, must be placed the cost of increasing the number of transmitting sites. The principle of small service areas has been developed most fully in cellular radio, where there is an additional advantage concerned with traffic density.

18.2.3 Frequency-reuse distance

In area-coverage planning, cochannel interference is usually the most serious constraint. Assuming a protection ratio greater than zero, transmissions at the same frequency cannot have contiguous service areas. Thus a number of different frequencies must be used to cover a given area.

In this situation a block of spectrum called a band will be allocated to a particular type of service, and divided in a regular manner into a number of channels which are numbered sequentially. Each channel has the same width, providing just enough bandwidth for the type of service.

A band organised in this manner is shown in Figure 18.3. The shaded area in each channel represents the spectrum of the radio service. For efficient frequency utilisation it is common for the spectral edges to overlap, this being taken into account in the calculation of adjacent-channel protection ratio. At each end of the band is a guard band, which accommodate the spectral edges of the first and last channels, plus similar edges from services in adjacent bands.

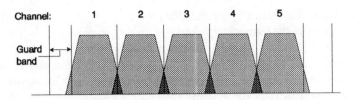

Figure 18.3   A band organised as five channels

It is obviously desirable to use the smallest number of channels without causing harmful interference, and this introduces the concept of frequency-reuse distance. This concerns the distance at which a given frequency can be used again.   In Figure 18.4 we have two transmitters providing the same type of service and with the same radiated power. The field strength due to Tx1 has fallen to E_1 at distance d. At the same point, the field strength of Tx2 has fallen to E_2. Distance d will be the interference-limited service area boundary for Tx1 if

$$E_1 - E_2 = R \qquad (18.1)$$

where R is the required protection ratio.

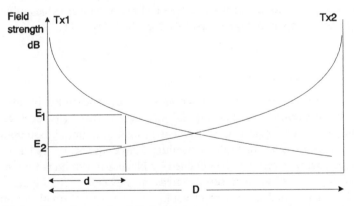

Figure 18.4   Frequency-reuse distance

The value of D/d determines how closely a given channel can be reused, and will depend on how rapidly field strengths diminish with distance. For free-space propagation, field strength E may be written as

$$E = K - 20 \log(d) \qquad (18.2)$$

where

K = a constant determined by the radiated power

d = distance from the transmitter.

For VHF and UHF propagation over land, measurements have shown that, for typical conditions, the rate of decay is substantially faster that '20 log (d)', and as a general approximation can be written

$$E = K - 40 \log(d) \tag{18.3}$$

Thus E_1 and E_2 in Figure 18.4 can be written

$$E_1 = K - 40 \log(d) \tag{18.4}$$

$$E_2 = K - 40 \log(D - d) \tag{18.5}$$

Equations 18.4 and 18.5 can be rearranged to give

$$\frac{D}{d} = 10^{R/40} + 1 \tag{18.6}$$

For conventional analogue services such as FM radio, typical values of cochannel protection ratio happen to be around 40 dB, for which $D/d = 11$.

18.2.4 Theoretical network planning

In practice the shape of a radio service area will be determined by topography, particularly the presence of hills and high-rise urban areas. However, as a basis for planning area-coverage it is convenient to make the initial assumptions that terrain is uniform and that all transmitters have the same radiated power. Continuous coverage can then be obtained by placing transmitters on a triangular grid, and approximating their service areas by hexagons. In Figure 18.5 the circles show the actual areas which would covered, over smooth terrain, for a given minimum field strength. Taking the shortest distance to a transmitter then gives hexagonal service areas

When the required reuse value of D/d is known, it is then necessary to choose an appropriate number of channels which can be assigned to a pattern of hexagons which can be repeated across the area to be covered. This is illustrated in Figure 18.6 for a network using 13 channels, with the bold lines delineating the reuse pattern.

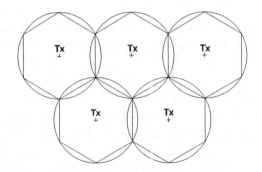

Figure 18.5   Continuous coverage by hexagonal service areas

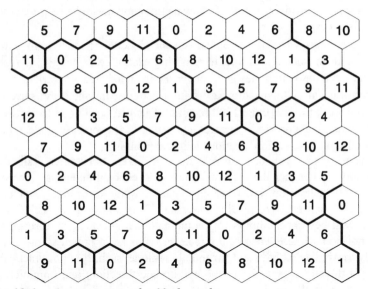

Figure 18.6   A reuse pattern for 13 channels

The pattern in Figure 18.6 maintains a D/d ratio for all cochannel transmitters of about six. In practice at least adjacent-channel separations also have to be considered, making the choice of reuse patterns more difficult. It will be seen in Figure 18.6 that all adjacent-channel transmitter pairs have a D/d of about four. A systematic approach can be taken to this form of planning [42], but it needs to be remembered that it is a preliminary stage; area planning must ultimately take account of actual topography.

18.3   Area-coverage related topics

Before describing some specific planning methods, it will be convenient to discuss a number of topics which are generally related to area coverage.

18.3.1 Antenna directional patterns

Where a transmitter, or mobile-radio base station, is in the centre of the area it is intended to serve, an omnidirectional radiation pattern is normally chosen. This is not the same as an 'isotropic' pattern, which refers to equal radiation into the surrounding sphere. The term 'omnidirectional' refers to a radiation pattern in the horizontal plane, where it radiates in all directions. To avoid wasting power up into the sky or down into the ground, such antennas usually form a beam in the vertical plane. Particularly for antennas well above ground level, this vertical pattern is sometimes given a 'tilt' of a few degrees downwards, to ensure coverage at short distances.

There are times when it is not convenient to serve an area from a central location, and in such cases the transmitting station can be given a shaped horizontal radiation-pattern design to cover the intended area. A variation on this is the sectoring of cellular-radio base-station patterns, in which the surrounding area is divided into, say, three 120° sectors. This can been done to increase the channel capacity of a previously unsectored cell as traffic increases, or can be used to economise on base-station locations in initial planning.

The area which can be covered from a given transmitter will depend considerably on whether the receivers can use directional antennas. The classic case is UHF television broadcasting, where highly directional antennas of moderate size and cost are readily available. This allows a service area to be extended well beyond the limits for omnidirectional antennas, by discriminating against noise and interference at the receiver. For radio broadcasting, which is generally viewed as mobile, and for mobile radio as such, it is not practicable to orientate a directional antenna in the appropriate direction, and omnidirectional antennas are normally used.

18.3.2 Shadow-fill-in and rebroadcast relays

In Figure 18.7 a television transmitter on the left serves house A well, but house B is shadowed by a nearby hill. When a pocket of shadow occurs like this within a service area, a common remedy is to fill it with a low-power transmitter. Where practicable, it is convenient to locate the fill-in station where it can receive the main station's signal, which it translates to a different frequency for retransmission into the shadow area. In this case the fill-in station is known as a rebroadcast relay (RBR). This saves having to provide a separate programme feed to the fill-in station.

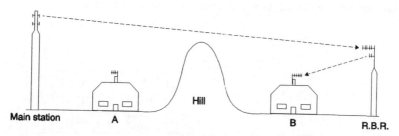

Figure 18.7   Shadow fill-in by rebroadcast relay

Directional television antennas can be used very effectively to select either the main or fill-in station. This can be important, since there are likely to be houses which receive approximately equal field strengths from the two transmitters. For this reason, fill-in stations are, if possible, sited to cover the shadow area from a substantially different direction than the main transmitter. What is not apparent in Figure 18.7 is that, in the UK UHF network, low-power fill-in stations normally use vertical polarisation whereas the main stations use horizontal. This provides additional discrimination between main and fill-signals.

Other types of area-coverage service can deal with shadow areas by similar techniques, but where directional antennas are not in use the design constraints are more severe.

18.3.3 Reflections, multipath and delay spread

In the VHF and UHF bands wavelengths are of the order of 1 m, and thus areas of concrete, roads, buildings etc. can readily provide flat surfaces with ample Fresnel clearance for reflection. Particularly for an omnidirectional receiving antenna, therefore, a substantial proportion of received power may come from reflections. For instance, in Figure 18.8 ray A may well provide a stronger signal than ray B. Although A undergoes both diffraction and reflection, its diffraction loss will be less than the diffraction loss for B, and this may more than compensate for the additional reflection. With slightly shorter buildings, A would not experience diffraction at all.

In a three-dimensional world such reflections can result in several significant ray paths between the transmitter and receiver. Because the paths will not have equal length, several versions of the radio modulation will be detected at different times, an effect known as delay spread. This is visible in television pictures as a 'ghost' image, usually slightly to the right of the main picture. In this case the reflected signal is weaker and arrives later than the direct signal. It is visible because the picture scanning is fast enough to show the small time difference. In television reception a directional antenna is often adjusted more to minimise such reflections than to maximise the direct signal. Delay spread affects an analogue sound channel equally, but is not normally noticeable to the ear.

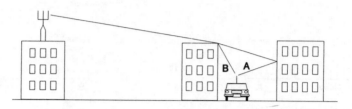

Figure 18.8   Multipath propagation in an urban environment

For a radio system using digital modulation, delay spread can be a serious source of error. In digital systems one or more information bits are modulated onto the carrier in what is called a symbol. If significant levels of received power are spread over a range of delays longer than the time interval from one symbol to the next, adjacent symbols will overlap. This is called intersymbol distortion. Digital area-coverage systems must take special measures against this effect. Considerable experimental work has been undertaken to collect statistics of delay spread in urban areas as a basis for countermeasures, particularly for mobile services.

Figure 18.9 shows a typical graph of received-signal strength plotted against delay for a multipath propagation channel. Delay is given as excess delay, i.e. the additional delay due to non-line-of-sight propagation. Signal strength is given in terms of power density. There are a number of parameters which can be obtained from such a graph which are used to characterise the channel, two of which are illustrated:

(*a*)   Delay interval is measured between two values of excess delay which mark the first time the power density exceeds a given threshold, and the last time it falls below it.

(*b*)   Delay window is the length of the middle portion of the power-delay profile containing a given percentage of the total energy.

In addition to delay spread, if the receiver or mobile is moving the signal will experience Doppler spreading due to the varying lengths of the ray paths. A surface plot of power density against excess delay and Doppler spread is sometimes, not very accurately, referred to as a 'scatter diagram'.

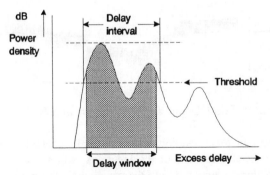

Figure 18.9   Delay-spread characteristics of a multipath channel

18.3.4 Location variability

Another effect of reflections is that the field strengths due to different radio paths can cause cancellations in the overall signal, resulting in minima or 'nulls'. In any multipath situation the field strength measured by a small omnidirectional antenna will show wide variations as it is moved over distances of the order of a wavelength. In effect this is a standing-wave pattern formed by coherent signals arriving over different paths from the same transmitter.

Superimposed on this fine structure of fading will be larger-scale variations of field strength due to shadowing by obstructions, such as terrain features, buildings, vegetation etc. Mobile services find railway and motorway cuttings a particular problem. Figure 18.10 shows typical results obtained by measuring field strength continuously as the receiver is slowly moved through the shadow of an obstruction. The 'rapid' fading due to multipath propagation causes an irregular sequence of variable-depth minima across the entire graph, but the slower reduction in signal as the receiver moves into a radio shadow is clearly superimposed.

For service planning it is impracticable to predict the detail of rapid fading, which thus must be allowed for on a statistical basis. Methods now exist for predicting shadow fading in some detail, but these require information on the position, size and electric properties of all significant obstacles. Although it is technically feasible to collect such information, the cost of doing so, and keeping it up to date, cannot usually be justified. Thus, at present location variability due to shadowing is also predicted on a statistical basis, and numerous measurement campaigns have been conducted by broadcasters and mobile-service operators to characterise various types of environment. These have shown that the spatial distribution of shadowing loss is approximately log-normal, and can thus be characterised by a mean and standard deviation.

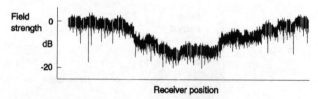

Figure 18.10   Rapid and slow fading due to multipath and shadowing

As a result of these factors, area-coverage is normally characterised in terms of the field strength exceeded at 50% of the locations within a given square area. This square must be large compared with the wavelength, but not so large that the unobstructed field strength would vary greatly across it. Squares of 100 m sides are a typical compromise for this purpose. Having obtained the '50% locations' field strength, the location variability of the square will be characterised statistically, usually by the standard deviation of an assumed log-normal distribution.

The introduction of digital modulation has placed greater demands on the assessment of location variability. Although digital systems can be engineered to operate satisfactorily at lower signal/interference ratios than analogue, the onset of degradation is much more abrupt. This is sometimes referred to colloquially as 'falling over a cliff'. With the introduction of digital mobile and broadcasting services it has been recognised that location variability needs to be defined more precisely, and in general larger allowances above the 50%-locations field strength are needed to obtain a given percentage coverage.

18.3.5 Height gain

An important type of spatial variation of signal strength is referred to as height gain. This is the increased signal level which is normally achieved by increasing the height of the receiving antenna. It is important to note that rapid fading will occur as an antenna is moved vertically, and that height gain generally refers to the variation of field strength with these rapid fluctuations averaged out. In open country, particularly where the ground is smooth, height gain is mainly due to ground reflections.   For urban environments height gain is dominated by shadowing due to buildings etc.

Height gain tends to increase by about 6 dB for each doubling of antenna height. This is illustrated by a proposed revision to ITU Recommendation P.370-7 [8] (see Section 18.4.1) which gives the height gain relative to a receiver height $h_2 = 10$ m, for h_2 in the range 1.5 to 40 m, by:

$$\text{height gain} = \frac{A}{6} 20 \log_{10} \left(\frac{h}{10} \right) \quad \text{dB} \qquad (18.7)$$

The parameter A is the change in dB for each doubling or halving of antenna height, and is given by

Location	VHF	UHF
Rural	4	4
Suburban	5	6
Urban	6	8

18.3.6 Height of base station

In mobile-radio planning, the height of the base station in relation to nearby ground cover is an important issue. In broadcasting, it is always assumed that the transmitter is sited above local ground cover. For conventional cellular radio it is assumed that the base station is above rooftop level. For 'microcells' it is assumed that the base station is below rooftop level, making a substantial difference to the radio path. When both ends of a radio path are near street level in an urban environment, propagation takes place mainly along what is called the 'street canyon', rather than over the rooftops. This can produce a waveguiding effect with a lower rate of loss with distance, compared with 'over-rooftop' propagation, until a corner is reached. In considering mobile planning methods it is important to know which type of propagation is assumed.

18.3.7 Topographic databases

Service-area planning requires information about the service area, and for automatic methods this needs to be in computer-readable form. Such collections of information are known as topographic databases, and with the increasing importance of mobile-radio services topographic data have acquired a considerable commercial value. A topographic database may consist of the following:

(i)   *Land height*: This is probably the oldest form of topographic data, and databases containing only such information are also known as digital elevation models, digital terrain models etc. They are normally produced by digitising the contours of existing maps, fitting a mathematical surface to the resulting data, and then sampling the surface at regular intervals to produce the height database. Cartographers are now starting to produce maps directly from digital data, and thus in the future the digitising of contours from papers maps should not be necessary.

(ii)   *Ground-cover categories*: This refers to the various obstacles to radio propagation which exist on top of the land surface. Since, to the radio engineer, buildings, vegetation etc. are generally viewed as obstructions, the term 'clutter' is often used in this context. In most topographic databases, ground cover is defined by allocating cover codes to squares of ground area to indicate, for instance, open land, suburban, urban or dense urban. These codes are then used to predict the statistics of location variability, height gain etc. expected at any location.

(iii)   *Digital street plans*: For microcellular applications, where propagation is primarily along the street canyons, digital maps giving the position of streets and buildings, but not building heights, can be used effectively. This assumes that propagation by diffraction over rooftops can be ignored compared with diffraction around and reflection/scattering from the building elevations. This approach is sometimes referred to as 'infinite building height'.

(iv)   *Explicit building data*: Owing to the increasing importance of radio services in urban environments, including indoor systems such as radio local-area networks, considerable interest is developing in the use of data giving detailed and explicit information on the environment, including position, shape, size and even materials of individual buildings. Propagation-prediction methods exist to use such information, and have been tested on detailed urban databases produced for limited areas. As computing power continues to increase, such methods will increasingly be used for routine planning operations.

At present most topographic data are collected with a large element of manual intervention. However, successful trials have been conducted to demonstrate the usefulness of remote sensing for this purpose, and this is likely to be used increasingly in the future.

18.3.8 Temporal variation

In tropospheric propagation the principal fading mechanism is variations of atmospheric refractivity, which can produce both reductions and enhancements of signal level. Broadcasting and mobile services are normally planned for an availability from 90% to 99%. Taking 90% as an example, it is thus necessary to predict the wanted field strength which will be exceeded for 90% of the time, and the unwanted potentially interfering field strengths exceeded for 10% of the time.

The complexity of this task, in conjunction with spatial variability, is considerable. The simplifying assumption is often made that no correlation exists between the various mechanisms causing variations. Also, over the distances involved in VHF and UHF coverage, fields strengths rarely fade far below the median value, although considerable enhancements occur for short time percentages at greater distances. Thus most service planning is based on median wanted field strengths, while for each unwanted signal the field strengths which will be exceeded for the relevant small time percentage are predicted. Propagation models for area-coverage thus normally predict field strengths exceeded from 50% down to a given small time percentage, typically 1%.

18.3.9 Selection of propagation mechanism

For the wanted signal, diffraction and possibly reflection and scattering must be evaluated in order to predict the median signal strength. Other mechanisms are not normally significant.

To predict unwanted field strengths exceeded for small time percentages, diffraction, troposcatter and ducting must all be evaluated, along possibly with reflection and scattering. Diffraction calculations should be performed with a value of k factor appropriate to the small time percentage of interest. Opinions are divided as to how far this process is valid. If the k factor is increased far enough, including passing through infinity to become negative, any terrain obstacle will in the end be depressed below the line of sight. This will result in a field strengths up to the free-space value, which down to about 1% time will rarely be exceeded. However, it is widely agreed that this process does not correspond to the physical principles involved.

It is better, and more common, to set an upper positive limit on k, and to employ a separate calculation to assess ducting field strengths. Troposcatter is important mainly because it sets an upper limit on the value of loss produced by a terrain obstruction. For instance, a potentially interfering transmitter might be behind a hill, and apparently screened off by a large diffraction loss from the required service area. However, it might be found that propagation over the hill by troposcatter can cause interference, and it is therefore essential to include the calculation.

18.3.10   OFDM and single-frequency networks

A technique to combat delay spread is orthogonal frequency-division multiplex (OFDM). This is a modulation scheme which uses a large number of closely spaced carriers, each digitally modulated at a symbol rate low enough to avoid intersymbol distortion. Although this results in low bit rates on each carrier, the aggregate bit rate over a large number of carriers is much higher. The term 'orthogonal' refers to a technique which avoids interaction between the modulations on adjacent carriers.

OFDM can provide wide-area-coverage to adjacent individual service areas using the same frequency for each transmitter. Provided that the modulation at all transmitters is synchronised, a receiver at the border of a service area will interpret signals from different transmitters in the same way as multipath due to reflections. This permits a single-frequency network of transmitters to provide continuous geographic coverage, greatly simplifying frequency planning and the filling-in of shadow areas.

18.3.11   Geographical information systems

Service areas are usually best described by a map, and area-coverage-planning tools can thus make good use of computerised geographical information systems (GIS).   The GIS is a general graphical user interface for a database of geographical information, such as might be used by water authorities, civil-engineering contractors etc. A GIS will allow maps and user-defined data, such as the position of underground pipes, to be stored and displayed in a flexible manner.

Several propagation-planning tools are now available commercially which incorporate a GIS, providing unprecedented facilities for displaying service area and related information. For instance, it would be practicable to store population figures, which would thus permit area-coverage in terms of percentage population to be computed and displayed.

18.4   Area-coverage propagation-prediction methods

There are two general types of propagation-prediction method used for area planning: empirical and deterministic. Although the distinction between these approaches is not always completely clear cut, the underlying methods are completely different:

(*a*)   An empirical method is based on a large number of measurements, from which averages are obtained for various conditions, such as frequency, antenna heights etc. The normal approach is to fit smooth curves to experimental data. The accuracy and usefulness of such methods depend greatly on how propagation conditions are classified.

(*b*)   A deterministic method determines propagation conditions by mathematically modelling the appropriate physical mechanisms. The success of such an approach is largely determined by the detail and accuracy of the necessary input data characterising the radio path, and explicit methods therefore make heavy demands upon topographic and radio-meteorological data.

A major complication in the deterministic approach is selecting the appropriate propagation mechanisms to model. It is known, for instance, that ducting occurs less than a certain percentage of time, and will not be effective for paths less than a certain length. Similarly, diffraction over irregular terrain can be modelled in various ways, and it may seem attractive to select a geometrical model according to the type of terrain. As a result of such considerations, the algorithms on which explicit methods are based often include conditional branches to select a calculation method. In practice it is difficult to ensure that no discontinuities are produced by such internal switching.

It is also worth noting that few deterministic methods are entirely free of empiricism. Best results are nearly always obtained by adjusting certain parameters to produce the best agreement with measured results. This is a perfectly valid technique if the parameters relate to appropriate physical phenomena, and in fact the ability to 'tune' a deterministic method to suit a particular type of location or climate can be a valuable advantage.

Explicit area-coverage-planning methods have benefited greatly from computer developments. A number of methods are now available commercially in computer form, making effective use of storage capability, processing power and colour graphics output.

18.4.1 *An empirical method: ITU Recommendation P.370-7 and associated texts*

A well-known empirical method is ITU Recommendation P.370-7 [8], which has been recommended for VHF and UHF broadcast planning by the ITU for many years. It contains a number of curves of field strength plotted against distance, obtained from extensive measurements, mainly in temperate climates including 'cold' and 'warm' seas, e.g. the North Sea and the Mediterranean. Similar curves suitable for mobile applications are contained in ITU Recommendation P.529-2 [15]. The curves are used in conjunction with a number of corrections.

The method does not necessarily require a complete path profile, but accuracy is improved if some topographic data are available. Terrain height information is needed for the following corrections:

(i)   The irregularity of the terrain is described by a parameter Δh, defined as the interdecile height of the terrain between 10 and 50 km from the transmitter.

(ii)   The height of the transmitting antenna should be the 'effective height', defined as height above the average land height between 3 and 15 km in the direction of the receiver.

(iii)   A correction for terrain clearance angle θ is defined.

The computation of Δh is illustrated in Figure 18.11. Two horizontal lines must be located. Between 10 and 50 km from the transmitter, 10% of the profile must be higher than the upper line, and 90% must be higher than the lower. Δh is the vertical distance between these two lines. Note that this calculation may ignore Earth curvature, with all heights measured from sea level. If required, Δh can be computed individually for each path during the prediction of a service area. However, since the method does not use the actual heights or positions of the hill tops, it is normal to use an average value of Δh for the terrain in question. The curves in Recommendation 370 are based on a Δh of 50 m, this being thought typical of much of the rolling terrain in Europe and North America.

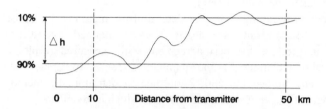

Figure 18.11   Terrain-irregularity parameter Δh

The calculation of terrain-clearance angle θ is illustrated in Figure 18.12. θ is the angle between the horizontal at the receiver and a straight line which just clears all terrain obstructions up to 16 km from the receiver in the direction of the transmitter. θ is positive if this line is below the horizontal, and negative if above. In this case Earth curvature should be taken into account for best accuracy.

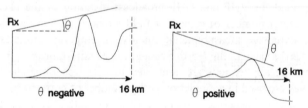

Figure 18.12   Terrain-clearance angle θ

All of the curves for field strength against distance in Recommendation 370 are for 1 kW radiated from a halfwave dipole, a receiver height of 10 m, for 50% locations, and a terrain irregularity factor Δh of 50 m for land paths. Each plot consists of a set of curves for effective transmitter heights of 37.5, 70, 150, 300, 600 and 1200 m. There are two sets of these plots, one for VHF (30 to 250 MHz) and one for UHF (450 to 1000 MHz). The other variables are given in Table 18.1

The 50% time curves are used to predict wanted field strengths, and the 10%, 5%, and 1% time curves for potentially interfering field strengths.   Thus availabilities of 90%, 95% and 99% time can be handled directly from the curves. In fact, the 5% time land curves are interpolated from the corresponding 10% and 1% curves using

$$E(5\%) = 0.653E(10\%) + 0.347E(1\%)$$ (18.8)

which assumes a log-normal distribution.   Thus UHF 5% time land data are available, even though no curve is given.

Table 18.1   *Variables used in ITU Recommendation 370 curves*

VHF*	UHF*	Path type	Percentage time
1a	9	Land	50
1b	13	Sea	50
2a	10	Land	10
2b	14a	Cold sea	10
2c	14b	Warm sea	10
3a		Land	5
3b	15a	Cold sea	5
3c	15b	Warm sea	5
4a	11	Land	1
4b	16a	Cold sea	1
4c	16b	Warm sea	1

* Figure numbers in Recommendation 370

Once the required field strengths have been read from the curves, the following corrections should be made:

Transmitter height: if the effective height of the transmitting antenna does not coincide with one of the six values used in the curves, linear interpolation on a dB scale is sufficiently accurate;

Δh: Recommendation 370 gives correction curves for values of Δh from 10 to 500 m as a function of distance from the transmitter (Figure 7 for VHF, and Figure 8 for UHF);

Terrain clearance: Recommendation 370 gives curves for the terrain-clearance-angle correction for θ computed as illustrated in Figure 18.12.

Percentage location: Recommendation 370 assumes a log-normal spatial distribution of field strength, and provides a correction curve from 1% to 99% locations (Figure 5 for VHF, and Figure 12 for UHF);

Receiver height: Recommendation 370 gives figures for the decrease in field strength when the receiver-antenna height is reduced from 10 m to 3 m for rural/urban and flat/hilly locations.

Recommendation 370 has a number of limitations, and in particular does not take explicit account of hilltops along a path (unless the dominant hilltop happens to determine the value of θ, in which case it will have a substantial effect). It has, however, been very widely used, and is still the basis for many bilateral agreements concerning service areas and interference.

18.4.2 BBC method

The BBC produced an area-coverage method intended mainly for the planning of the UHF television broadcasting bands in the UK. It has not been made available commercially, but the algorithms have been published [68]. It is essentially a deterministic method, although some parameters can be set according to local conditions which might be determined by measurements.

The method uses a variety of diffraction calculations according to the type of path, plus troposcatter and ducting. It also requires a number of parameters to be adjusted according to local conditions. It is subject to discontinuities, particularly due to the switching of diffraction models, and these are particularly noticeable if field strength is plotted against distance over fairly smooth terrain. However, the method has been highly successful in achieving a very high coverage for four television channels over the UK.

18.4.3 Mobile planning methods

With the commercial success of cellular radio, considerable effort has been put into planning methods intended specifically for mobile services, particularly for the urban environment. Many of these are empirical methods based on statistical topographic data.

Measurements reported by Okumura *et al.* in 1968 [141] have been very influential in the development of these methods. Okumura presented the data graphically, but Hata [102] provided a mathematical formulation based on the same data, which facilitates computerised methods. Hata's model is often a reference point when comparing mobile planning methods:

$$L_b = 69.55 + 26.16\log(f) - 13.82\log(h_b) - A$$
$$+ \{44.9 - 6.55\log(h_b)\}\log(d) \qquad \text{dB} \tag{18.9}$$

where

$$A = \{1.1\log(f) - 0.7\}h_m - \{1.56\log(f) - 0.8\} \tag{18.10}$$

and

f	=	frequency in MHz in the range $150 - 1000\,\text{MHz}$;
d	=	distance of the mobile from the base station in km in the range $1 - 20\,\text{km}$;

h_b = height of the base station in m in the range $30 - 200$ m, assumed to be above rooftop level;

h_m = height of the mobile in m in the range $1 - 10$ m.

Two other well known papers on mobile planning are those by Walfish and Bertoni [184] and Ikegami [109].

Because many towns are built on a fairly regular plan, mobile planning methods have had some success with characterising urban ground cover statistically, but with rather more detail than is normal in broadcasting. Typical parameters are illustrated in Figure 18.13, where:

d = horizontal distance from base station to mobile;

α = angle to the horizontal of the ray from base station to the diffraction edge E local to the mobile;

w = width between building elevations;

b = centre-to-centre spacing of buildings;

Δb = height of the base station above rooftop level;

Δm = height of the rooftops above the mobile;

m = height of the mobile above ground level.

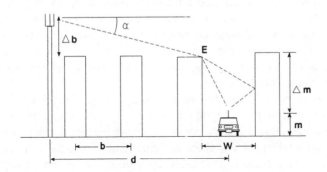

Figure 18.13   Parameters for urban-mobile-propagation models

The importance of α, Δb, Δm and w is that they permit the effect of multiple diffraction over successive roof tops between the base station and E to be evaluated. Some models also take into account changes in terrain height, and also the angle between direction of arrival of the radio signal and the characteristic street direction at the mobile.

Chapter 19

A brief review of some applications of millimetre waves

C.J. Gibbins

19.1 Introduction

Millimetre waves, i.e. frequencies in the range from 30 to 300 GHz, though long regarded as being in the realm of exotica, are now finding many applications in remote sensing, radar and communications. Indeed, in remote sensing especially, millimetre waves have attained a maturity not considered possible 10–20 years ago, by virtue of their facility to provide measurements either not feasible or not practicable in any other waveband.

Today, however, millimetre-wave technology is no longer confined to the experimental laboratory. Prompted largely by the requirements of radio astronomers and for military purposes, considerable developments in the technology have taken place over the last decade or two, to the extent that a wide range of very sophisticated devices and systems is now commercially available throughout the millimetre-wave region of the electromagnetic spectrum, and, while the costs for some specialised components and systems may remain rather high for some time to come, due largely to their limited application and demand, for many other systems and equipments the costs are expected to fall rapidly, especially as mass markets for millimetre-wave systems develop, particularly in the domestic sector.

This chapter discusses just some of the very wide range of applications of millimetre waves currently being used or considered, in three broad areas: remote sensing, radar and communications. In the latter two applications, in particular, there is now a real prospect of developing mass consumer markets for millimetre-wave products.

19.2   Applications in remote sensing

Millimetre waves have been, and are, widely used to study our environment — the Earth — using the technique of remote sensing, in ways not possible or practicable by direct (i.e. *in situ*) measurement. Two broad areas of remote sensing include studies of the atmosphere (its composition and dynamics, for example) and studies of the Earth's surface, including near-surface phenomena such as the weather. Atmospheric studies can be carried out either from the ground or from elevated platforms such as balloons, aircraft or satellites, while operational measurements of surface parameters are generally performed only from elevated platforms.

19.2.1 Attenuation and emission by atmospheric gases

Remote sensing of atmospheric constituents, either from the ground or by observations from satellites, relies on the fact that molecules of a number of atmospheric gases possess dipole moments and thus interact with the electromagnetic field. Specifically, such molecules, when in local thermodynamic equilibrium, absorb incident electromagnetic energy and then re-emit that same energy as radiation in the form of spectral lines, arising from quantum-mechanical transitions between energy levels within the molecule [189]. The atmospheric gases which dominate the spectrum at millimetric wavelengths are oxygen and water vapour, while many other minor constituents also have spectral lines in this region. Figure 19.1 shows the spectrum of atmospheric thermal emission (or brightness temperature, see Section 19.2.2) of oxygen and water vapour, together with the positions of spectral lines from high-altitude ozone, as an example.

The frequency, magnitude and shape of these spectral lines contain information on the types and densities of gases that are present at different altitudes in the atmosphere. Such spectral lines are collision or pressure broadened, i.e. the higher the pressure, the wider the line (in the frequency domain). The millimetre-wave region, in fact, is particularly useful, more so than the more conventional infra-red bands, for remote sensing, because spectral lines remain pressure broadened to much higher altitudes than at infra-red wavelengths, as indicated in Figure 19.2, which shows typical half-widths of spectral lines as a function of altitude. In the pressure-broadened region, spectral linewidths depend on pressure, which can be related to altitude through the hydrostatic equation. Thus, measuring the width of emission lines provides information on the altitude at which the emitting molecules are situated. Ultimately, the lines become Doppler broadened at the highest altitudes, with the linewidth depending primarily on temperature and frequency. The effects of Doppler broadening are therefore greater at infra-red wavelengths, while at millimetric wavelengths, information on the vertical distribution of gases can be obtained up to much higher altitudes than at the shorter wavelengths.

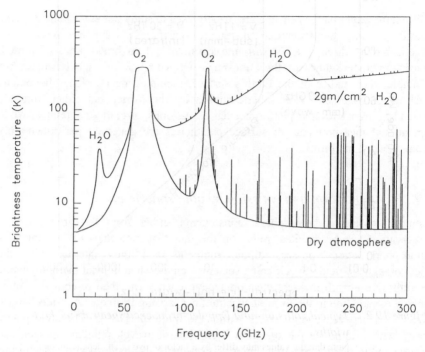

*Figure 19.1   Spectrum of the atmospheric thermal emission brightness
temperature at the zenith, showing the emission lines of oxygen,
water vapour and upper-atmospheric ozone*

The upper curve is for a water-vapour atmosphere of $10\,g/m^3$ at ground level,
decreasing exponentially with a scale height of $2\,km$; the lower curve is for a dry
atmosphere

Figure 19.1 shows the atmospheric spectrum in terms of the thermal emission of
gases. Since the atmosphere is, in general, in thermodynamic equilibrium, then
Kirchhoff's Law states that the amount of energy absorbed by the molecules will
also be emitted, or reradiated. Hence, measurement of the radiated thermal
emission enables the absorption (or attenuation) of the atmosphere to be
determined.

Measurement of the intensity and spectral shape of this emitted thermal
radiation will enable the level of attenuation, or absorption, to be determined as a
function of frequency, and thence, with one or two assumptions, the number and
distribution of the particular gases can be derived. Measurement of the thermal
radiation is generally accomplished with radiometers.

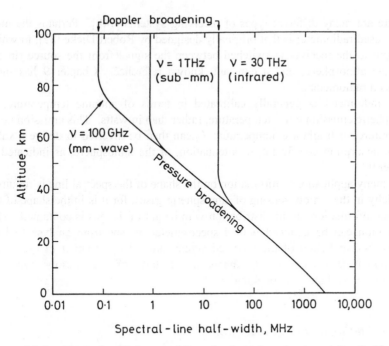

Figure 19.2   *Typical collision- and Doppler-broadened spectral-line half-widths*
Note that the values may differ by a factor of two or so, depending on the gas

19.2.2 Radiometry

The power level in thermal radiation from an absorbing medium is given by the equation

$$P = \alpha kTB \qquad (19.1)$$

where α is the (fractional) absorption coefficient of the medium ($0 \le \alpha \le 1$), k is Boltzman's constant, T is the temperature of the black body (the absorbing medium, i.e. the atmosphere) and B is the bandwidth in which the power is measured. The quantity αT is generally referred to as the emission noise temperature, or brightness temperature T_B of the atmosphere. For thermal radiation emitted by the Earth's atmosphere, the power is typically less than 10^{-15} W, in a 1 MHz bandwidth. Such power levels can be smaller than the noise power generated internally within many receiving systems, and thus radiometric receivers must be designed to introduce a minimum of noise and to be linear in output if they are to provide accurate reproduction of the spectral characteristics of the input signal.

There are many different types of receiving systems [148]. Perhaps the most widely used radiometer is that originally designed by Robert Dicke [82], in which the input to the receiver is switched between the signal from the source (in our case, the atmosphere) and a constant, known reference. Chapter 9 has more details on radiometers.

The radiometer is generally calibrated in terms of antenna temperature, or atmospheric emission noise temperature, rather than in watts. This emission noise temperature, or brightness temperature T_B can then be inverted to derive a value for the absorption coefficient, or attenuation, of the atmosphere, as indicated in Chapter 9.

For many applications, information on the shape of the spectral line is required, especially in the remote sensing of atmospheric gases, for it is in the shape of the line that information on the distribution with height of the gas is contained. This information can be deduced with a spectrometer, or spectrum analyser, which divides the bandwidth of the received signal into a large number of (generally contiguous) narrow-bandwidth channels. Details of the various types of spectrometer available can be found in [132].

19.2.3 Methods of data inversion

Deriving information on the vertical distribution of the densities of gases invokes use of the theory of radiative transfer, which takes a particularly simple form in the millimetre-wave region, on the low-frequency tail of the Planck radiation function, i.e. when $h\nu \ll kT$, where ν is the frequency (in spectroscopic notation) and T is the temperature of the atmosphere. In this region, indeed, at all radio frequencies, the classical Rayleigh–Jeans law holds, whereby the intensity (or power) of radiation is linearly proportional to the temperature of the absorbing/emitting medium, when in local thermodynamic equilibrium, i.e. when Kirchhoff's law applies. For a black body, i.e. one which is completely absorbing (characterised by $\alpha = 1$ in eqn. 19.1), the emitted radiation intensity is simply kTB, where B is the bandwidth over which the power is measured.

Now, consider an element of the atmosphere at an altitude h, as shown in Figure 19.3. If the absorption coefficient at this altitude is given by $\kappa(\nu,h)$, then this atmospheric layer will emit an amount of radiation $\kappa(\nu,h)T(h)dh$ per unit bandwidth. When this radiation is received at the ground, it will have been attenuated by the atmosphere below altitude h by an amount $\exp\{-\tau(\nu,h)\}$, where τ is the optical depth given by

$$\tau(\nu,h) = \int_0^h \kappa(\nu,h')dh' \tag{19.2}$$

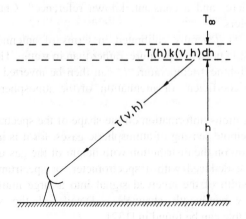

*Figure 19.3   Schematic diagram of the radiative transfer scenario, illustrating
the emission from an element dh of the atmosphere at an altitude h
received by a ground-based radiometer after being attenuated by
an amount τ(v,h)*

Integrating over the entire atmosphere, then, the total received power, expressed in terms of brightness temperature (or equivalent black-body temperature), will be

$$T_B(v) = T_\infty \exp\{-\tau(v,\infty)\} + \int_0^\infty \kappa(v,h)T(h)\exp\{-\tau(v,h)\}dh \qquad (19.3)$$

where T_∞ is the brightness temperature of any source external to the atmosphere, for example the 3 K cosmic background.

If we now define a weighting function $W(v,h)$, such that

$$W(v,h) = \frac{\kappa(v,h)T(h)\ \exp\{-\tau(v,h)\}}{n(h)} \qquad (19.4)$$

where $n(h)$ is the number density at altitude h of the molecule being measured, then, ignoring the first term in eqn. 19.3, we obtain

$$T_B(v) = \int_0^\infty W(v,h)n(h)dh \qquad (19.5)$$

These weighting functions describe the probability of emission at a given altitude, or, in other words, the amount of radiation emitted per molecule and received at the ground, and can readily be calculated. In general, the weighting functions for single-frequency observations are ill-conditioned and nonorthogonal, i.e. not statistically independent, and it has become customary to define weighting functions for measurements of the difference in brightness temperatures at two frequencies, one of which usually corresponds to the centre of the emission line v_0 and another frequency v_i to one side of the centre frequency. Figure 19.4 shows the typical form of such un-normalised weighting functions at various frequency differences $v_i - v_0$ from the line centre, for vertical (zenith or nadir) observations, and illustrates the effective probability that the radiation contributing to the difference in emission noise temperatures at the two frequencies originates from a given altitude. The weighting functions have their maximum values at the higher altitudes for frequencies v_i close to the centre of the emission line, and peak at lower altitudes further away from the centre frequency.

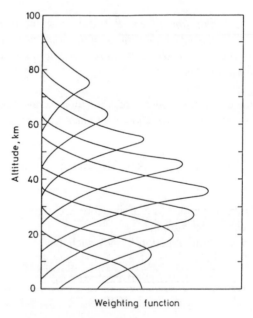

Figure 19.4   *Examples of typical weighting functions (arbitrary scale) for zenith or nadir remote sensing*

The higher-altitude weighting functions are for frequencies close to the centre of the emission line, while the lower-altitude weighting functions correspond to frequencies further away from the line centre

If a series of brightness-temperature measurements is made over a range of frequencies, then a number of mathematical techniques exist to invert this integral equation (eqn. 19.5) to yield an estimate of $n(h)$, the vertical density profile of the gas [174], which is the primary objective of the measurement, for input to photochemical models of the atmosphere, e.g. to understand climate change, or to deduce such parameters as attenuation for calculating link budgets in satellite-communications systems. As can be seen from the weighting functions, the probability of emission extends over a range of altitudes, which indicates that the vertical resolution of the estimate of the density profile from such measurements is not very high. Typically, vertical resolutions of 7–10 km are realistically achieved.

19.2.4 Ground-based measurements of upper-atmospheric water vapour

As an illustration of the technique, which has been used by a number of workers for a range of different gases, we shall consider measurements of upper-atmospheric water vapour using the emission line at 22.235 GHz. The spectrum of emission from the 22 GHz line has been measured with a Dicke radiometer and spectra of the kind shown in Figure 19.5 were obtained [164]. The shape of these spectra contains the information on the number of water-vapour molecules and at what height they are distributed in the atmosphere; the spectra can then be inverted mathematically to yield vertical profiles of water-vapour density, or mixing ratio, of which Figure 19.6 shows an example.

The 22 GHz water-vapour line is very weak, of course, and it is preferable to make use of stronger lines if possible, to reduce integration times. However, tropospheric water vapour introduces considerable attenuation at higher frequencies where lines are stronger and measurements from the ground (of water vapour and many other molecules with emission lines at higher frequencies) are simply not feasible. In such cases, elevated platforms such as aircraft or balloons have been used by a number of groups, using spectral-line receivers or radiometers, to measure a range of different gases.

Ground-based systems can provide continuous measurements in time, but only at one location, while those using aircraft or balloons can only provide information for a very short period of time. In most studies of atmospheric chemistry, however, global measurements are required, and these can only be achieved from orbiting satellites.

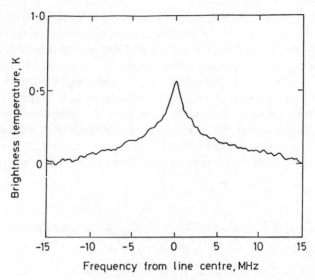

Figure 19.5   Example of the spectrum of brightness temperatures measured at the ground from the 22.235 GHz water-vapour emission line [164]
Reproduced by permission of Macmillan Magazines Ltd.

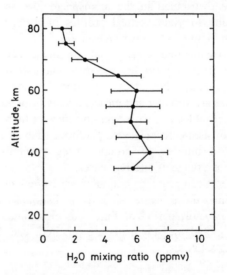

Figure 19.6   Example of the vertical profile of water-vapour mixing ratios, in parts per million by volume (ppmv), deduced from spectra of the kind shown in Figure 19.5 [164]
Reproduced by permission of Macmillan Magazines Ltd.

19.2.5 Limb-sounding from satellites

Remote sensing of the atmosphere from satellites can use one of two configurations: nadir observations or limb-sounding. Nadir measurements, in which relatively small antennas are directed towards the Earth's surface, have weighting functions similar to those shown in Figure 19.4. They are quite broad in the vertical, as a function of altitude, and can only provide a limited number of statistically-independent measurements over the altitude range. However, by using a large antenna to measure radiation tangentially from the limb of the atmosphere, as shown in Figure 19.7, the shape of the antenna beam can be used to restrict the altitude range from which radiation is received, enabling much narrower weighting functions to be defined, of the form shown in Figure 19.8. With such weighting functions and a vertically scanning antenna, much higher resolution can be obtained in the vertical profiles of molecular densities.

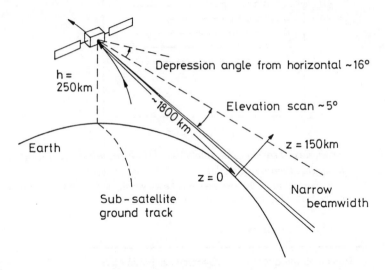

Figure 19.7   Geometry of the limb-sounding technique

One such experiment operating at present is the Microwave Limb Sounder (MLS) [119], on board the Upper Atmospheric Research Satellite (UARS), which was launched in 1991. Table 19.1 lists the frequencies of the radiometers which are on the MLS, together with the constituents and parameters which are measured.

The MLS project is currently providing much valuable data on the global distributions of ozone, water vapour, chlorine monoxide (which results from the photodissociation at high altitudes of refrigerants and aerosol propellants and which catalytically destroys the ozone which protects the biosphere from harmful solar ultra-violet radiation) and even sulphur dioxide emitted from recent volcanic eruptions.

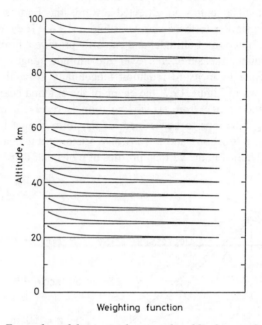

Figure 19.8   Examples of theoretical, normalised limb-sounding weighting functions (arbitrary scale)
The sharp cut-off at the lower region of each weighting function is determined by the beamwidth of the antenna; see also the caption to Figure 19.4

Table 19.1   MLS-instrument parameters

Radiometer frequency	Geophysical parameters
GHz	
63	Pressure
183	H_2O, O_3
205	O_3, ClO, (SO_2)

19.2.6 Satellite observations of the Earth's surface

Satellite-borne millimetre-wave radiometers are able not only to measure atmospheric constituents, but, in nadir or ground-scanning modes, can provide a wide range of information on surface and near-surface phenomena. Indeed, surface-measuring systems have been flown on satellites for much longer than have limb-scanning instruments.   For example, the Electronically Scanning Microwave Radiometer (ESMR), flown on *Nimbus-6* which was launched in 1975, was a 37 GHz radiometer which studied cloud liquid, sea ice and open sea coverage, surface composition and soil type together with surface features and moisture.   The Scanning Multispectral Microwave Radiometer (SMMR), flown on *Nimbus-7/Seasat* in 1978 carried radiometers operating at 6.6, 10.7, 18.0, 21.0 and 37.0 GHz to measure sea-surface temperature, near-sea-surface winds, water vapour, liquid-water content, cloud-droplet size and rainfall rate.

The measurement technique uses an antenna which scans from side to side to receive radiation from the surface across a 'swathe' along the ground track, measuring a temperature 'image' of the surface.   The technique relies on the fact that surface brightness temperatures at different elevation angles depend on the frequency, the polarisation and the nature of the surface, or near-surface, parameters to be studied.   An example of such differences is shown in Figure 19.9, for bare soil of different moisture contents, illustrating the way in which different polarisations behave as a function of elevation angle.   These results are for a frequency of about 1.4 GHz, but demonstrate the principle of the measurement technique.

One of the more recent instruments is the Special Sensor Microwave/Imager (SSM/I) [105], which was flown in 1987 and which includes radiometers operating at frequencies of 19.35, 22.235, 37 and 85.5 GHz, to measure a wide variety of parameters, including ocean wind speed, ice coverage, age and extent, precipitation intensity, cloud-water content and land-surface moisture.

Of particular interest in these millimetre-wave instruments is their ability to measure rainfall on a global scale, and especially over oceans, for which data are very sparse indeed.   Such information is particularly important in monitoring and understanding the global water budget, especially in tropical regions where more than two-thirds of worldwide precipitation falls.   The energy thereby released helps to power the global atmospheric circulation, influencing both weather and climate.   Thus, improving our knowledge of tropical rain is clearly of paramount importance to increasing our understanding of the world's climate and especially of potential changes in climate.

Unfortunately, measuring tropical rain by conventional surface techniques is very difficult, because of its high variability and because surface observations are not feasible over the vast jungle and ocean regions in the tropics.   Such regions are really only accessible for measurement by spacecraft; however, a major problem with conventional, near-Sun-synchronous, polar-orbiting satellites, of the type generally in use to date, is that they are not able to provide coverage of diurnal cycles in tropical regions, a knowledge of which is very important to understanding the climatology of tropical rain.

To provide more detailed sampling of tropical rain, the SSM/I is planned to be flown, together with other instruments, including a microwave radar and visible/infrared radiometers, on the Tropical Rainfall Measuring Mission

(TRMM), due for launch in 1997. This satellite will be dedicated to measuring tropical regions of the Earth and will have a low-inclination orbit of 35° from the equator; the daily orbital track will include South America, Africa, Australia, India and China, together with southern regions of Japan and the USA, for example. TRMM will provide very extensive sampling in the tropics together with coverage of diurnal cycles of tropical rain not provided by sun-synchronous, polar-orbiting satellites.

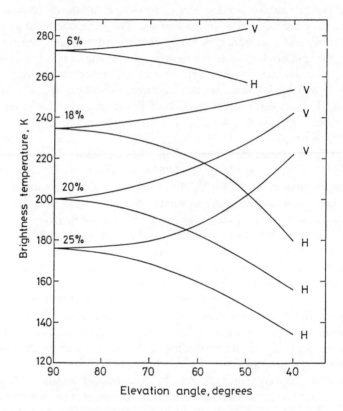

Figure 19.9   Typical examples of microwave brightness temperatures calculated at different angles of elevation, from the surface of bare soil, at different percentage moisture contents, for vertical (V) and horizontal (H) polarisations.

19.3 Millimetre-wave radars

19.3.1 Types of radar

A comprehensive review of meteorological and other radar systems is not feasible in this Chapter; more detailed information can be obtained from [167], for example. The most common form of radar is the pulsed system, in which the transmitter sends out a train of short radio-frequency pulses. Other systems include the continuous-wave (CW) radar and the frequency-modulated continuous-wave (FM-CW) radar. In the former, designed to detect moving targets, the received signal can be mixed with the transmitted signal to determine the Doppler frequency by homodyne detection, from which the radial target velocity can be obtained. This type of radar cannot indicate the range of the target, but by modulating the transmitter frequency, the FM-CW radar can be used to infer target range from the difference or beat frequency between the instantaneous transmitter frequency and the frequency of a received echo.

19.3.2 Advantages of millimetre waves for radar

Millimetre waves can provide a number of advantages for radar systems, primarily because, for a given resolution, sensitivity or precision, the antennas can be very much smaller than those used at more conventional microwave frequencies. Table 19.2 lists some of the frequency bands which have allocations for radar, together with typical antenna sizes for a beamwidth of 1°.

Table 19.2   1°-beamwidth antenna sizes at different frequencies

Frequency	1° beamwidth antenna size
GHz	cm
9	200
33	70
80	25
94	20

These small antenna sizes, coupled with the generally smaller dimensions (compared with those in the microwave band) of millimetric components, result in compact radar systems which are highly desirable for mobile applications, such as in missiles, satellites, aircraft and even automobiles.

Narrow beamwidths provide greater resolution and precision in target tracking and discrimination; an immunity to ground multipath and clutter down to low elevation angles; high antenna gain resulting in lower peak and average power requirements; the capability to detect and locate small objects such as cables and poles; and good resolution of closely spaced targets together with a high immunity to jamming.

A second advantage of the millimetric bands lies in the wide bandwidths available, the benefits of which include wideband spread-spectrum capability to reduce multipath and clutter; a high information-rate capability to yield fine-structure detail of target signatures; very high range resolution for precision tracking and target identification; greater sensitivity in radiometers; large Doppler-frequency shifts from low radial-velocity targets resulting in improved detection and recognition capability of slowly moving objects, and a high immunity to jamming and interference, owing to the large number of frequencies which can be used.

There are, of course, a number of disadvantages of the millimetric wavebands for radar applications. These include the need for high precision in component manufacture, resulting in high costs; narrow beamwidths which can make large-volume searches impractical, and which are thus not particularly suitable for target search and acquisition; severely reduced range capability in adverse weather conditions, poor foliage penetration and large Doppler shifts which can be outside the receiver bandwidth.

19.3.3 Examples of millimetre-wave radars

Most applications of radar involve achieving as long a detection range as possible and, as a result, research and development of millimetre-wave radar systems have concentrated in the regions of low atmospheric attenuation, i.e. near the window frequencies of 35, 94, 140 and 220 GHz (see Figure 1.2), with very little development at frequencies higher than 220 GHz. The majority of such applications fall into one of four generic groups:

(i)   surveillance and target acquisition;
(ii)   instrumentation and measurements;
(iii)   guidance and seekers;
(iv)   fire control and tracking.

From this list, it is clear that many, if not most, millimetre-wave radars have been developed for military applications, and it will not be the purpose of this chapter to discuss in detail the wide variety of such applications. However, some interesting civil applications will be discussed briefly, as examples of the advantages of the millimetre-wave region.

19.3.3.1 Automotive radars
One application of particular relevance to the development of mass markets for millimetre-wave systems, however, is that of automotive radar, where, at the

shorter wavelengths, at frequencies near 94 GHz, for example, good resolution can be achieved with antennas small enough to be easily integrated into automobiles.

Automotive radar has the potential to provide drivers with a variety of safety features [170]. Collision avoidance is an obvious benefit, for example; when driving in conditions of low visibility, radar can provide an early warning of slower-moving vehicles ahead, either passively by visual or audible indication, or actively by reducing speed. An extension of this principle involves headway measurement, where the radar measures the distance to the next vehicle in front; such a system can be made even more sophisticated to provide intelligent cruise control. In these headway-measurement systems, the radar output is processed to interact with the electronic engine- and brake-management systems now being fitted to the latest generation of cars, thus allowing the system automatically to maintain a constant and safe headway.

Another category of automotive radars currently being investigated is that of obstacle-warning radars, which are intended to give drivers warning of any possible hazards which might be encountered on the road ahead. Such systems will be even more sophisticated than collision-avoidance radars or intelligent cruise-control systems, since they will involve a scanning antenna to provide the driver with a perspective view of the area in front of the vehicle. Considerable signal processing will, of course, be required to extract only the very basic and essential information on possible hazards ahead for presentation to the driver, via a suitable 'man–machine interface'. It is not envisaged at present that obstacle-warning radars should exercise any control over the vehicle; their purpose will be to aid drivers to recognise potential hazards, in conditions of poor visibility, using some kind of 'head-up' display system.

In addition to facilitating integration into automobiles, by virtue of the small size of equipment, millimetre waves provide a further advantage over lower frequencies in that radar cross-sections, i.e. the effectiveness of the target as a radar reflector, can be much greater than at longer wavelengths. This means that small targets can be detected more easily. As an example, Table 19.3 lists some typical radar cross-sections of various targets at frequencies near 12 GHz and near 94 GHz, from which it can be seen that a small car has roughly the same equivalent size when illuminated at 94 GHz as has an airliner at 12 GHz.

These figures are somewhat oversimplified, however, since radar cross-sections are notoriously variable because of the complexity of target topography (and in target material) and can change by an order of magnitude if the aspect angle changes by only a few degrees. The situation is also further complicated by the fact that automotive radars will generally be required to operate in the near field. Nevertheless, Table 19.3 serves as an illustration of the greater sensitivity (and target detectability) of the shorter wavelengths for small targets.

The ranges required for automotive radar will typically be of the order of hundreds of metres, rather than the 100 km or so of more conventional radar systems and, since the received power in the signal reflected from a target is proportional to the fourth power of range, it is thus clear that the power requirements for automotive radar will be rather modest, and therefore compatible with a mass market. For example, it has been estimated that an obstacle-warning radar will require a transmitter power of about 10 mW, while the simpler intelligent cruise-control system will need less than 1 mW. Such power levels are readily achievable with current solid-state-oscillator technology.

Table 19.3   Typical radar cross-sections for different targets

Frequency	Target	Area
GHz		m^2
12	Man	1
	Jet fighter	1
	Airliner	10
95	Man	1
	Small car	10
	Van	15
	Debris	>0.1
	Road surface	<10^{-4} m²/square metre of road surface

Much work has already been carried out on various automotive radar systems. Intelligent cruise control has been tested on motorways (with further trials still in progress) and has been demonstrated to be very reliable, missing no small targets and not being subject to false alarms. Trials of obstacle-warning radar using scanning antennas have yielded promising results, though the effectiveness of such systems will clearly depend heavily on the way in which information is presented to the driver, i.e. on the 'man–machine interface'. The evidence from this work is that millimetre-wave technology is now capable of providing effective automotive radar systems which can make driving in poor weather conditions significantly safer and less tiring.

19.3.3.2 Adverse-weather aircraft-landing systems
Another application in which the millimetre-wave region is currently being explored concerns instrument landing systems for use by aircraft in adverse weather conditions [95]. Adverse-weather imaging sensors are being developed in both the infra-red and the millimetre-wave regions, the latter using imaging radars and radiometers. Millimetre-wave systems offer the prospect of being usable in rain and fog, the latter being particularly limiting at infra-red wavelengths. Systems have been developed to operate in the 33 GHz and 94 GHz bands, the higher frequency providing three times the resolution of radars operating at 33 GHz using antennas of the same size. By scanning the ground in

front of the aircraft as it approaches the landing field, the imaging system provides a picture of the runway which may not be visible because of the adverse weather.

19.4   Applications of millimetre waves in communications

19.4.1 Advantages and disadvantages of the millimetre-wave region

Spectrum congestion at the lower frequencies, the ever-increasing demand for more communication channels and the requirement for higher-data-rate transmissions have inevitably led to active consideration and application of radiocommunication systems operating at frequencies above 30 GHz.

At such frequencies, however, the Earth's atmosphere and the prevailing weather, especially precipitation, have a dramatic effect on the transmission of millimetre waves.   Oxygen and water vapour in the troposphere impose fairly severe levels of attenuation in certain frequency bands, as shown in Figure 1.2. These effects are omnipresent, can be readily calculated and can even be put to advantageous use, for example in using the high levels of atmospheric attenuation in the 60 GHz band to enable multiple reuse of the same frequencies over relatively short distances, thus ensuring efficient spectrum utilisation.   The second effect, that of precipitation attenuation, also shown in Figure 1.2 for different kinds of rain, from light drizzle to intense thunderstorms, is more problematic, since rain, being a stochastic process, is less easily predictable.   Much effort is being devoted to quantify more precisely the effects of rain at millimetric wavelengths and to facilitate the prediction of such effects statistically, in order that the planning and design of future millimetre-wave communication systems can more easily and accurately be accomplished.

Many of the advantages (and disadvantages) of the millimetric wavebands mentioned in Section 19.3 for radar systems apply equally well to communication-systems applications.   Small, compact equipment facilitates system installation, small antennas can provide high gains, reducing the need for high-power transmitter oscillators, or increasing the range over which transmissions can reliably be achieved, very wide bandwidths (and thence high datarates) are readily achievable and realisable; the higher frequencies have (at present) an immunity to interference (and jamming) because of the low population of transmissions, and also because operation in certain frequency bands (e.g. near 60 GHz) can make use of the additional atmospheric attenuation to reduce the possibility of interference.   This high atmospheric attenuation can also increase the efficiency of spectrum utilisation, as noted above, through multiple frequency reuse over quite short distances.   The facility for covert systems operation in the 60 GHz band is also clearly evident.

19.4.2 Examples of applications in communications

A number of communications systems have been proposed, or planned, to operate in the millimetric wavebands. As examples, typical and representative (though by no means exhaustive) applications include:

(*a*) short-range transmissions with intensive frequency reuse near 60 GHz, for example in urban microcellular systems;

(*b*) applications where cable wayleave is either difficult or not possible to obtain, such as communication between buildings separated by a major road, a railway track or a river;

(*c*) short-range (up to 20 km) feeder links at 38 GHz to longer-distance optical-fibre networks such as the personal communications network (PCN);

(*d*) self-trunking, nonregulated, point-to-point links operating near 60 GHz

(*e*) electronic news gathering and outside-broadcast services; and

(*f*) high-capacity, point-to-point and point-to-multipoint transmissions of data, speech and video, including final connections to customers from a cable or optical-fibre network.

19.4.3 Multipoint Video Distribution Service

This last application is, in fact, currently receiving considerable interest, because of the potential for a mass-market delivery of entertainment services — the so-called Multipoint Video Distribution Service (MVDS) — in which a large number of television channels (20 – 30 are being envisaged) are broadcast at frequencies near 40 GHz to those towns and communities considered not to be economically large enough for the more conventional cable-television network [153]. The principle, illustrated in Figure 19.10, invokes the use of a local studio which gathers together a range of material — different channels and programmes from off-the-air UHF broadcast television, from satellite television, from video tapes and even from local community live broadcasts — and transmits these to the local community, over a radius of perhaps 2 – 3 km.

Small rooftop antennas, no larger than about 15 cm, receive the 40 GHz transmissions, while the front-end receiver down-converts the signals to an intermediate frequency compatible with current satellite television systems, thus enabling existing 'set-top boxes' to be used for MVDS reception.

Millimetre waves are thus seen as an exciting new opportunity for delivering mass-market multichannel television entertainment to the home. MVDS is considered to be economical and inexpensive to install, especially when compared with the more-conventional cable systems, for the smaller towns and communities, and it has the capability of bringing some 20–30 channels of television to millions of viewers in small to medium-sized towns quite soon, long before cable television (if ever) reaches them. Indeed, such multichannel entertainment may only be possible for the smaller towns through MVDS, which can offer, via one small antenna, a single access point to multiple television programming from all possible sources.

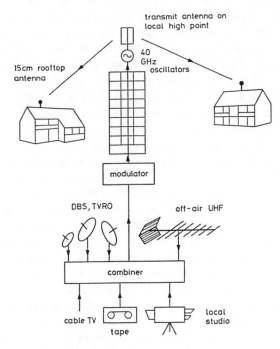

Figure 19.10   Schematic diagram illustrating the concept of the Multipoint Video Distribution Service (MVDS) at 40 GHz

MVDS is a comparatively late entry into the entertainment and television field, primarily because the technology has, until recently, been very expensive. Recent developments in gallium arsenide (GaAs) monolithic integrated-circuit technology, at these high frequencies, have, however, enabled active components to be developed to the point where mass-market consumer applications have become both realistic and realisable. Furthermore, the installation of MVDS does not require the expense of laying cables, while the higher cost of receivers will be offset by the lower cost of small antennas and their installation, compared with direct-broadcast satellite systems. Finally, the environmental advantage of the small MVDS antennas should also not be overlooked.

19.4.4 Mobile Broadband System (MBS)

The Mobile Broadband System (MBS) is a project within the RACE (Research and development of Advanced Communication in Europe) framework to develop a third-generation digital microcellular service to make the Broadband Integrated Services Digital Network (B-ISDN) available to mobile users [108]. Two 1 GHz bands have provisionally been allocated at 62 and 65 GHz, for the down and up links, respectively. By operating within the oxygen absorption bands, the microcells will have a coverage typically of about 100 m, although the use of directional antennas along highways, for example, could extend this range. With data rates of up to 34 Mbit/s per channel, MBS is being developed to provide full multimedia services of data, voice and video to the mobile user. The system is designed to have application in a wide variety of areas, for example in the emergency services, for mobile local-area networks (LANs) and for television outside broadcasts such as electronic news gathering.

Applications in the emergency services could, for example, provide live audio and video information from the scene of an accident to doctors at the hospital, who could then give advice to medical staff in attendance at the accident. Fire services could employ small helmet-mounted cameras to transmit video to temporary control centres to enable guidance to be given to firefighters. Although wireless LANs are currently employed in the exchange of data, applications being considered for the future include the need for video, for example in the use of robots in hazardous locations, where the robot may be operated by remote control using high-definition video from a robot-mounted camera. In outside broadcasts of television, a number of different television cameras at different locations in a sports arena, for example, could be connected to a local control centre using MBS links. This would provide a high level of flexibility and, further, enable the system to be installed very quickly. Other applications envisaged for MBS include such diverse areas as traffic management and city guidance, traffic advice, pictorial data for travel, access to banking services for mobile users, interconnection of mobile LANs and surveillance applications.

Like MVDS, the Mobile Broadband System will depend critically on the development of low-cost millimetre-wave components and systems, but it is seen as having enormous potential as the next generation multimedia communication service for a very wide range of user applications in a whole host of fields.

19.5 Summary

This chapter has considered some of the many applications now in use or being considered which exploit the millimetre-wave spectrum. These applications include extensive remote-sensing experiments aimed at studying and understanding our environment, i.e. the Earth's atmosphere, climatology and weather, and studies of the Earth's surface and its resources. Millimetre waves are furthermore being actively explored for mass-market consumer-oriented applications such as automotive radar systems to improve safety on the roads, and for multichannel television entertainment services, while a whole new generation of multimedia communication is becoming available for mobile users as a result of the exploitation of the millimetric wavebands. Such applications are only now being realised because of extensive developments in millimetre-wave technology, which will have the effect of bringing a dramatic reduction in the hitherto high costs of such technology.

Finally, while millimetre waves have until very recently been considered rather esoteric, and confined to the experimental, high-budget laboratory, it should be remembered that the investigation of millimetre waves is by no means a new or even a recent endeavour. The very first 60 GHz signals were generated and used for studies of scattering from and polarisability of various objects by Acharya Jagdish Chandra Bose, at the University of Calcutta, as far back as 1897 [60]. Indeed, J.C. Bose also developed semiconductor detectors, rectangular waveguides and horn and lens antennas, all operating at 60 GHz, and this nearly 100 years ago!

Chapter 20

References

20.1 ITU publications

Sources of ITU-R texts are outlined in the Preface of this book.

1   Radio Regulations, edition of 1990, revised 1994
2   Recommendation BS.412-7 'Planning standards for FM sound broadcasting at VHF', 1995
3   ITU-R Recommendation F.339-7 'Bandwidths, signal-to-noise ratios and fading allowances in complete systems, RF series, fixed service at frequencies below about 30 MHz', 1994
4   ITU-R Recommendation IS.847-1: 'Determination of the co-ordination area of an Earth station operating with a geostationary space station and using the same frequency band as a system in the terrestrial service', 1994
5   ITU-R Recommendation P.114-7 'Prediction of sky-wave field strength at frequencies between 150 and 1700 kHz', 1995
6   ITU-R Recommendation P.341-4 'The concept of transmission loss for radio links', 1995
7   ITU-R Recommendation P.368-7 'Ground-wave propagation curves for frequencies between 10 kHz and 30 MHz', 1994
8   Recommendation P.370-7 'VHF and UHF propagation curves for the frequency range from 30 MHz to 1000 MHz', 1995
9   ITU-R Recommendation P.372-6 'Radio noise', 1994
10   ITU-R Recommendation P.373-7 'Definitions of maximum and minimum transmission frequencies', 1995
11   ITU-R Recommendation P.434-6 'Reference ionospheric characteristics and methods of basic MUF, operational MUF and ray-path prediction', 1995
12   ITU-R Recommendation P.452-7 'Prediction procedure for the evaluation of microwave interference between stations on the surface of the Earth at frequencies above about 0.7 GHz', 1995
13   ITU-R Recommendation P.453-5 'The radio refractive index: its formula and refractivity data', 1995

14   ITU-R Recommendation P.526-4 'Propagation by diffraction', 1995
15   Recommendation P.529-2 'Prediction methods for the terrestrial land mobile service in the VHF and UHF bands', 1995
16   ITU-R Recommendation P.530-6 'Propagation data and prediction methods for the design of terrestrial line-of-sight systems', 1995
17   ITU-R Recommendation P.531-3 'Ionospheric effects influencing radio systems involving spacecraft', 1994
18   ITU-R Recommendation P.533-5 'HF propagation prediction method', 1994
19   ITU-Recommendation P.617-1 'Propagation prediction techniques and data required for the design of trans-horizon radio-relay systems', 1994
20   ITU-R Recommendation P.618-3 'Propagation data and prediction methods required for the design of Earth-space telecommunications systems', 1995
21   ITU-R Recommendation P.619-1 'Propagation data required for the evaluation of interference between stations in space and those on the surface of the Earth', 1994
22   ITU-R Recommendation P.620-2 'Propagation data required for the calculation of coordination distances in the frequency range 1–40 GHz', 1995
23   ITU-R Recommendation P.676-2 'Attenuation by atmospheric gases in the frequency range 1–350 GHz', 1995
24   ITU-R Recommendation P.684-1 'Prediction of field strength at frequencies below about 500 kHz', 1994
25   ITU-R Recommendation P.832 'World atlas of ground conductivities', 1994
26   ITU-R Recommendation P.836 'Surface water vapour density', 1994
27   ITU-R Recommendation P.837-1 'Characteristics of precipitation for propagation modelling', 1994
28   ITU-R Recommendation P.838 'Specific attenuation model for rain for use in prediction methods', 1994
29   ITU-R Recommendation P.840-1 'Attenuation due to clouds and fog', 1994
30   ITU-R Recommendation P.841 'Conversion of annual statistics to worst-month statistics', 1994
31   ITU-R recommendation P.842-1 'Computation of reliability and compatibility of HF radio systems', 1994
32   ITU-R recommendation P.845-2 'HF field strength measurement', 1995
33   ITU-R Recommendation RBS.705 'HF transmitting antenna characteristics and diagrams', 1994
34   ITU-R Recommendation SF.558-2 'Maximum allowable values of interference from terrestrial radio links to systems in the fixed-satellite

service employing 8-bit PCM encoded telephony and sharing the same frequency bands', 1994

35   ITU-R Recommendation SF.615 'Maximum allowable values of interference from the fixed-satellite service into terrestrial radio-relay systems which may form part of an ISDN and share the same frequency band below 15 GHz', 1994

36   ITU-R Recommendation SF.1006 'Determination of the interference potential between Earth stations of the fixed-satellite service and stations in the fixed service', 1994

37   ITU Report 250-6 'Long-distance ionospheric propagation without intermediate ground reflection', 1990

38   ITU-R Report 879-1 'Methods for estimating effective electrical characteristics of the surface of the Earth', 1990

39   ITU-R Report 895-2 'Radio propagation and circuit performance at frequencies below about 30 kHz', 1990

40   ITU-R Report 1010 'Propagation data for bi-directional coordination of Earth stations', 1990

41   ITU-R Resolution 25 'Computer programs for the prediction of ionospheric characteristics, sky-wave transmission loss and noise', 1994

20.2 CCIR publications

42   Report 944 'Theoretical network planning', Annex to Vol. X, pt 1, 1990

43   Supplement to Report 252-2 'Second CCIR computer-based interim method for estimating sky-wave field strength and transmission loss at frequencies between the approximate limits of 2 and 30 MHz', 1978

20.3 COST project publications

44   COST 205: 'Influence of the atmosphere on radio propagation on satellite Earth paths at frequencies above 10 GHz', Commission of the European Communities report EUR 9923, 1985, ISBN 92 825 5412 0 [also in *Alta Freq.*, 1989, **54** (3)]

45   COST 207: 'Digital land mobile radio communications. Management committee final report 14th March 1984 – 13th September 1988', Commission of the European Communities report EUR 12160 (catalogue no. CD-NA- 12160-EN-C), 1989, ISBN 92 825 9946 9

46   COST 210: 'Influence of the atmosphere on interference between radio communications systems at frequencies above 1 GHz. Final report', Commission of the European Communities report EUR 13407 (catalogue no. CD-NA- 13407-EM-C), 1991, ISBN 92 826 2400 5

20.4 Other references

47   ABRAMOWITZ, M. and STEGUN, J. A.: 'Handbook of mathematical functions' (Dover, 1965, 1972)

48   ALLPRESS, S. A.: 'Optimising signal rate and internal diversity order for the mobile cellular DS-CDMA systems'. PhD thesis, University of Bristol, UK, 1993

49   ALLPRESS, S. A. and BEACH, M. A.: 'Measurement and characterisation of the wideband DS-CDMA radio channel'. Proceedings of ICAP 93, Edinburgh, UK, 1993. *IEE Conf Publ.* 370, Part 2, pp. 429–432

50   ALLPRESS, S. A. and BEACH, M. A.: 'On the optimum DS- CDMA channel bandwidth for personal communication systems'. Proceedings of IEEE 43rd VTC, New Jersey, USA, 1993, pp. 436-439

51   BARCLAY, L. W.: 'Statistical modelling of HF links'. AGARD conference proceedings 238, Ottawa, Canada, 1978

52   BASHIR, S. O.: 'Three years statistics of refractive index gradient and effective Earth radius factor for the state of Bahrain'. Proceedings of ICAP 89, *IEE Conf. Publ. 301*, Part 2, pp. 220–223

53   BATTAGLIA, M. R.: 'Modelling the radar evaporative duct'. RANRL technical note 3/85, Department of Defence, DSTO, RAN Research Laboratory, Australia, 1985

54   BEACH, M. A., *et al.*: 'An evaluation of direct sequence CDMA for future mobile personal communication networks'. Proceedings of IEEE VTC, St. Louis, USA, 1991, pp. 73–70

55   BEAN, B. R. and DUTTON, E. J.: 'Radio meteorology'. US National Bureau of Standards, monograph 92, 1966, chap. 1

56   BELLO, P. A.: 'Characterisation of randomly time- variant linear channels', *IEEE Trans.*, 1963, **CS–11**, (4), pp. 360–393

57   BILITZA, D., RAWER, K., BOSSY, L. and GULYAEVA, T.: 'International reference ionosphere – past, present and future: 1 Electron density'. *Adv. Space Res.*, 1993, **13**, (3), pp. 3–13

58   BOITHIAS, L.: 'Radio wave propagation' (McGraw-Hill, New York, 1987) (also North Oxford Publishers, 1987)

59   BOOKER, H. G. and WALKINSHAW, W.: 'The mode theory of tropospheric refraction and its relation to wave-guides and diffraction' *in* 'Meteorological factors in radio wave propagation'. Physical Society report, London, UK, 1946, pp. 80–127

60   BOSE, J. C.: 'Collected physical papers' (Longman Green, New York, 1927)

61   BRACEWELL, R. N., *et al.*: 'The ionospheric propagation of low- and very-low-frequency radio waves over distances less than 1000 km', *Proc. IEE*, 1951, **98**, Part III, pp. 221–236

62   BRADLEY, P. A. and HOWARD, D. R.: 'Transmission loss at high frequencies on 3260 km temperate-latitude path', *Proc. IEE*, 1973, **120**, (2), pp. 173–180

63   BREMMER, H.: 'Terrestrial radio waves' (Elsevier, 1949)

64   BRENNAN, D. G.: 'Linear diversity combining techniques', *Proc. Inst. Radio Eng.*, 1959, **47**, pp. 1075–1102

65   BRUSSAARD, G.: 'A meteorological model for rain-induced cross-polarisation', *IEEE Trans.*, 1976, **AP–24**, (1), pp. 5–11

66   BUDDEN, K. G.: 'Radio waves in the ionosphere' (Cambridge University Press, 1961)

67   BUDDEN, K. G.: 'The waveguide mode theory of wave propagation' (Logos Press, 1961)

68   CAUSEBROOK, J. H.: 'Computer prediction of UHF broadcast service areas'. BBC Research Department report 1974/4, 1974

69   CHEUNG, J. C. S., *et al.*: 'On the impact of soft handoff on cellular direct sequence code division multiple access system'. Commission of the European Communities RACE PLATON/UOB/037/WPD2/07/93, issue 1, July 1993

70   CLARKE, R. H.: 'A statistical theory of mobile radio reception', *Bell Syst. Tech. J.*, 1968, **47**, pp. 957–1000

71   COLLINS, R. E.: 'Antennas and radiowave propagation' (McGraw-Hill, International Students Edition, 1985)

72   COMMISSION OF THE EUROPEAN COMMUNITIES: RACE mobile communications workshop, DG XIII/B, RA129-93, Metz, France, 16-18 June 1993

73   COX, D. C. and LECK, R. P.: 'Correlation bandwidth and delay spread multipath propagation statistics for 910 MHz urban mobile radio channels', *IEEE Trans.*, 1985, **COM–23**, pp. 1275–1280

74   CRAIG, K. H.: 'Roots of the mode equation for propagation in an elevated duct'. Proceedings of ICAP 85, *IEE Conf. Publ. 248*, 1985, pp. 274–278

75   CRAIG, K. H.: 'Propagation modelling in the troposphere: parabolic equation method', *Electron. Lett.*, 1988, **24**, pp. 1136–1139

76   CRAIG, K. H. and LEVY, M. F.: 'Parabolic equation modelling of the effects of multipath and ducting on radar systems', *IEE Proc. F*, 1991, **138**, (2), pp. 153–162

77   CROFT, T. A.: 'A review of oblique ray tracing and its application to the calculation of signal strength' *in* JONES, T. B. (Ed.): 'Oblique ionospheric radiowave propagation at frequencies near the lowest usable high frequency', *AGARD Cont. Proc.*, 1969, **13**, Technivision, Slough, UK, pp. 137–169

78   CTIA Technology Forum, Qualcomm, 5 December 1991

79   D'AMICO, M.: 'Melting layer modelling and scatter by raindrops'. ESTEC internal working paper N.1618, 1991

80   DEBYE, P.: 'Polar molecules' (Dover, 1957), pp. 89–90

81   DEYGOUT, J.: 'Multiple knife-edge diffraction of radio waves', *IEEE Trans.*, 1966, **AP–14**, (4), pp. 480–489

82   DICKE, R. H.: 'The measurement of thermal radiation at microwave frequencies', *Rev. Sci. Instrum.*, 1946, **17**, pp. 268–275

83   DIXON, R. C.: 'Spread spectrum systems' (John Wiley, 1984), 2nd ed.

84   DOCKERY, G. D.: 'Modeling electromagnetic wave propagation in the troposphere using the parabolic equation', *IEEE Trans.*, 1988, **AP–36**, pp. 1464–1470

85   ECKERSLEY, P. P.: *Proc. Inst. Radio Eng.*, 1930, **18**, p. 1160

86   EIA/TIA proposed interim standard, Qualcomm document 80-10201, rev. DCR03651, 15 May 1992

87   EPSTEIN, J. and PETERSON, D. W.: 'An experimental study of wave propagation at 850 MC', *Proc. Inst. Radio Eng.*, 1953, **41**, (5), pp. 595–611

88   FERGUSON, J. A.: 'Status of the naval ocean systems long wave propagation capability "LWPC"'. Paper presented at *Third Workshop on ELF/VLF Radio Noise*, Stanford University, USA, 1988

89   GANS, M. J.: 'A power spectral theory of propagation in the mobile radio environment', *IEEE Trans.*, 1972, **VT–21**, pp. 27–38

90   GILHOUSEN, K. S.: 'Mobile power control for CDMA', *Communications*, 1992, pp. 34–38

91   GILHOUSEN, K. S., *et al.*: 'On the capacity of a cellular CDMA system', *IEEE Trans.*, 1991, **VT–40**, pp. 303–312

92   GIOVANELLI, C. L.: 'An analysis of simplified solutions for multiple-knife-edge diffraction', *IEEE Trans.*, 1984, **AP–32**, (3), pp. 297–301

93   GODDARD, J. W. F. and BAPTISTA, J. P. V.: 'Multi- parameter radar applied to propagation studies'. Proceedings of *Olympus Utilisation Conference*, Seville, Spain, 1993, ESA publication WPP-60, pp. 609–613

94   GOLOMB, S. W.: 'Shift register sequences, revised edition' (Aegean Park Press, California, 1982)

95   GREENWOOD, S. W.: 'The application of imaging sensors to aircraft landings in adverse weather', *Microwave J.*, September 1992, pp. 80–89

96   GRIFFITHS, J.: 'Radiowave propagation and antennas: an introduction' (Prentice Hall International (UK), 1987)

97   GRIFFITHS, J. and MCGEEHAN, J. P.: 'Inter-relationship between some statistical distributions used in radiowave propagation', *IEE Proc. F*, 1982, **129**, (6), pp. 411–417

98   HAGN, G. H., GIBSON, A. J. and BRADLEY, P. A.: 'Propagation on frequencies above the basic maximum usable frequency'. Proceedings of the *7th International Ionospheric Effects Symposium*, Alexandria, VA, USA, 1993, paper 5A-1 (published 1994)

422   *Propagation of radiowaves*

99   HALL, M. P. M.: 'Effects of the troposphere on radio communications'. IEE Electromagnetic Waves Series 8 (Peter Peregrinus, London, 1979)
100   HARDEN, B. N., NORBURY, J. R. and WHITE, W. J. K.: 'Use of a log-normal distribution of raindrop sizes in millimetric radio attenuation studies'. *IEE Conf. Publ. 169*, 1978, Part 2, pp. 87–91
101   HASHEMI, H.: 'Simulation of the urban radio propagation channel', *IEEE Trans.*, 1979, **VT–28**, (3), pp. 213–225
102   HATA, M.: 'Empirical formula for propagation loss in land mobile radio services', *IEEE Trans.*, 1980, **VT–29**, pp. 317–325
103   HAWORTH, D. P., MCEWAN, N. J. and WATSON, P. A.: 'Relationship between atmospheric electricity and microwave radio propagation', *Nature*, 1977, **266**, pp. 703–704
104   HERRING, R. N.: 'General raytracing techniques for predicting fields in an arbitrarily varying troposphere'. Proceedings of ICAP 85, *IEE Conf. Publ. 248*, 1985, pp. 279–282
105   HOLLINGER, J. P., PIERCE, J. L. and POE, G. A.: 'SSM/I instrument evaluation', *IEEE Trans.*, 1990, **GRS–28**, pp. 781–790
106   HOLT, A. R., UZUNOGLU, N. K. and EVANS, B. G.: 'An integral equation solution to the scattering of electromagnetic radiation by dielectric spheroids and ellipsoids', *IEEE Trans.*, 1976, **AP–26**, (5), pp. 706–712
107   HOLTZMAN, J. M. and JALLOUL, L. M. A.: 'Rayleigh fading effect reduction with wideband DS-CDMA signals'. Rutgers University, technical report, August 1991
108   HUISH, P., ZUBRZYCKI, J. T. and LURDES LOURENCO, M.: 'Mobile broadband system'. IEE colloquium on *Exploiting the millimetric wavebands*, 7 January 1994, *IEE Colloquium Digest 1994/002*
109   IKEGAMI, F., YOSHIDA, S., TAKEUCHI, T. and UMEHIRA, M.: 'Propagation factors controlling mean field strengths on urban street', *IEEE Trans.*, 1984, **AP–32**, pp. 822–829
110   JAMES, G. L.: 'Geometrical theory of diffraction for electromagnetic waves', IEE Electromagnetic Waves Series 1 (Peter Peregrinus, London, 1976)
111   JAPANESE MINISTRY OF POSTAL SERVICES: 'Atlas of radio wave propagation curves for frequencies between 30 and 10 000 Mc/s', Radio Research Laboratory, Tokyo, 1957, pp. 172–179
112   JESKE, H.: 'Die Ausbreitung elecktromagnetischer Wellen im cm- bis m-band uber dem Meer unter besonderer Berucksichtigung der meteorologischen Bedingungen in der maritimen Grenzschicht' *in* 'Hamburger Geophysikalische Einzelschriften' (De Gruyter & Co, Hamburg, 1965)
113   JORDAN, E. C. and BALMAIN, K. G.: 'Electromagnetic waves and radiating systems' (Prentice Hall, New York, 1968), chap. 16

114   KIFT, F., BRADLEY, P. A., MARTIN, L. T. J. and BRAMLEY, E. N.:
       'HF oblique sounding measurements over a 6700 km temperate-latitude
       path', *Proc. IEE*, 1969, **112**, (12), pp. 1985–1991
115   KLAASSEN, W.: 'Attenuation and reflection of radio waves by a melting
       layer of precipitation', *IEE Proc. H*, 1990, **137**, pp. 39–44
116   KNIGHT, P., *in* HALL, M. P. M. and BARCLAY, L. W.: 'Radiowave
       propagation', IEE Electromagnetic Waves Series 30 (Peter Peregrinus,
       London, 1989)
117   KO, H. W., SARI, J. W., THOMAS, M. E., HERCHENROEDER, P. J.
       and MARTONE, P. J.: 'Anomalous propagation and radar coverage
       through inhomogeneous atmospheres'. AGARD CP-346, 1984, pp. 25.1–
       25.14
118   KRAUS, Z. Z. and CARVER, X. X.: 'Electromagnetics' (McGraw-Hill,
       International Students Edition, 1973), 2nd ed.
119   LAHOZ, W. A.: 'The 183 GHz water and ozone channels on the upper
       atmosphere research satellite microwave limb sounder', *Int. J. Remote
       Sensing*, 1991, **12**, pp. 33–53
120   LAWS, J. O. and PARSONS, D. A.: 'The relation of rain-drop size to
       intensity', *Trans. Am. Geophys. Union*, 19??, **24**, pp. 432–460
121   LAWTON, M. C., DAVIES, R. L. and McGEEHAN, J. P.: 'A
       ray-launching method for the prediction of indoor radio channel
       characteristics'. IEEE international symposium on *PIMRC*, Kings College
       London, UK, 1991, pp. 104–108
122   LEE, W. C. Y.: 'Overview of cellular CDMA', *IEEE Trans.*, 1991,
       VT–40, pp. 291–301
123   LIEBE, H. J.: 'An updated model for millimeter wave propagation in
       moist air', *Radio Sci.*, 1985, **20**, pp. 1069–1089
124   LIVNEH, N., *et al.*: 'Frequency hopping CDMA for future cellular radio'.
       Proceedings of IEEE 42nd VTC, Denver, USA, 1992, pp. 400–404
125   LORENZ, R. W.: 'Field strength prediction method for mobile telephone
       system using a topographical data bank'. *IEE Conf. Publ. 188*, 1980, pp.
       6–11
126   LUEBBERS, R. J.: 'Finite conductivity uniform-GTD knife edge
       diffraction in prediction of propagation path loss', *IEEE Trans.*, 1984,
       AP–32, (1), pp. 000–000
127   MANABE, T., IHARA, T., AWAKA, J. and FURUHAMA, Y.: 'Rain
       attenuation statistics of millimeter waves and raindrop size distribution'.
       Proceedings of URSI Commission F symposium on *Wave propagation:
       remote sensing and communications*, University of New Hampshire, USA,
       1986, pp. 10.5.1–10.5.4
128   MARSHALL, J. S. and PALMER, W. McK.: 'The distribution of
       raindrops with size', *J. Meteor.*, 1948, **5**, pp. 165–166

129   MASON, B. J.: 'The physics of clouds' (Oxford University Press, 1971), 2nd ed.

130   MCLINTOCK, R. W. and KEARSEY, B. N.: 'Error performance objectives for digital networks', *Radio Electron. Eng.*, 1984, **54**, pp. 79–85

131   MEDEIROS FILHO, F. C., COLE, R. S. and SARMA, A. D.: 'Millimetre-wave rain induced attenuation: theory and experiment', *IEE Proc. H*, 1986, **133**, pp. 308–314

132   MEEKS, M. L. (Ed.): 'Methods of experimental physics. Vol. 2 Part B Astrophysics' (Academic Press, New York, 1976)

133   MELFI, S. H. and WHITEMAN, D.: 'Observation of lower-atmospheric moisture structure and its evolution using a Raman lidar', *Bull. Am. Meteorol. Soc.*, 1985, **66**, (10), pp. 1288–1292

134   MILLER, I. and FREUD, J.: 'Probability and statistics for engineers' (Prentice Hall, 1985), 3rd ed.

135   MILLINGTON, G.: 'Ground-wave propagation over an inhomogeneous smooth Earth. Part 1', *Proc. IEE*, 1949, **96**, pp. 53–64

136   MILLINGTON, G. and ISTED, G. A.: 'Ground-wave propagation over an inhomogeneous Earth. Part 2 Experimental evidence and practical implications', *Proc. IEE*, 1950, **97**, pp. 209–222

137   MORFITT, D. G., *et al.*: 'Numerical modelling of the propagation medium at ELF/VLF/LF'. AGARD conference proceedings 305, 1981, paper 32

138   NORTON, K. A.: 'The propagation of radio waves over the surface of the Earth and in the upper atmosphere. Part 1', *Proc. Inst. Radio. Eng.*, 1936, **24**, pp. 1367–1387

139   NORTON, K. A.: 'The propagation of radio waves over the surface of the Earth and in the upper atmosphere. Part 2', *Proc. Inst. Radio. Eng.*, 1937, **25**, pp. 1203–1236

140   NORTON, K. A.: 'The calculation of ground-wave field intensity over a finitely conducting spherical Earth', *Proc. Inst Radio Eng.*, 1941, **29**, pp. 623–639

141   OKUMURA, Y., OHMORI, E., KAWANO, T. and FUKUDA, K.: 'Field strength and its variability in VHF and UHF land-mobile radio service', *Rev. Electr. Commun. Lab.*, 1968, **16**, pp. 825–873

142   Olympus Utilisation Conference, Seville, Spain, 1993, ESA publication WPP-60

143   OMURA, J. K.: 'Spread spectrum radios for personal communication services'. IEEE international symposium on *Personal, Indoor and Mobile Radio Comms.*, November 1992

144   PARSONS, J. D.: 'The mobile radio propagation channel' (Pentech Press, 1992)

145   PARSONS, J. D. and BAJWA, A. S.: 'Wide band characterisation of fading mobile radio channels', *IEE Proc. F*, 1982, **129**, (2), pp. 95–101

146   PATTERSON, W. L.: 'Comparison of evaporation duct and path loss models', *Radio Sci.*, 1985, **20**, pp. 1061–1068

147   PICQUENARD, A.: 'Radio wave propagation' (Macmillan, 1974)

148   PRICE, R. M.: 'Radiometer fundamentals' *in* MEEKS, M. L. (Ed.): 'Methods of experimental physics. Vol. 12 Part B Astrophysics' (Academic Press, New York, 1976)

149   PROAKIS, J. G.: 'Digital communications' (McGraw-Hill, International Student Edition, 1989), 2nd ed., chap. 7

150   PRUPPACHER, H. R. and PITTER, R. L.: *J. Atmos. Sci.*, 1971, **28**, pp. 86–94

151   PURLE, D. J., *et al.*: 'Investigation of propagation through a frequency hopped channel', *Electron. Lett.*, 1992, **29**, (9), pp. 785–786

152   PURSLEY, M. B.: 'Performance evaluation of phased-coded spread spectrum multiple-access communications. Part 1 System analysis', *IEEE Trans.*, 1977, **COM–25**, pp. 795–799

153   RADIOCOMMUNICATIONS AGENCY: 'Performance specification for analogue multipoint video distribution systems (MVDS) transmitters and transit antennas operating in the frequency band 40.5–42.5 GHz'. Report of the 40 GHz MVDS Working Group, September 1993, document MPT 1550, November 1993

154   RAMO, S., WHINNERY, J. R. and VAN DUZER, T.: 'Fields and waves in communication electronics' (J Wiley & Sons, 1984), 2nd ed.

155   RATCLIFFE, J. A.: 'The magnetoionic theory and its applications to the ionosphere' (Cambridge University Press, 1959)

156   RAY, P. S.: 'Broadband complex refractive indices of ice and water', *Appl. Optics*, 1972, **11**, pp. 1836–1844

157   RODA, G.: 'Troposcatter radio links' (Artech House, 1988)

158   ROTHERAM, S.: 'Radiowave propagation in the evaporation duct', *Marconi Rev.*, 1974, **37**, pp. 18–40

159   ROTHERAM, A. S.: 'Ground-wave propagation. Part 1 Theory for short distances', *IEE Proc. F*, 1981, **128**, pp. 275–284

160   ROTHERAM, A. S.: 'Ground-wave propagation. Part 2 Theory for medium and long distances and reference propagation curves', *IEE Proc. F*, 1981, **128**, pp. 285–295

161   ROTHERAM, S.: 'Ground-wave propagation', *Marconi Rev.*, 1982, **45**, (1), pp. 18–48

162   RUSH, C. M. and GIBBS, J.: 'Predicting the day-to-day variability of the midlatitude ionosphere for application to HF propagation predictions'. AFCRL technical report 73-0335, Defense Documentation Centre, Alexandria, VA, USA, pp. 2–9

163   SCHILLING, D. S.: 'Broadband CDMA overlay'. Proceedings of IEEE 43rd VTC, New Jersey, USA, 1993, pp. 452–455

164   SCHWARTZ, P. R., CROSKEY, C. L., BEVILACQUA, R. M. and OLIVERO, J. J.: 'Microwave spectroscopy of H_2O in the stratosphere and mesosphere', *Nature*, 1983, **305**, pp. 294–295

165   SEN, H. K. and WYLLER, A. A.: 'On the generalisations of the Appleton–Hartree magnetoionic formulas', *J. Geophys. Res.*, 1960, **65**, p. 3931

166   SIMON, M. K., *et al.*: 'Spread spectrum communications' (Computer Science Press, Rockville, 1985), vols. 1–3

167   SKOLNIK, M.: 'Radar handbook' (McGraw-Hill, New York, 1990), 2nd ed.

168   SMITH, E. K. and WEINTRAUB, S.: 'The constants in the equation for atmospheric refractive index at radio frequencies', *Proc. Inst. Radio Eng.*, 1953, **41**, pp. 1035–1037

169   SOMMERFELD, A.: 'The propagation of waves in wireless telegraphy', *Ann. Phys.*, 1909, **28**, p. 665

170   STOVE, A. G.: 'Obstacle detection radar for cars', *Electron. Commun. Eng. J.*, 1991, **3**, pp. 232–240

171   SWALES, S. C., *et al.*: 'A comparison of CDMA techniques for third generation mobile radio systems'. Proceedings of IEEE 43rd VTC, New Jersey, USA, 1993, pp. 424–427

172   TATARSKI, V. I.: 'Wave propagation in a turbulent medium' (McGraw-Hill, New York, 1961)

173   TURIN, G., CLAPP, F. D., JOHNSTON, T., FINE, S. and LARRY, D.: 'A statistical model of urban multipath propagation', *IEEE Trans.*, 1972, **VT–21**, (1), pp. 1–9

174   TWOMEY, S.: 'Introduction to the mathematics of inversion in remote sensing and indirect measurements' (Elsevier, Amsterdam, 1977)

175   ULBRICH, C. W.: 'Natural variations in the analytical form of the raindrop size distribution', *J. Clim. Appl. Meteorol.*, 1983, **22**, pp. 1764–1775

176   VAN DE HULST, H. C.: 'Light scattering by small particles' (Dover, New York, 1981)

177   VAN DER POL, B. and BREMMER, H.: 'The diffraction of electromagnetic waves from an electrical point source round a finitely conducting sphere', *Philos. Mag. Ser. 7*, 1937, **24**, pp. 141–176

178   VAN DER POL, B. and BREMMER, H.: 'The diffraction of electromagnetic waves from an electrical point source round a finitely conducting sphere', *Philos. Mag. Ser. 7*, 1937, **24**, pp. 825–864

179   VAN DER POL, B. and BREMMER, H.: 'The diffraction of electromagnetic waves from an electrical point source round a finitely conducting sphere', *Philos. Mag. Ser. 7*, 1938, **25**, pp. 817–834

180 VAN DER POL, B. and BREMMER, H.: 'The diffraction of electromagnetic waves from an electrical point source round a finitely conducting sphere', *Philos. Mag. Ser. 7*, 1939, **26**, pp. 261–275

181 VINCENT, W. R., DALY, R. F. and SIFFORD, B. M.: 'Modeling communications systems' *in* FOLKESTAD, K. (Ed.): 'Ionospheric radio communications' (Plenum Press, New York, 1968), p. 321

182 WAIT, J. R.: 'Electromagnetic waves in stratified media' (Pergamon Press, New York, 1962)

183 WAIT, J. R. and SPIES, K. P.: 'Internal guiding of microwave by an elevated tropospheric layer', *Radio Sci.*, 1969, **4**, pp. 319–326

184 WALFISH, J. and BERTONI, H. L.: 'A theoretical model of UHF propagation in urban environments', *IEEE Trans.*, 1988, **AP–38**, pp. 1788–1796

185 WATERS, J. W.: 'Absorption and emission by atmospheric gases' *in* MEEKS, M. L. (Ed.): 'Methods of experimental physics. Vol. 12 Part B Astrophysics' (Academic Press, New York, 1976)

186 WATT, A. D.: 'VLF radio engineering' (Pergamon Press, 1967)

187 WIJAYASURIYA, S. S. H., *et al.*: 'Rake decorrelating receiver for DS-CDMA mobile radio networks', *Electron. Lett.*, 1992, **29**, pp. 395–396

188 YILMAZ, U. M., KENNEDY, G. R. and HALL, M. P. M.: 'A GaAs FET microwave refractometer for tropospheric studies'. AGARD CP-346, 1983, pp. 9.1–9.5

Index